QH438.5 .P48 2004
Phenotypic plasticity :
Northeast Lakeview Colleg
33784000123240

PHENOTYPIC PLASTICITY

PHENOTYPIC PLASTICITY

Functional and Conceptual Approaches

Edited by
Thomas J. DeWitt
Samuel M. Scheiner

2004

OXFORD
UNIVERSITY PRESS

Oxford New York
Auckland Bangkok Buenos Aires Cape Town Chennai
Dar es Salaam Delhi Hong Kong Istanbul Karachi Kolkata
Kuala Lumpur Madrid Melbourne Mexico City Mumbai Nairobi
São Paulo Shanghai Taipei Tokyo Toronto

Copyright © 2004 by Oxford University Press

Published by Oxford University Press, Inc.
198 Madison Avenue, New York, New York 10016

www.oup.com

Oxford is a registered trademark of Oxford University Press

All rights reserved. No part of this publication may be reproduced,
stored in a retrieval system, or transmitted, in any form or by any means,
electronic, mechanical, photocopying, recording, or otherwise,
without the prior permission of Oxford University Press.

Library of Congress Cataloging-in-Publication Data
Phenotypic Plasticity : functional and conceptual approaches /
edited by Thomas J. DeWitt, Samuel M. Scheiner.
p. ; cm.
Includes bibliographical references and index.
ISBN 0-19-513896-1
1. Genotype-environment interaction. 2. Phenotype.
[DNLM: 1. Phenotype. 2. Adaptation, Biological. 3. Evolution. 4.
Genotype. QH 438.5 P541 2003] I. DeWitt, Thomas J. II. Scheiner,
Samuel M., 1956–
QH438.5 .P48 2003
576.5'3—dc21 2002153341

9 8 7 6 5 4 3 2 1

Printed in the United States of America
on acid-free paper

To my wife, Heather Wilkinson, and daughter, Maretta DeWitt.
One passed my genes to the other, and now both enhance my environment.

T.J.D.

To my brother, Avram Scheiner, and sister, Harriet Scheiner,
proof of the interaction between genetics and environment.

S.M.S.

Preface

Our intent in assembling this book was to produce a conceptually based overview of the large and flourishing field of phenotypic plasticity—where we have been and where we are going. Plasticity research has burgeoned for more than a decade with surprisingly little effort to provide such a general summary and synthesis. The present volume is intentionally broad in coverage so that it will be of interest both to specialists broadening their understanding and to those seeking an entry into the subject. Our choice of authors yields a mix of received wisdom and novel insights.

Besides fleshing out major concepts, this volume emphasizes the functional aspects of plasticity. The value of plasticity to organisms depends on how often the correct phenotype–environment matches are made. Thus, plasticity is a major factor influencing how organisms relate to their environments and how populations evolve. Even our chapters on the proximate mechanisms and underpinnings of plasticity are directed toward these ultimate concerns of function and how function determines evolution. Although many authors of diverse backgrounds have contributed to the volume, the book retains a unified feeling because we have selected authors who themselves stress the functional importance of plasticity.

In chapter 1, we introduce major concepts, issues, and terms, attempting to provide enough basic information to orient nonresearchers or those beginning research in the field. In chapter 2, Sahotra Sarkar, a historian of science, orients our readers to the sequence of major developments in the field. The inclusion of a professional historian in such a volume is highly unusual but will provide unique insights for practicing biologists that they might otherwise fail to obtain.

Chapter 3 begins our mechanistic coverage. Jack Windig, Carolien de Kovel, and Gerdien de Jong examine the linkage from genotype to phenotype, the information needed to make that link, and where the gaps in our knowledge lie. In chapter 4, Jean David, Patricia Gibert, and Brigitte Moreteau look at patterns of evolution of plasticity and reaction norms using the model genus *Drosophila*. A developmental perspective is provided Tony Frankino and Rudy Raff in chapter 5, examining how plasticity unfolds as an epigenetic process and the importance of considering plasticity in a developmental context.

The middle portion of the book delves into plasticity as an adaptation. Chapters 6 and 7 are theoretical. In chapter 6, David Berrigan and Sam Scheiner provide a comprehensive review of the theory of plasticity evolution. In chapter 7, Thom DeWitt and Brian Langerhans look in more depth at how plasticity can be integrated with other solutions to environmental heterogeneity. Andy Sih, in chapter 8, contrasts and compares the many topics that fall under the general rubric of plasticity, especially addressing animal behavior and how traditional emphases in that field can be informed by mainstream plasticity studies, and vice versa. The functional ecology and plasticity adaptations of animals and plants are presented, respectively, by Paul Doughty and Dave Reznick in chapter 9 and Susan Dudley in chapter 10.

Finally, we switch gears again to look at wider issues. In chapter 11, Jason Wolf, Butch Brodie, and Mike Wade show how genotype–environment interaction can be considered a special case of context-dependent evolution. Then Carl Schlichting puts the evolution of plasticity in a macroevolutionary context in chapter 12. Last, in chapter 13, we highlight recent advances and suggest future directions for plasticity research. We expect this field to be just as dynamic with as many exciting new advances in the next twenty years as it has been for the past twenty.

We are grateful to those colleagues who reviewed chapters: Kathleen Donohue (Harvard University), Richard Gomulkiewicz (Washington State University), Ary Hoffman (La Trobe University), Curt Lively (Indiana University), Trudy Mackay (North Carolina State University), Andrew McCollum (Cornell College), Lisa Meffert (Rice University), Beren Robinson (University of Guelph), Bernard Roitberg (Simon Fraser University), and Anthony Zera (University of Nebraska).

Contents

Contributors

David Berrigan
National Cancer Institute
Executive Plaza North MSC 7344
Bethesda, Maryland 20892

Edmund D. Brodie III
Indiana University
Department of Biology
Jordan Hall 142
1001 East Third Street
Bloomington, Indiana 47405

Jean R. David
Laboratoire Populations, Génétique et Evolution
Avenue de la Terrasse
CNRS
91198 Gif sur Yvette Cedex
France

Gerdien de Jong
Evolutionary Population Biology
Padulaan 8
NL-3584 CH Utrecht
The Netherlands

Carolien G. F. de Kovel
Human Genetics Department (120)
UMC St Radboud
P.O. Box 9101
6500 HB Nijmegen
The Netherlands

Thomas J. DeWitt
Department of Wildlife and Fisheries Sciences
Texas A&M University
2258 TAMU
College Station, Texas 77843

Paul Doughty
Department of Zoology and Entomology
University of Queensland
Brisbane, Queensland 4072
Australia

Susan A. Dudley
Department of Biology
McMaster University
1280 Main Street West
Hamilton, Ontario L8S 4K1
Canada

W. Anthony Frankino
Section of Evolutionary Biology
Institute for Evolutionary and Ecological Sciences
Leiden University
Kaiserstraat 63
P.O. Box 9516
2300 RA Leiden
The Netherlands

Patricia Gibert
Laboratoire Populations, Génétique et Evolution
Avenue de la Terrasse
CNRS
91198 Gif sur Yvette Cedex
France

R. Brian Langerhans
Department of Wildlife and Fisheries Sciences
Texas A&M University
2258 TAMU
College Station, Texas 77843

Brigitte Moreteau
Laboratoire Populations, Génétique et Evolution
Avenue de la Terrasse
CNRS
91198 Gif sur Yvette Cedex
France

Rudolf A. Raff
Biology Department
Center for the Integrative Study of Animal Behavior
Indiana Molecular Biology Institute
Indiana University
Bloomington, Indiana 47405

David N. Reznick
Department of Biology
University of California
Riverside, California 92521

Sahotra Sarkar
Department of Philosophy
University of Texas at Austin
Waggener Hall 316
Austin, Texas 78712

Samuel M. Scheiner
Division of Environmental Biology
National Science Foundation
4201 Wilson Boulevard
Arlington, Virginia 22230

Carl D. Schlichting
Department of Ecology and Evolutionary Biology
University of Connecticut
75 N. Eagleville Road, U-43
Storrs, Connecticut 06269

Andrew Sih
Department of Environmental Science and Policy
University of California—Davis
One Shields Avenue
Davis, California 95616

Michael J. Wade
Indiana University
Department of Biology
Jordan Hall 142
1001 East Third Street
Bloomington, Indiana 47405

Jack J. Windig
Institute for Animal Science and Health
Department of Animal Genetics and Genetic Diversity
P.O. Box 65
8200 AB Lelystad
The Netherlands

Jason B. Wolf
Department of Ecology and Evolutionary Biology
University of Tennessee
Knoxville, Tennessee 37996

PHENOTYPIC PLASTICITY

1

Phenotypic Variation from Single Genotypes

A Primer

THOMAS J. DEWITT
SAMUEL M. SCHEINER

The Breadth of the Topic

A plant senses competitors for light and elongates its stem, accepting and elevating the challenge. A zooplankter senses diurnal predators and alters its daily vertical movements. A great breadth of ideas fall under the rubric of phenotypic plasticity, and this book is designed to express that breadth as a broad historical and conceptual review, bringing together a variety of (sometimes conflicting) viewpoints. In this chapter, we set the stage for the rest of the book by reviewing these diverse ideas under an intentionally broad definition of plasticity: *environment-dependent phenotype expression.* We emphasize the value of combining proximate, ultimate, and historical views. We hope to convey the need to understand how trait values are influenced by the environment, how individuals vary in this ability, and what such variations imply for how organisms live and reproduce.

Why Study Phenotypic Plasticity?

Phenotypic plasticity, as a paradigm, has broad significance and appeal because it unites perhaps all of biology. Phenotypic plasticity embraces genetics, development, ecology, and evolution and can include physics, physiology, and behavioral science. Although observations of environmentally induced phenotypes were once met blankly or with umbrage (chapter 2), the same observations today provoke biologists to ask how and why plasticity occurs. The change of interest reflects our new understanding that plasticity is a powerful means of adaptation. Alternative alleles or their products react differently to the environment; those with favorable reactions persist while others go extinct. This mechanism produces flexible organisms that respond to environmental shifts with beneficial phenotypic changes (chapters 8–10).

Although the view of plasticity as provider of elaborate adaptations has stimulated a recent boom in research, plasticity also can be a liability for organisms. If a single phenotype is best in all circumstances, then environmentally induced deviation away from the best phenotype only reduces fitness. For example, *Eurosta solidaginis* larvae producing an intermediate gall size on goldenrod plants are least vulnerable to predators (Abrahamson and Weis 1999). Yet the gall size produced by the fly depends on the plant genotype, an aspect of the fly's environment (Weis and Gorman 1990). Thus, environmental influences interfere with a fly's ability to consistently produce the best gall size. Often genetic changes are required to compensate for maladaptive plastic responses to the environment (i.e., countergradient variation).

Even when plasticity can potentially help organisms solve the problem of alternative phenotypic optima in different environments, costs and limits of plasticity may make plasticity suboptimal compared with a compromise level of plasticity that is more economical (Van Tienderen 1991; reviewed in DeWitt et al. 1998). And once plasticity has evolved, it may obviate the need for alternative adaptations to environmental variation. So, as either adaptation or constraint, or as part of an integrated set of strategies, plasticity is a key element in the functioning of organisms in variable environments.

Defining Plasticity

A common definition of phenotypic plasticity is *the environmentally sensitive production of alternative phenotypes by given genotypes* (for a semantic review, see Stearns 1989). This definition leaves considerable flexibility in deciding what types of traits exhibit plasticity, because the word "phenotype" is left for individuals to define for themselves. Some scientists prefer to restrict the concept of phenotypic plasticity to developmental processes (chapter 5) rather than other labile means of expressing phenotypes, such as physiological or behavioral shifts (chapter 8). Another view holds plasticity to be any environment-dependent gene expression, which can include gene regulatory processes that may have no gross phenotypic effects.

The danger of too broad a definition is that all biological processes are to some extent influenced by the environment. Thus, everything falls in the realm of plasticity. We fail to see this as a problem, as long as the point of addressing these diverse phenomena as plasticity is to focus on the genotype–environment interaction. Such breadth of scope reinforces the idea that a particular trait value as observed in a given environment always is a special case of a potentially more complex relationship. That is, specific phenotype–environment observations are a fraction of a multidimensional space. This view promotes in our thinking the constant and useful caveat that given phenotype distributions may only apply for the environment in which observation is conducted. Extrapolation beyond given conditions must be justified rather than assumed.

Often there is resistance to use of the language and conceptual framework of plasticity to describe phenomena outside of development. Is suppression of vertical migration in plankton, based on chemical cues from predators, really plasticity? Yet the wealth of concepts and unique analytical tools in the field of plasticity research might inform tangent fields of inquiry with different traditions. Likewise, these other fields can similarly inform mainstream studies in phenotypic plasticity research (e.g., chapters 5, 7, 8, and 11). Thus, each of the following examples, with increasingly liberal definitions of plasticity, can be considered plasticity: (1) development of alternative leaf types in high versus low light, (2) induced chemistry in response to herbivory, (3) production of lactase enzymes in bacteria triggered by the presence of lactose, (4) suppression of the vertical migration instinct in zooplankton by chemical cues from fish, (5) production of fever in endotherms upon infections, (6) buildup of muscles with use, and (7) animal learning. We suggest you draw the line where you wish, but be prepared to learn from those who draw their line elsewhere.

We also distinguish between this broad definition of plasticity as a trait, and a more narrow scope of evolutionary theory of plasticity (chapter 6). Theories of plasticity evolution all deal, in some fashion, with evolution and adaptation in uncertain environments, where uncertainty can be a function of a variety of ecological and biological mechanisms. Other aspects, such as costs of plasticity, are treated as constraints that are fixed by aspects of the organism's biology and are external to the evolutionary dynamic. Plasticity as a paradigm for evolutionary studies encompasses a much wider set of questions (chapter 13) than are covered by evolutionary models of plasticity.

Environmental or Genetic?

Although the definition of plasticity can expand or contract to accommodate various traits, it is important to keep the definition narrow with respect to which aspects of trait variation we refer to as plasticity. Biologists commonly partition total phenotypic variance (V_P) into that due to genetic effects (V_G) and that due to environmental effects (V_E). The equation $V_P = V_G + V_E$ can be found in most introductory biology textbooks. Typical in this gross view is that all deviations from genotype values (= breeding values) are deemed "environmental." But the environmental component is not accorded any functional or breeding value and is not distinguished from developmental or stochastic noise. The recognition of phenotypic plasticity, systematically induced variation attributable to specific environmental states, has allowed us to refine and go beyond the simple dichotomy (Via and Lande 1985).

With recognition of the importance of phenotypic plasticity, we expand the variance partition to $V_P = V_G + V_E + V_{G \times E} + V_{\text{error}}$ (Scheiner and Goodnight 1984; Via and Lande 1986), which includes explicit recognition of a systematic environmental effect (V_E) and, perhaps more important, a genotype–environment interaction ($V_{G \times E}$). This interaction specifies that the environment's effect is different for some genotypes relative to others (figure 1.1). So persistence of one subset of genotypes over others can change the average effect of the environment. For example, if divergent natural selection favors some genotypes over others based on genotypic reactions to the environment, then adaptive evolution of plasticity will occur.

Finally, a typological dichotomy we must disintegrate involves the question, "Is variation plastic or genetic?" This question is enduring and perennially misleading. The query often reflects the incorrect view of environmental and genetic effects being exclusive entities. Besides the obvious fact that genes and environment can interact ($G \times E$ variation), plastic responses are underlain by genes even when plasticity exhibits no

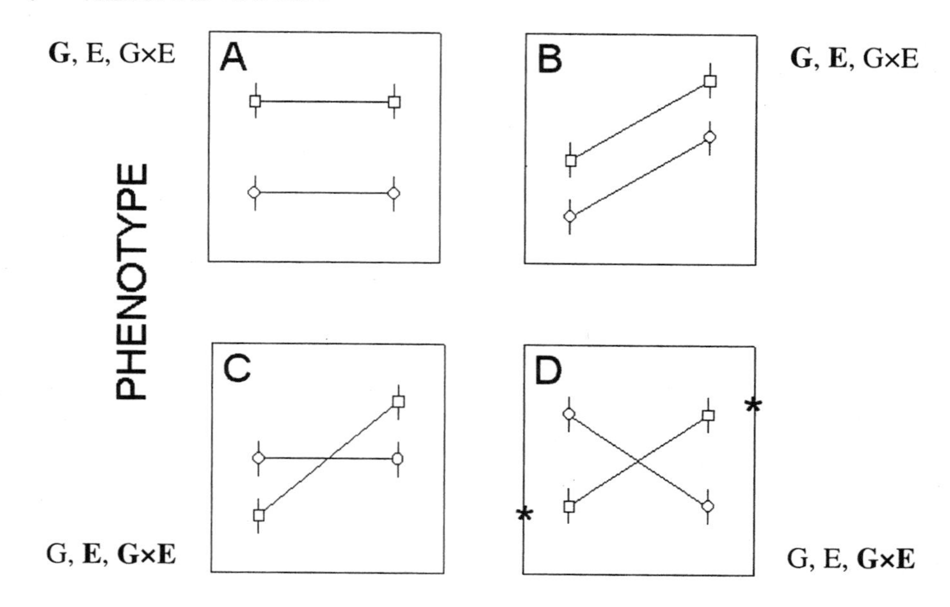

Figure 1.1. A diversity of reaction norms. Mean phenotype and variance in two environments for a family are denoted with circles, and those for another family are denoted with squares. Family means across environments are connected with a line, the *reaction norm*, to indicate familial identity. Boldface terms adjacent to the panels (*G*, *E*, *G*×*E*) indicate significant genetic, environmental, or gene–environment interaction variance, respectively. (A) Flat reaction norms (i.e., no phenotypic plasticity) with consistent genetic differences between families in both environments. (B) Sloped, parallel reaction norms, indicating plasticity and additive genetic variation for trait means but no interaction variance. That is, both genotypes are similarly plastic. (C) Differently sloped reaction norms, indicating genetic variation for plasticity (i.e., interaction variance). Because the overall slope is positive, there is also an effect of environment. Because the marginal mean phenotypes do not differ between families, there is no primary genetic effect (i.e., no genetic main effect). (D) As in C, family marginal means are the same for both families, but the families have opposite reactions to the environment. Therefore, only interaction variance is illustrated here. Asterisks are placed in D to illustrate adaptive optima. Because the family represented by squares would perform better than the other family, and families differ genetically in their degree of plasticity, plasticity would evolve.

additive genetic variance. Such responses are still genetic in the sense that they represent a range of reaction that could be subject to modification if suitable mutations arose and were favored by selection. That all organisms in a population have alleles responding similarly to an observed portion of an environmental gradient does not imply the reaction norm is nongenetic. [Conversely, just because a response is plastic still requires that we distinguish between active and passive plasticity (chapter 10).] The number of fingers on human hands is not heritable in the strict sense, yet it would be hard to contend that the trait is not genetic. These points may seem obvious, but we frequently see plasticity cited as being "nongenetic," a tradition going back at least to Wright (1931). Put another way, such traits are actually *perfectly* heritable, in the sense that you will express the *exact* phenotype of your parents. Such traits are not heritable in the breeding-value (or evolutionary) sense that the inherited trait distinguishes you from random individuals in the population. Yet there are genetically fixed responses to the environment, genetically fixed plasticity.

Therefore, the answer to the question, "Is variation plastic or genetic?" is simple—it's genetic. Sometimes it is not heritable in the narrow sense (i.e., additive genetic variance), however.

Plasticity versus Developmental Noise

Phenotypic variation from single genotypes can be produced by phenotypic plasticity or developmental noise. Developmental noise consists of random fluctuations that arise during development that alter the phenotypic product of development (Lynch and Gabriel 1987; Scheiner et al. 1991). The effect of developmental noise on fitness could be good, bad, or neutral. For example, developmental noise is costly under stabilizing selection because organisms cannot consistently produce the optimal phenotype (Yoshimura and Shields 1992; DeWitt and Yoshimura 1998). Conversely, developmental noise can be good in fluctuating environments because genotypes with noisy development have broader environmental tolerance (*sensu* Lynch and Gabriel 1987), so the geometric mean fitness among generations is increased.

Scheiner et al. (1991) raised the question of whether phenotypically plastic genotypes are by necessity "developmentally noisy." Scheiner et al. (1991) found no clear relationship between developmental noise and plasticity in bristle number, wing length, or thorax length in *Drosophila melanogaster*. Likewise, DeWitt (1998) failed to find evidence that developmental noise is associated with plasticity for shell shape in a freshwater snail.

In the equation, $V_P = V_G + V_E + V_{G \times E} + V_{\mathrm{error}}$, V_E is plasticity and V_{error} is developmental (or behavioral, etc.) noise. The designation of "noise" implies only that phenotypic deviations from a mean are random in direction but not necessarily random in magnitude. For example, in some environments it may be useful to hedge one's bets by producing variable offspring (Kaplan and Cooper 1984). In such environments, selection favors random phenotypic deviations from a mean, whereas other environments perhaps present strictly stabilizing selection for a mean with no variance. Therefore, it is reasonable to expect interesting evolutionary patterns where both trait means and variances, and potentially higher moments of phenotype distributions (e.g., skewness), vary across environments in an adaptive manner (chapter 7).

To illustrate this point, consider the freshwater snail *Physa*, in which chemical cues from fish induce crush-resistant shells. When fish cues are detected by snails it is certain that fish are present, yet the absence of cues does not always imply the absence of the predator. Thus, noninducing environments (absence of inducing cues) can be inscrutable (*sensu* Leon 1993). Inscrutable environments favor bet hedging, so we can expect snails in the noninducing environment to produce a moderate shell shape with greater variance than that produced in the inducing (fish) environment. Empirical data match this prediction (figure 1.2; T.J.D., unpublished data). Imagine the simultaneous optimization that selection conducts for all these moments along an environmental gradient. It is perfectly reasonable to speak both of reaction norms for trait means and of reaction norms for developmental noise (DeWitt and Yoshimura 1998) and potentially higher moments (chapter 7).

Genetics or Ecology?

There is a point at which the mathematical abstractions about genetic correlations and phenotypic variance components lose intuitive value to ecologists. What do they tell us about the ecology of the organism? Similarly, geneticists may lose interest in studies

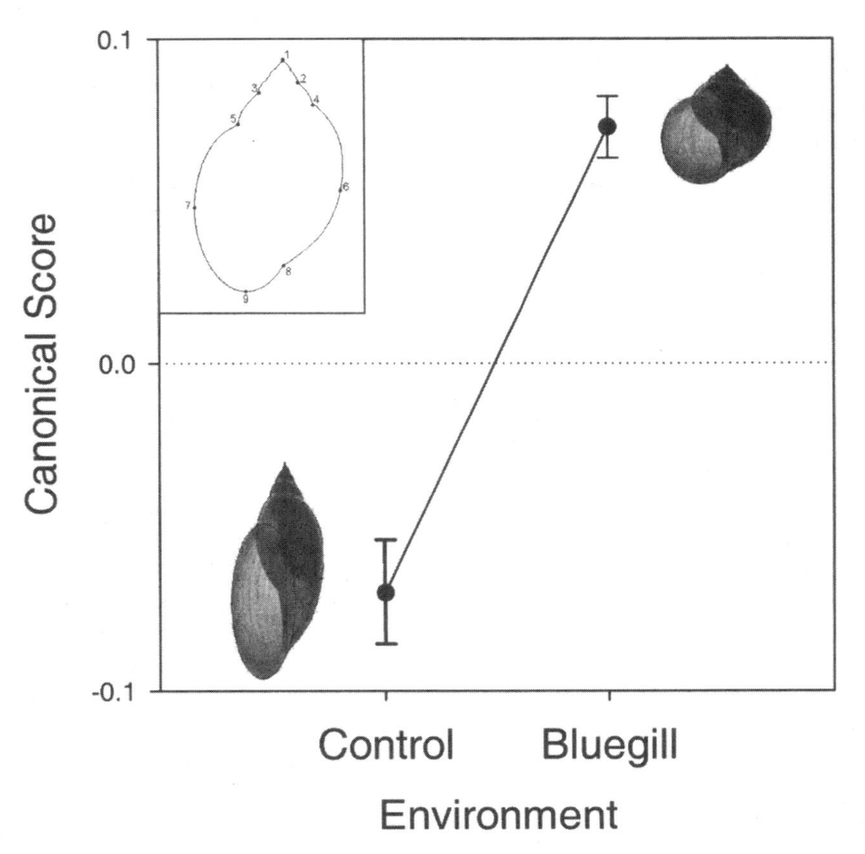

Figure 1.2. Reaction norm for shell morphology in the freshwater snail *Physa virgata*. The reaction norm depicts treatment means and variances of the first canonical axis (from a MANCOVA of partial warp scores; Bookstein 1991). The canonical y-axis describes elongate shells in a noninducing (control) environment and rotund shells in an inducing (bluegill sunfish) environment. Landmarks used in the analysis are indicated in the inset. Shell images illustrate shape differences between treatments magnified 10× using tpsSuper software (Rohlf 2000).

that do not report genetic variation for the traits. We sympathize with both views. It is unfortunate that genetic and functional aspects of plasticity are often studied in isolation of one another; quantitative genetic and functional ecology approaches should be complementary. Consider the traits a snail uses to avoid its many predators. One of the traits (growth rate) has high heritability within the environments, but little correlation (r_G) between environments (DeWitt 1996), which suggests that different genes contribute to the trait in alternative environments (Falconer and Mackay 1996). Although this suggestion is a mathematical deduction, the ecology of the animals indicates the likely cause. Behavior (hence, behavior genes) seems to determine growth rate in one environment—snails raised with fish perform antipredator behaviors that restrict feeding, so growth suffers (DeWitt 1998; Langerhans and DeWitt 2002). In an environment with an alternative predator, snails reduce allocation to reproduction (determined by a different suite of genes), seemingly to reduce time spent in a vulnerable size class (Crowl and Covich 1990; DeWitt 1998). Thus, we have a functional (and adaptive) basis for understanding the mathematical deduction. This example illustrates the sort of multiple-trait, multiple-environment approach that is needed to address plasticity evolution.

In many plasticity studies, fitness itself is frequently treated as a trait (e.g., Schmitt 1993; Stratton 1994). Environmental tolerance curves can be thought of as reaction norms for fitness over an environmental gradient. Conversely, reaction norms for fitness can be thought of as parts of environmental tolerance curves. Each view yields insight into the evolutionary ecology of organisms. However, we think that it is more fruitful to focus on individual traits and their reaction norms, not fitness *per se*, because this allows one to ask questions about plasticity as an adaptation.

Plasticity as Adaptation

The Problem

Consider the Olympic hopeful with two performance passions: to be a great sumo wrestler and to be a great pole-vaulter. He has an obvious problem. He needs a large body mass to increase performance in sumo, but large mass is a liability for pole vaulting. Where performance in one environment is inversely related to performance in another, a functional trade-off exists. Assuming that performance translates into fitness (Arnold 1992), then the functional trade-off creates divergent natural selection. The theoretical maximum fitness under divergent natural selection is achieved only by expressing the best phenotype in each environment (i.e., "perfect plasticity"). That is, there has to be perfect phenotype–environment matching for greatest adaptive value. Perfect plasticity is an insuperable strategy. Obviously, such perfection is never actually achieved in nature when the phenotypes diverge greatly. Yet a critical issue in plasticity studies is to define this maximum—to know what the best phenotype is in each environment. Knowing the functional ecology of an organism, we can adequately assess the degree to which an organism increases its fitness by facultative adjustments to the environment. Once the functional ecology and reaction norms are defined, we can think about the constraints that prevent perfection. Typically, however, this functional approach only proceeds about halfway, as detailed below.

The Benefits of Plasticity

The role of phenotypic plasticity in adapting to natural environments has been the focus of considerable work for decades. However, despite the large volume of work in this field (see figure 13.1), adaptive plasticity has not often been documented thoroughly (Scheiner 1993a; Gotthard and Nylin 1995)—but see chapters 8 and 9 for several examples. This problem persists because proof of adaptive plasticity requires analysis of fitness in multiple environments.

Consider the following example: Dodson (1988) showed that several cladoceran zooplankton (*Daphnia* spp.) produce small bodies when raised with fish, and that small size reduced predation risk in the presence of fish. Is the plasticity in body size adaptive plasticity? Probably. However, all that the preceding information tells us about adaptation is that small body size is favored in the presence of fish. We cannot make conclusions on the adaptive value of plasticity without further information. Three alternatives come to mind:

1. Perhaps small size is also favored in the absence of fish. In this case, nonplastic individuals strictly canalized for small bodies in all environments are more fit than are plastic types.
2. Perhaps there is no selection for body size in the absence of fish, but plasticity exhibits costs or limitations that make it a less suitable strategy than always producing small bodies. In this case, less plasticity might be the optimal strategy.
3. Perhaps plasticity in antipredator behavior is what is really adaptive, but performing the behaviors results in a correlated response in body size (DeWitt 1998). In this scenario, genotypes that are behaviorally responsive to predators survive better and their small body is merely a by-product.

Therefore, to show adaptive plasticity in the face of the first alternative, we have to show the induced phenotype to be adaptive in each environment being considered. Said another way, one must show that the canalized phenotype is inferior in the alternative environment. Examples of one-sided functional ecology are common, where authors demonstrate increased fitness of the induced character state in the inducing environment without testing for higher fitness of the alterative phenotype in other environments (e.g., Appleton and Palmer 1988; Dodson 1989; Parejko and Dodson 1991). Despite this rather stringent litmus test for demonstrating adaptive value, however, the cost of an induced defense is often easy to imagine, and so most cases of one-sided functional ecology are reasonably informative about adaptation. That said, documenting the absolute adaptive value of plasticity requires documentation of functional trade-offs, not merely selection within single environments.

As for alternative 2, we should be mindful of constraints and actively seek them out. The topic of costs of plasticity recently has been in vogue (reviewed in DeWitt et al. 1998). The current tests for costs (e.g., DeWitt 1998; Scheiner and Berrigan 1998; Donohue et al. 2000; Agrawal 2002; Johansson 2002; Relyea 2002) indicate that costs are certainly not pervasive, and probably are rare.

Thus, the major constraint on the evolution of plasticity likely involves limits: logistic constraints that prevent the evolution of perfect plasticity. Obviously, lack of genetic variation for plasticity may be an important constraint. Yet almost all studies find at least some genetic variation for plasticity (Scheiner 1993a). Other limits are probably more important, at least in the long term. Cue reliability is probably extremely important. In the freshwater snails mentioned above, for example, a severe constraint is that the snails cannot tell predatory sunfish from nonpredatory sunfish (Langerhans and DeWitt 2002). The snails therefore end up responding inappropriately to nonpredatory fish. The responses (reduced growth and altered shell shape) are maladaptive because they make snails vulnerable to common alternative predators and limit fecundity. So, research emphases need to shift to include as much (or more) work on limits as is currently being directed to costs.

The third alternative, correlations among traits masking the true object of selection, illustrates a topic rarely addressed in plasticity studies. For more general studies of natural selection, this issue was brought to the fore 20 years ago (Lande and Arnold 1983). Recent advances show how these methods can be applied to trait plasticities (Scheiner and Callahan 1999; Scheiner et al. 2000, 2002b). We must consider the entirety of the organism in order to determine patterns of selection and parse functional adaptations.

Concluding Remarks

A central accomplishment of the Modern Synthesis (Provine 1971) has been the gain of a deep understanding about how natural selection shapes phenotypes. Up to now, that understanding has been confined almost entirely to fixed traits or traits in only a single environment. We are now ready to achieve a similar deep understanding divergent natural selection and the evolution of trait plasticity. To achieve this understanding, we need to understand the nature of plasticity, its evolution, and its effects on diversification. We need to know more about the mechanistic underpinnings of plasticity (chapters 3–5). We need to understand how plasticity is optimized and integrated with other strategies for dealing with variable environments (chapters 6 and 7). We need comprehensive (i.e., two-sided) studies of functional ecology, and we need to address constraints more often than has been common (chapters 9 and 10). We need to define the relevant trait space of plasticity (chapters 8 and 11). Finally, we need to discern how plasticity participates in evolutionary diversification at levels above populations (chapter 12).

Central to all these aspirations is not only that we expand among topics studied under the rubric of plasticity, but also that we integrate them. Our goals for this book are to present a diverse collection of ideas that embrace the breadth of concepts surrounding plasticity and to provide a pathway toward that integration.

2

From the *Reaktionsnorm* to the Evolution of Adaptive Plasticity

A Historical Sketch, 1909–1999

SAHOTRA SARKAR

This chapter gives a brief history of the norm of reaction (NoR) from its initial introduction to current disputes about the evolutionary etiology of adaptive plasticity, focusing on conceptual issues. The following section describes how the NoR fared in the West until the 1940s and how a conceptual framework for genetics that had no place for the NoR was constructed (see also Sultan 1992). I then describe the fate of the NoR in the Soviet Union in the 1920s, how it was used to deflate claims of the inheritance of acquired characters, how it was incorporated into theories of organic selection, and how it was repatriated to the West. These two sections are short; a more detailed discussion can be found in Sarkar (1999). Then follows a description of how the evolutionary etiology of adaptive NoRs emerged as a topic of controversy between those who espoused specific plasticity genes and those who viewed the adaptation of NoRs as a by-product of selection for trait means. In the final section, I optimistically explore the possibility that molecular genetic data may not only resolve this controversy but may also yield new insights into the genesis and evolution of the NoR and phenotypic plasticity.

The Origin of the Concept

After the rediscovery of Mendel's work around 1900, the discrete nature of the Mendelian factors led to a temporary popularity of discrete jumps or "saltations" as the primary mechanism of evolutionary change. Hugo de Vries (1901, 1903), one of the first to reproduce Mendel's results, proposed a "mutation theory" of evolution. Wilhelm Johannsen, best known for introducing the genotype–phenotype distinction, argued that evolution consisted of discontinuous changes between "pure lines" (Johannsen

1909). Among those who resisted these non-Darwinian moves was the zoologist Richard Woltereck. At a June 1909 meeting of the German Zoological Society, held to commemorate the centenary of Darwin's birth, Woltereck (1909) interpreted years of work on *Daphnia* and *Hyalodaphnia* species to support the Darwinian view that evolution occurred through natural selection acting on small, continuous variations.

Woltereck studied morphologically distinct strains of *Daphnia* and *Hyalodaphnia* from different German lakes. These were pure lines that maintained their form through several generations of parthenogenesis. His focus was on continuous traits such as head height at varying nutrient levels (figure 2.1), because only these could shed light on the question of whether underlying evolutionary processes were continuous. For head height, Woltereck found that the phenotype (1) varied between different pure lines, (2) was affected by some environmental factors such as nutrient levels, (3) was almost independent of other factors such as temperature, and (4) showed cyclical variation with such factors as the time of year. However, the phenotypic response to a given environmental change was not identical for different lines. Woltereck drew "phenotypic curves" to depict this phenomenon (figure 2.2). These curves changed for every new variable that was considered. Thus, there potentially were almost infinite numbers of them, and Woltereck (1909, p. 135) coined the term "*Reaktionsnorm*" to indicate the totality of the relationships embodied in them.

Woltereck argued that what was inherited was this *Reaktionsnorm* and that hereditary change consisted of modification of that norm. He identified the *Reaktionsnorm* with the genotype: "*Der 'Genotypus' . . . eines Quantitativmerkmals is die vererbte Reaktionsnorm*" ("The genotype of a quantitative trait . . . is the inherited reaction norm"; 1909, p. 136). Thus, because the reaction norm consisted of continuously varying phenotypic curves, Darwinism was saved from the saltationist challenge because selection would act on small changes in the genotype (*qua Reaktionsnorm*). Johannsen (1911, p. 133) endorsed the concept of the reaction norm and thought it to be "nearly synonymous" with the concept of "genotype." But, contrary to Woltereck, he argued that the latter had not shown any inconsistency between the varying curves and the existence of constant genotypes. Rather, these curves depicted the phenotypes that arose from the "reactions of the genotypical constituents" (p. 145). Under Johannsen's interpretation, Woltereck's contribution consisted of a quantitative picture of what Nilsson-Ehle (1914) a little later called "plasticity" and interpreted as having general adaptive significance. In the West (i.e., the U.S. and Europe outside what became the Soviet Union), where Johannsen's distinction between genotype and phenotype became part of the standard picture of genetics, the subsequent decades witnessed a general trend to emphasize the constancy and causal efficacy of the genotype at the expense of the complexity of its interactions. The NoR remained a relatively unknown concept during this period.

Under Woltereck's interpretation, the genotype was less a deterministic force than an enabling agent in phenogenesis. This view found resonance in the Soviet Union where the NoR emerged as a concept of central importance. The two points of view were partly integrated in the 1950s, and the evolution of the NoR became a topic of further research starting in the 1960s and 1970s. In both traditions, contrary to Woltereck's own use, the NoR came to indicate each individual phenotypic curve rather than the totality of relationships depicted by such curves.

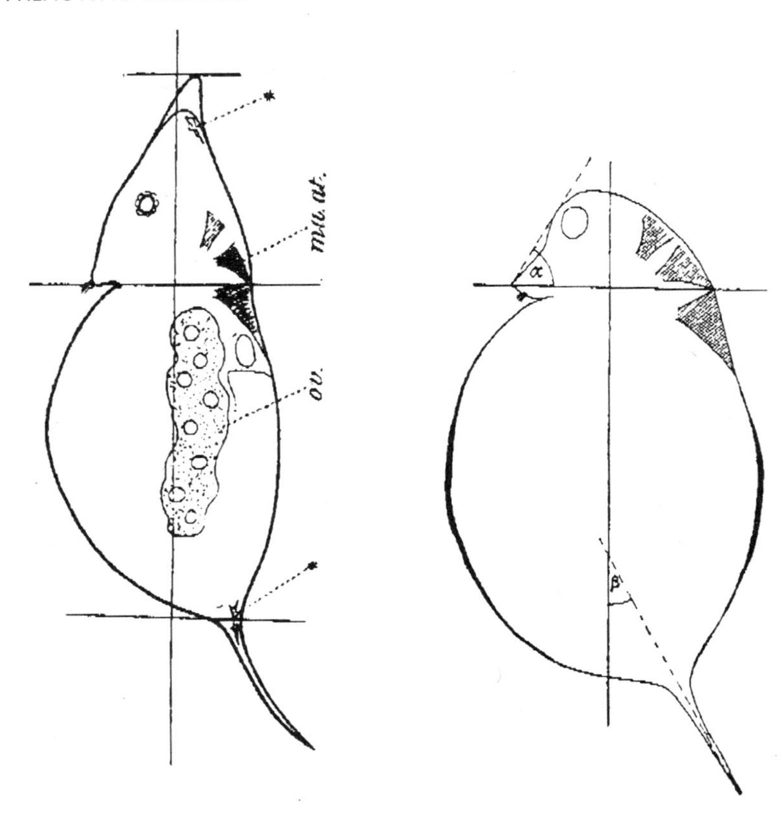

Figure 2.1. Side views of *Hyalodaphnia cucullata* female (left) and *Daphnia longispina* (right). In *Hyalodaphnia*, head height is measured along the vertical axis between the topmost and middle horizontal lines. The relative head height (which was the phenotype that Woltereck considered) is the head height divided by the distance between the topmost and lowest horizontal lines (and multiplied by 100 to be expressed as a percentage). (*mu. at.*: muscles of the antenna; *ov.*: ovary). In *Daphnia*, α and β are examples of quantitative characters. [After Woltereck (1909, p. 114).]

The Primacy of the Gene

Although Woltereck's experiments were widely discussed, at least in Germany (e.g., Baur 1922; Goldschmidt 1920, 1928), the NoR was ignored elsewhere in the West until 1950. Krafka (1920) published what was probably the first graphical depiction of NoRs. The phenotype was eye facet number in *Drosophila melanogaster* and its dependence on temperature (figure 2.3). Krafka referred to Woltereck's work and must clearly have been directly influenced by it, although he did not explicitly invoke the NoR. In 1930, using data from Krafka and many other sources, Hersch (1930) produced several graphical representations of NoRs. He tried to provide a mathematical description of the curves (see also Hersh 1934). Driver (1931) produced similar figures. Neither Herch nor Driver mentioned Woltereck or referred to the NoR.

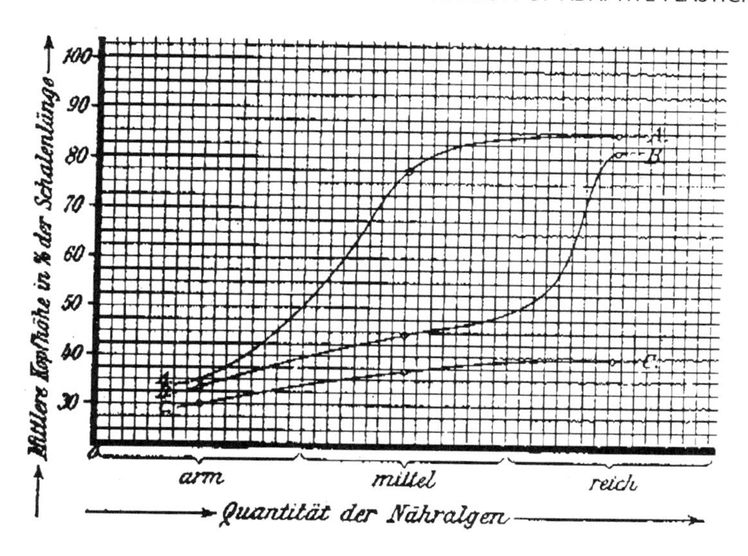

Figure 2.2. Phenotypic curves for females of three pure lines of *Hyalodaphnia cucullata*. Abscissa, nutrient level; ordinate, relative head height. (A) Strain from Moritzburg. (B) From Brosdorf. (C) From Kospuden. All strains were grown at a constant intermediate temperature and were from a "middle generation" of parthenogenesis. The curves show nonuniform variation between pure lines. [From Woltereck (1909, p. 139).]

Using Krafka's (1920) data, Hogben (1933) drew NoRs for eye facet number (figure 2.4) also without acknowledgement of Woltereck's work. Because the NoRs were not all parallel, Hogben argued for the "interdependence" of nature and nurture and against facile genetic reductionism: the claim that phenogenesis can be entirely explained from a genotypic basis (Sarkar 1998). That is, phenotypic variation cannot be additively decomposed into genotypic and environmental parts because of an interaction between the genotype and environment. It formed part of Hogben's critique of eugenic proposals to improve allegedly desirable human phenotypes by genetic intervention through selective breeding. Hogben's argument constitutes the first of three major shifts in the conceptualization of the NoR. Instead of being a tool to rescue Darwinism from saltationism, it became a standard-bearer for the complexity of nature–nurture interactions during phenogenesis.

Eventually this argument became a standard use of the NoR in debates over the origin of complex human traits such as IQ (e.g., Lewontin 1974). But in the 1930s and 1940s, the most influential arguments for the significance of genotype–environment interactions were Haldane's (1936, 1946) algebraic analyses. Meanwhile, genetics and evolutionary biology coalesced around a Mendelian core during this period (see Sapp 1987). This involved attempts (1) to reticulate the structure of Mendelian genetics to encompass the complexity of phenogenesis and (2) to delineate exactly the genotypic contribution to phenogenesis and, in the spirit of (1), impute as much causal efficacy to the genotype as possible.

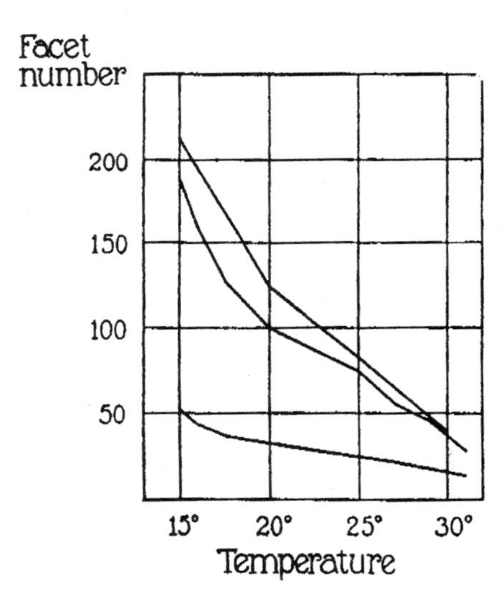

Figure 2.3. NoRs for dependence of eye facet number on temperature of *Drosophila melanogaster* females. Abscissa, temperature; ordinate, average facet number. The top curve corresponds to the unselected stock; the other, lower curves correspond to two mutants (*Low Selected* and *Ultra-bar*, respectively). Krafka drew two conclusions from this graph: "(1) The mean facet number at any given temperature is not the same for all stocks. (2) The difference in the mean number of facets between any two temperatures is not a constant for all three stocks. In other words, the number of facets is determined by a specific germinal constitution plus a specific environment" (p. 419). [From Krafka (1920, p. 419).]

The Reticulation of Mendelian Genetics

The first experimental program to address explicitly the complexity of phenogenesis and phenotypic variability emerged in the Soviet Union. In Moscow in the 1920s, an active genetics research group formed around the pioneering population geneticist S. Chetverikov (Adams 1980). In 1922, D. D. Romaschoff discovered the "abdomen abnormalis" mutation in *Drosophila funebris* that resulted in a degeneration of abdominal stripes. There was individual variability in the mutant phenotype, which Romaschoff (1925) interpreted as a difference in the strength of the mutation's effect. The manifestation of the mutation depended on environmental factors, in particular, on the dryness and liquid content of food. But Romaschoff could not rule out the possible influence of other genes. N. W. Timoféeff-Ressovsky studied the recessive "radius incompletus" mutation of *D. funebris*. In mutant flies, the second longitudinal vein did not reach the end of the wing. Timoféeff-Ressovsky (1925) created different pure lines, each homozygous for this mutation. Descendants included phenotypically normal flies. The proportion of normals was fixed for each pure line but varied between lines. External factors had little influence; the differences between the lines were apparently under the control of genetic factors. Some lines gave a large proportion of mutants but manifested the mutation weakly; the converse was also true.

Rather than accept that these results manifestly demonstrated a systematic indeterminacy in the genotype–phenotype relation, Vogt (1926) introduced two new concepts to interpret them from a genocentric point of view: a mutation's "expressivity" was the extent of its manifestation; its "penetrance" was the proportion of individuals carrying it that manifested any effect at all. In Vogt's definitions differences between lines were ignored. Expressivity and penetrance became properties of the mutation itself (and, eventually, the allele) rather than relative to the genetic background. Timoféeff-Ressovsky enthusiastically endorsed the new concepts (Timoféeff-Ressovsky and Timoféeff-Ressovsky 1926). The terms were introduced into the English literature by Waddington (1938), who incorrectly attributed them to Timoféeff-Ressovsky. Waddington's book, along with Timoféeff-

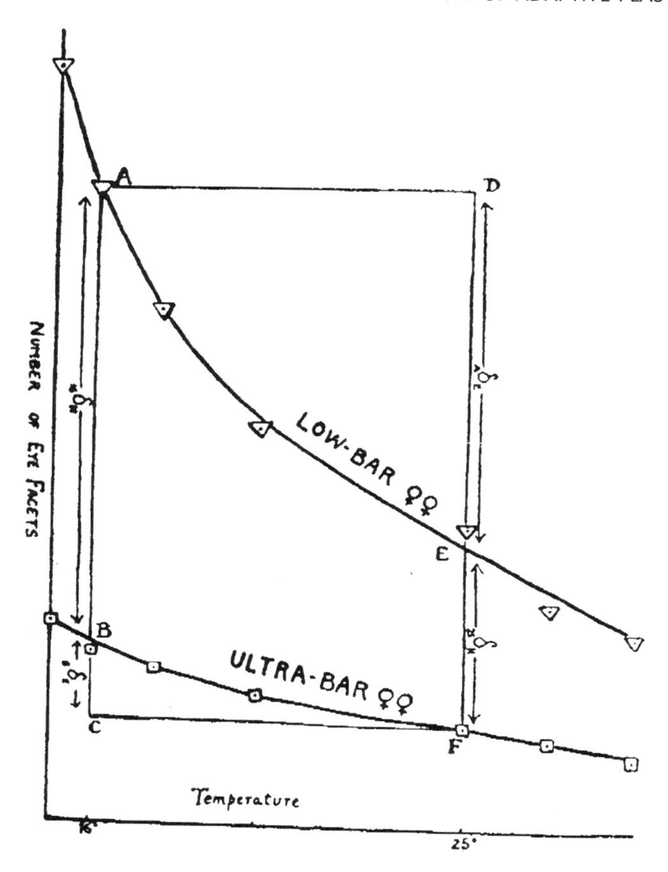

Figure 2.4. NoRs for eye facet number dependence on temperature of *Drosophila melanogaster* females. Hogben's innovation was to point out that the nonparallelism of the NoRs showed the "interdependence" of nature and nurture. [After Hogben (1933, p. 384).]

Ressovsky's increasing prominence within the genetics community, made the terms common currency by the 1950s.

What the original results of Romaschoff and Timoféeff-Ressovsky showed was a predictable complexity in the genotype–environment interaction. Both data sets permitted the construction of NoRs, although Vogt's reinterpretation made such a move moot. Two related aspects of that reinterpretation deserve emphasis: (1) Vogt, and those who used "penetrance" subsequently, ignored the systematic differences between pure lines; and (2) Vogt explicitly introduced expressivity and penetrance as properties of genes on par with, although different from, dominance. Given (1), it was now possible, as textbooks do to this day, to define penetrance as the conditional probability that a phenotype will manifest, given a particular gene. What this ignores is that because the genotypic background is ignored, there is no reason to expect measured empirical frequencies to converge to a definite probability.

The introduction of expressivity and penetrance constituted a convoluted reticulation of the structure of Mendelian genetics by ad hoc extension of the concept of the gene. Besides having their standard transmission properties, genes were no longer only recessive or dominant (or displaying varying degrees of dominance); they also had degrees of expressivity and penetrance. There was no clear distinction between expressivity and dominance: expressivity is indistinguishable from the degree of dominance. The purpose of the new concepts was to maintain a complete genetic etiology in the face of phenotypic plasticity induced by genotype–environment interactions. Variability in the phenotypic manifestation of a trait became a result of a gene's expressivity and (indirectly) its penetrance.

Heritability Analysis

A second development, initiated in the United States, also contributed to the neglect of the complexity of genotype–environment interactions during this period. In 1918 R. A. Fisher introduced the analysis of variance that permitted the decomposition of phenotypic variability within a population into genotypic and environmental components and their interaction (Fisher 1918). Independently, Sewall Wright (1920) introduced almost equivalent methods. Wright (1921) distinguished between three genotypic components of variability of a continuous trait: (1) additive effects of alleles at all loci, (2) dominance at each locus, and (3) interactions between loci (epistasis). Although Wright had implicitly been using a concept of heritability since the 1920s in analyzing breeding designs, Lush (1943) finally explicitly defined and distinguished between "narrow-sense" and "broad-sense" heritability (for a history, see Bell 1977).

Although Fisher (1931) criticized heritability for being too simplistic a statistic for the effective analysis of breeding problems, its use spread (see Lerner 1950). The problem lies in the interpretation of the statistic. Conventionally, following Lush (1943), broad-sense heritability is interpreted as the fraction of the phenotypic variance that is due to genotypic variation, with the rest attributed to environmental variation. However, this interpretation is valid only if the genotype–environment interaction is the same for all genotypes and all environments. This is where the NoR enters the story. The conventional interpretation requires that all NoRs be parallel to each other, as Hogben (1933) implicitly realized but was only later made explicit by Layzer (1974) and Lewontin (1974) (see also Moran 1973; Feldman and Lewontin 1975; Jacquard 1983). Experimental data on a wide variety of species almost never give rise to parallel reaction norms. However, these interactions are notoriously hard to detect (Wahlsten 1990; Windig 1997). The continued use of heritability analysis constitutes yet another way in which a commitment to genocentrism permits an avoidance of the complexity of phenogenesis that the explicit use of NoRs would demonstrate.

The Soviet Espousal

In sharp contrast to the West, the NoR emerged as a potent conceptual tool in Soviet genetics in the 1920s. There it was deployed primarily to deflate claims of the inheritance of acquired characteristics (Blacher 1982). In 1926, Theodosius Dobzhansky argued that what was inherited was not a trait but an NoR. Returning to the "abdomen abnormalis" mutation of *D. funebris*, Dobzhansky pointed out that the mutant phenotype did not manifest for generations if the food was dry. However, it reappeared if the offspring were supplied with moist food. He argued that this and other such examples showed that, even when environmental factors induced a trait, an unchanged NoR was inherited according to Mendel's rules. Dobzhansky's interpretation is the second of the three major innovative shifts in the conceptualization of the NoR: it was now a Mendelian unit of inheritance.

This interpretation, which was close to Woltereck's original one, made the NoR a heritable factor. It is not accidental that it found popularity in a communist state with an official scientific and socioeconomic ideology, dialectical materialism, which emphasized environmental influences on individuals. Dialectical materialism also promoted an ontology of processes over entities that was in dissonance with the Western ideology of genes as almost infinitely stable entities that determined traits. Most Soviet geneticists, until their suppression during the Lysenko era, attempted to find an interpretation of genetics that was less static and mechanistic than in the West. The NoR, by emphasizing the inheritance only of a capacity subject to environmental modulation, helped provide such an interpretation of Mendelism.

Stabilizing Selection

The most important conceptual innovation from Soviet genetics—and this constitutes the last of the three major conceptual shifts of the NoR—was a sharp distinction between adaptive and nonadaptive NoRs. The former were incorporated into models of "organic selection," originally proposed by Baldwin (1896, 1902), Osborn (1897a), and Lloyd Morgan (1900) but ignored subsequently (Huxley 1942, p. 524), and then independently formulated in the Soviet Union by E. J. Lukin and others around 1936. In Gause's (1947) exposition, organic selection was based on four principles:

> (1) Organisms frequently respond to environmental changes by adaptive phenotypic modifications. (2) Similar adaptive characters may be genotypically fixed in races normally living in the corresponding environments. (3) It is proved that conversion of modifications into mutations is not possible. (4) Hence modifications can only be *substituted* by coincident mutations, if the latter are associated with some advantages in the process of natural selection. (p. 22)

The most influential version of this theory is found in I. I. Schmalhausen's (1949) *Factors of Evolution*. Originally published in Russian in 1947, that book may have remained unknown in the West had not Dobzhansky arranged for its translation into English in 1949. [In 1948, with the completion of Lysenko's takeover of Soviet "genetics," Schmalhausen was removed from his professorship and forced to return to purely descriptive embryological work (Wake 1987).] Schmalhausen's term for organic selection was "stabilizing selection." Stabilization consisted of the replacement of an adaptive phenotypic response by an identical genotypic one, ensuring its transmission to future generations.

The NoR was central to this analysis. Assuming that it was "hereditary," Schmalhausen (1949) distinguished "adaptive norms" from "morphoses":

> [E]very genotype is characterized by its own specific "norm of reaction," which includes adaptive modifications of the organism to different environments. When expression of the adaptive modifications is so complete that it transforms the entire organization, the genotype is said to possess adaptive norms, which are particular expressions of the gen-

> eral norm of reaction. . . . Nonadaptive modifications are of an entirely different character. They arise as new reactions which have not yet attained a historical basis. Either the organism encounters new environmental factors with which it never had to deal before or its norm of reaction is changed (disturbed) as a result of mutation. (pp. 7–8)

A mutation was a "change in . . . reaction norm" (p. 10). Expressivity and penetrance were properties of mutations. Schmalhausen explicitly restricted the scope of the latter concept to "a group of mutants of similar origin" (p. 16), undoing some of the harm that Vogt had wrought.

Once distinguished, adaptive norms moved to center stage. They were buffered against environmental changes. Using the capacity for growth in the shrimp *Artemia salina* and the fish *Periophthalmus variabilis* as examples, Schmalhausen argued, "[I]n the process of evolution there have arisen definite optimum norms of growth that are determined by the ecologic position of the organism, especially by its relationships with other organisms. Modifications are possible thus only within the relative narrow limits of this norm. Therefore, it is not the *modification itself* but its confinement within definite limits that should be regarded as an *adaptation*" (pp. 184–185). Because adaptive norms were ipso facto selected for, their persistence (unlike that of morphoses) would be ensured both by having a genotypic basis and by selective spread in future generations.

Gause (e.g., 1941, 1942, 1947) developed similar ideas but emphasized that "the possibilities of genotypic response to changed environment are much wider than phenotypic ones. Hence with extreme alteration of conditions direct selection of genoadaptations takes place, but when the environment changes less directly the work of organic selection can be observed with certainty" (Gause 1947, pp. 32–33). Primarily using several *Paramecium* species, Gause established an experimental program to study the effects of organic selection. This led to even more caution:

> It is unlikely on *a priori* grounds that the resemblance of genoadaptations to phenoadaptations which they substitute will be a far-reaching one. As far as substituting or organic selection is based upon a greater adaptive value of possible genovariations as compared to modifications, and not upon the inheritance of the latter, it is probable that the likeness between these two types of acquirement will be limited to a superficial resemblance only. It is hardly reasonable to expect that genoadaptation will imitate all particulars of a physiological response. (p. 37)

Experiments with the fly *Fannia canicularis* and with *Paramecium bursaria* confirmed these suspicions. Gause emphasized that specialization and plasticity would have an inverse relation to each other. He ended his 1947 paper with a plea: "We hope that this line of thoughts and investigations will not come to a dead end again. . . . It can perhaps be considered as an important beginning of the new trend in biology" (p. 65). The ascendancy of Lysenko ended that possibility within the Soviet Union.

Meanwhile, apparently unaware of the Soviet work, Waddington (1940a,b, 1942) developed similar ideas although in a more rudimentary form. Waddington's (1940a) proposal that development was "canalized" was identical to Schmalhausen's claim of "autoregulation" of NoRs. However, Waddington (1942), unlike Schmalhausen, placed his ideas centrally within the contemporary framework of Western genetics by pointing out that the evolution of trait canalization seemed similar to the possible evolution of dominance. If dominance was an evolved phenomenon (Fisher 1928a,b, 1931), then the evolution of dominance provided a model for the evolution of a buffered phenotype. Waddington thought that this argument could subsume other forms of buffering, including canalization. However, unlike Fisher, he produced no quantitative analysis to buttress his claim. In an attempted synthesis of Schmalhausen's theory of stabilizing selection and Waddington's ideas on canalization with conventional population genetics, Lerner (1954) developed a more general model of "genetic homeostasis."

Schmalhausen, in his preface to *Factors of Evolution*, noted that his manuscript was completed in 1943 before he became aware of Waddington's (1942) paper. Waddington's later work on the genetic assimilation (e.g., Waddington 1953) explicitly drew on Schmalhausen's book. Meanwhile, Gause has priority in the establishment of a systematic research program to study plasticity and organic selection.

Adaptive Norms

If Schmalhausen had already brought adaptive norms to center stage, Dobzhansky directed attention almost exclusively to them. Dobzhansky moved from the Soviet Union to the United States in 1927 (Coe 1994). One consequence of this move was the repatriation of the NoR to the West. In 1937, in *Genetics and the Origin of Species* he introduced the NoR to the Anglophone world. That account reflected the consensus view from the Soviet Union: "[O]ne must constantly keep in mind the elementary consideration which is all too frequently lost sight of in the writings of some biologists; what is inherited in a living being is not this or that morphological character, but a definite norm of reaction to environmental stimuli. . . . [A] mutation changes the norm of reaction" (Dobzhansky 1937, p. 169).

On occasion, Dobzhansky used the NoR in this general sense that included both adaptive norms and morphoses. In 1950, in the fourth edition of *Principles of Genetics*, "reaction range" was used to refer to NoRs (Sinnott et al. 1950, p. 22). Phenotypic variability in the reaction range showed how phenogenesis depended on heredity–environment interactions. "Environmental plasticity" was emphasized; it could vary between traits in the same organism (p. 22) and between different environmental ranges (p. 23). The reaction range could only be incompletely known because no genotype could be experimentally exposed to all possible environments. Hence, any trait (including IQ) could be modified beyond known values by appropriate environmental intervention (p. 23).

Dobzhansky developed these themes in his 1955 book, *Evolution, Genetics, and Man*: "[T]he norm of reaction of a genotype is at best only incompletely known. . . . The existing variety of environments is immense and new environments are constantly produced. Invention of a new drug, a new diet, a new type of housing, a new educational system, a new political regime introduces new environments" (Dobzhansky 1955a, p. 75). There was no sharp distinction between hereditary and nonhereditary diseases (p. 76). Schmalhausen's sharp distinction between adaptive norms and morphoses had begun to fade. Except in very rare cases, a radical change of environment could well make a morphosis adaptive. The rare cases were those in that a morphosis was a result of internal disharmony during phenogenesis. But Dobzhansky, unlike Schmalhausen, and like most geneticists from that period, generally ignored embryology.

For Dobzhansky, by 1955, the origin of the adaptive norm had become a basic problem of population genetics. In his opening address to the 20th Cold Spring Harbor Symposium on Quantitative Biology, Dobzhansky (1955b, p. 3) redefined "adaptive norm" and, therefore, the NoR as a population-level entity. Gathering data on adaptive norms was already an important part of Dobzhansky's program to elucidate the genetics of natural populations of *Drosophila*. Dobzhansky and Spassky (1963) gave yet another new twist to the definition of an adaptive norm:

> Natural populations of *Drosophila*, man, and presumably of all sexual, diploid, and outbreeding organisms contain a multitude of genotypes. A majority of these genotypes make their carriers tolerably well adapted to survive and to reproduce in environments which the population frequently encounters in its natural habitats. The array of such genotypes constitutes the adaptive norm of the species or population. Some genotypes yield, however, low fitness in the habitual environments; these compose the genetic load of a population. And finally, some genotypes confer a fitness distinctly above the mean of the adaptive norm; these are the genetic elite of the population. (p. 1467)

These definitions incorporate three innovations: (1) The adaptive norm now consisted of the set of genotypes itself rather than phenotypic manifestations of those genotypes. Dobzhansky thus accommodated the Soviet articulation of the NoR to the genocentrism of the West. (2) The genetic elite did not constitute the adaptive norm despite having the highest fitness. The reason for this was that Dobzhansky and Spassky (1963) found that "an intensive study . . . revealed that they behave as 'environmental narrow specialists'; they produce superior homozygotes only in a certain environment, and when the environment is altered turn out to act as subvitals which reduce the viability more or less strongly" (p. 1482). Adaptive norms consist of generalists. (3) The definition of the genetic load was radically different from the standard one due to Crow (1958): the proportion by which the population fitness is decreased from that of an optimum genotype. Dobzhansky and Spassky argued that no such optimum exists in nature except, possibly, as "a single individual in a single environment" (p. 1482). Of Dobzhansky's definitional innovations, only one survived in a very mitigated form. This was the idea that an adaptive norm should be defined not only for individual genotypes but also for an entire population. As the next section shows, when the etiology of NoRs became the focus of research, mean values of a trait for a population in different environments were connected to form NoRs.

The Evolution of Reaction Norms

If NoRs—individual or populational—are adaptive, there is presumably an evolutionary story to be told of their etiology. In particular, they are likely to be either direct targets of selection or indirect results of selection for traits. Neither Dobzhansky nor his collaborators broached this problem. However, in 1965, Bradshaw published a seminal review of phenotypic plasticity in plants that suggested a model for the evolution of NoRs based on genetic control of plasticity. Bradshaw did not refer to NoRs but framed his discussion in terms of plasticity that he explicitly defined: "*Plasticity* is . . . shown by a genotype when its expression is able to be altered by environmental influences" (Bradshaw 1965, p. 116). As Bradshaw recognized, this definition did not restrict plasticity to be adaptive; however, he chose to put most of his emphasis on adaptive plasticity. Positing a strong dichotomy between plasticity and stability of a trait, he associated the latter with Waddington's "canalization" and Schmalhausen's occasional use of "stabilizing selection" as autoregulation. Homeostasis, for Bradshaw, signified the opposite of plasticity (p. 117).

Bradshaw's most important contribution was to propose that plasticity was under specific genetic control. Three arguments supported this claim. (1) Plasticity was not a property of the entire genome but "a property specific to individual characters in relation to specific environmental influences" (p. 119). (2) The plasticity of a trait varied among different species of the same genus and varieties of the same species. "Such differences are difficult to explain unless it is assumed that the plasticity of a character is an independent property of that character and is under its own specific genetic control" (p. 119). (3) Waddington's (1959) and others' (e.g., Bateman 1959) work on canalization and the genetic assimilation of acquired characters showed that stability was under genetic control. Because plasticity was the opposite of stability, plasticity must also be under genetic control.

If plasticity was under genetic control, it was subject to selection. Bradshaw (1965) argued that Waddington's (1959) selection experiments, which increased the capacity of a strain of *Drosophila melanogaster* to react to increased salinity by developing larger papillae, showed that plasticity could be altered through selection (p. 124). Drawing on Levins's (1962, 1963) pioneering theoretical work on evolution in changing environments, Bradshaw listed four selection regimes favoring plasticity. (1) *Disruptive selection in time*: Genotypic changes through mutation could only take place on time scales of at least a generation. Therefore, "[i]f the duration of the environmental fluctuation is much less than the generation time, any adaptation that occurs can only take place by plasticity" (p. 127). (2) *Disruptive selection in space*: If spatial heterogeneity occurred on scales of only 10 m or more, genetically different plant populations were known to form. However, if environmental heterogeneity occurred on smaller spatial scales, this option may not be available; plasticity may be the only available adaptive response. (3) *Directional selection*: "If directional selection is very severe and the normal, directly adaptive, genetic variation is limited, further adaptation may be afforded by plasticity" (p. 136). This was supported by experiments of Thoday (1955, 1958). (4) *Stabilizing selection*: More clearly than Waddington, Bradshaw noted the connection between such selection and evolutionary models of the origin of dominance. However, little experimental data provided support for stabilizing selection, "although common sense suggests that it could be found commonly if appropriate measurements were made" (p. 137).

Bradshaw's 1965 paper concluded with a discussion of conditions disfavoring plasticity (e.g., any condition that favored canalization) and possible mechanisms for generating plasticity: "The mechanisms involved are varied. At one extreme the character may show a continuous range of modification dependent on the intensity of the environmental stimulus. At the other the character may show only two discrete modifications" (p. 150). In the latter case, the environmental stimulus directed phenogenesis along one of two canalized developmental pathways. The former had been called "dependent morphogenesis" by Schmalhausen; the latter, "autoregulatory dependent morphogenesis."

Optimization Models

Starting in the 1970s, but only gathering momentum in the 1980s, models for the evolution of NoRs were constructed. This theoretical interest reflected growing experimental investigations of plasticity. The reasons for the neglect of plasticity before 1980 are unclear. As Schlichting (1986) observed:

> Until 1980, theoretical work on plasticity was limited; and empirical research, with the notable exception of Subodh Jain's efforts [Marshall and Jain 1968; Jain 1978, 1979], was largely unfocused. The reasons for such neglect are puzzling, especially considering the clarity of Bradshaw's review. Surely part of the problem was the growing fascination with the detection and measurement of "genetic" variation, of which plasticity must have seemed the antithesis. Another problem was that environmentally induced variability in an experiment is typically avoided at all costs. Experimental complexity and the problem of measuring plastic responses also retarded progress. (p. 669)

Consistent with the view that the NoR was a target of selection, but with less concern for genetic detail than what Bradshaw (1965) had shown, optimization models began to be constructed in the 1970s (e.g., Huey and Slatkin 1976) and became a cottage industry in the 1980s. With the exception of Fagen (1987), who considered fitness, other optimization models only dealt with life history traits. Lively (1986a), Fagen (1987), Houston and McNamara (1992), and Moran (1992) modeled discrete environments with spatial variation; Moran (1992) also included temporal variation. Stearns and Koella (1986), Perrin and Rubin (1990), Clark and Harvell (1992), and Caswell (1983) modeled continuous environmental variation: the first two of these papers dealt with spatial variation; the last two, as well as Huey and Slatkin (1976), considered temporal variation (see chapter 5 for a detailed review).

These models were of limited interest because they were restricted to life history traits. Besides, they suffered from two standard problems with the assumptions of optimization models: (1) unlimited genetic variation and (2) no genetic or other constraints on evolution (Scheiner 1993a). These problems reflect the assumption that selection alone explains the feature of interest (Gould and Lewontin 1979). But even if a general skepticism about optimization is not justified, these models contributed little to the debate about the etiology of NoRs that emerged in the 1980s.

Plasticity as a By-product of Selection

As Via (1994) noted, when Bradshaw published his review in 1965, "there were no population genetic models for the evolution of quantitative traits under natural selection, even in single environments" (p. 51). By the 1980s that situation had changed. As an offshoot of an attempt to elucidate the evolutionary dynamics of quantitative characters in spatially heterogeneous environment, Via and Lande (1985) produced several models for the evolution of adaptive—as well as nonadaptive—NoRs. This spawned further work, some consisting of quantitative genetic models and others of "gametic models," that is, models with explicit genetic bases, by Via (1987), de Jong (1988, 1989, 1990, 1995), Van Tienderen (1991), Gabriel and Lynch (1992), Gomulkiewicz and Kirkpatrick (1992), Gavrilets and Scheiner (1993a,b), and Van Tienderen and Koelewijn (1994), among several others.

Following Bull (1987), Via (1994) noted that two types of plasticity could be distinguished. The first type results when genotypes are exposed to different repeatable and predictable environments. Via (1994) explicitly—and others implicitly—restricted the use of the NoR to this situation. "Central to the concept of the reaction norm is the idea that the environments involved are repeatable and predictable aspects of the organism's habitat" (p. 37). The second type is "noisy plasticity" that is a response to "largely unpredictable variability within environments" (p. 37). The plastic response to the latter situation is "different from an adaptive norm, because the result is variability in the phenotype within the current environment rather than the production of a particular phenotype" (p. 37). Bull (1987) modeled this latter situation; I do not consider it any further in this chapter.

The central disputed issue became the question whether adaptive NoRs emerged as direct targets of selection or as by-products of selection acting on traits themselves. This issue is connected with that of the existence of specific "plasticity genes," that is, genes that through either their expression or a regulatory role modify a trait's expression. As Bradshaw (1965) had pointed out, if there are such genes (loci), they (the alleles) would be subject to selection. Contrary to Bradshaw, the models described in this subsection make no reference to specific plasticity genes; rather, all genes effectively act as plasticity genes (see Scheiner 1998).

In these models, adaptive NoRs emerge while selection acts on the means of quantitative traits. Selection is "stabilizing": it acts to drive populations toward optima that may or may not be accessible. The analytic technique used came to be called the "character state" approach (Via et al. 1995). The crucial technical insight went back to Falconer (1952), who showed that a character expressed by the same genotype in two different environments could be modeled as two distinct, genetically correlated characters. The expression of a character in each environment was a "character state." Falconer's insight allowed the co-option of standard techniques from quantitative genetics for the analysis of the evolution of NoRs. In this approach, the set of phenotypic means constituted the NoR of the genotype.

Via and Lande (1985), Via (1987), Van Tienderen (1991), and Gomulkiewicz and Kirkpatrick (1992) considered environmental variation due to spatial heterogeneity. In this scenario—"coarse-grained" environments, in Levins's (1968) terminology—an individual completed its development in a single environmental type. Following Christiansen (1975), Via and Lande (1985) distinguished between "soft" and "hard" selection: soft selection involves the regulation of each population in each environment; in hard selection, the population is regulated globally. Via and Lande (1985) constructed four models assuming two (discrete) environments: (1) a panmictic population subject to soft selection, (2) the same population subject to hard selection, (2) a subdivided population subject to soft selection, and (4) the same population subject to hard selection. They assumed weak selection, a population large enough for mutation to replenish any variation lost because of selection, and intermediate optima for traits.These models also assumed—and, except for epistasis, all the others considered in this chapter make these assumptions—that nonoverlapping generations, autosomal inheritance, and the effects of mutation, drift, epistasis, and recombination were negligible compared with that of selection. If the genetic correlation between character states was not ±1, and there was no cost to plasticity, in models (1) and (2) a joint optimum for both environments was achieved. If there was adequate gene flow between subpopulations, models (3) and (4) led to similar conclusions.

Via (1987) extended model (1) to a situation with four environments. Van Tienderen (1991) introduced a cost to plasticity. For soft selection, he found that the equilibrium population mean of the trait was a compromise between that which belonged to the optimal NoR and that which had the lowest cost. It was also closest to the optimal mean in the most frequent habitat. For hard selection, high costs generated an adaptive landscape with multiple peaks. Which peak attracted the population depended on the initial conditions, that is, the history of the population.

Using techniques developed by Kirkpatrick and Heckman (1989), Gomulkiewicz and Kirkpatrick (1992) modeled soft and hard selection in continuously varying environments. In contrast to Via and Lande (1985), their interest was as much in constraints on adaptive evolution as on the emergence of adaptive NoRs. Under soft selection, if additive genetic variance existed for all the conceivable evolutionary changes of the population mean of the trait, the NoR evolved to its optimum. Otherwise, genetic constraints prevented optimization. Under hard selection—as with Van Tienderen's (1991) model—the population's history determined which equilibrium was reached.

Gomulkiewicz and Kirkpatrick (1992) also modeled temporal variation. These were of two types: (1) within-generation variation—the same individual experienced a range of environments; and (2) between-generation variation—each individual experienced a constant environment but this changed between generations. In the former case, a distinction between two types of trait (also due to Schmalhausen) became important: labile traits for which the individual adjusted phenotypic expression throughout its life, and nonlabile traits, the expression of which was fixed during development. For labile traits, Gomulkiewicz and Kirkpatrick found that NoRs evolved to optima. For nonlabile traits, in the absence of genetic constraints, selection took NoRs to compromise equilibria that took into account the intensity and duration of the environmental fluctuations. In the latter case, the order in which a population experienced different environments was critical to its evolution. Genotypically identical populations could diverge in their evolutionary futures depending on the sequence of environmental encounters. No general results were possible, and Gomulkiewicz and Kirkpatrick were content to develop a framework that could be adapted to specific experimental situations.

Via (1994) summarized the insights from these theoretical explorations:

> Though the details of the evolutionary trajectories followed by populations under each of these selection scenarios differ, as can the equilibrium reaction norms, they have three important features in common: (1) Selection is assumed to act only on the phenotypic character states expressed in the environment in which an individual finds itself at the moment; (2) within each environment, selection acts to move the population towards the optimum phenotype . . . ; and (3) evolution in variable environments requires that populations respond to selection . . . in a "quasi-simultaneous" way, as parts of the population experience each environment or as the population experiences different environments in a sequence. . . . In all these models, the optimum joint phenotype—and thus the optimal reaction norm—will not be attained unless there is sufficient genetic variation for all the character states and their combinations. . . . In a few of the models, the optimum phenotype cannot be attained. . . . In these . . . models, the population evolves until it arrives at the best compromise among different selective forces. (p. 43)

The last remark is misleading. What Van Tienderen (1991) and Gomulkiewicz and Kirkpatrick (1992) had shown was that evolution toward the optimum could halt in the presence of constraints and, especially, that the particular optimum reached might depend on evolutionary history. The important point, however, is Via's first one: the evolution of adaptive NoRs could occur through selection acting on the traits themselves. Selection need not act on plasticity modeled as a distinct trait with its own specific genetic etiology.

The Return of the Gene

Meanwhile, the 1980s also saw the revival by many workers of Bradshaw's (1965) hypothesis of the specific genetic control of plasticity. For example, Scheiner and Goodnight (1984) argued that five populations of the grass *Danthonia spicata* showed highly significant differences in the total phenotypic variance for 11 out of 12 plastic traits. Although there were significant genotypic differences between the populations, these were "small relative to the ability of all populations to grow and reproduce under a wide range of experimental conditions" (p. 848). They even suggested that the genotypic differences were the result of drift rather than selection. More important, Schlichting (1986) argued that "the amount and pattern of plastic response can evolve independently of the character mean" (p. 677). For instance, he noted, the character shoot:root ratio had a similar mean in two *Portulaca* species (*P. grandiflora* and *P. oleraca*) but showed marked differences in the amount and direction of plasticity. These data were supposed to show that "the genetic control of the plasticity of a trait is distinct from that of the trait itself, [and] evolutionary forces can act independently on characters and the plasticities of these characters" (p. 677).

De Jong (1988, 1989, 1990a) began quantitative modeling of variation in plasticity. She did not explicitly endorse the existence of "genes for plasticity," but her model—and all the others discussed in this subsection—allowed for the existence of specific plasticity genes and independent evolution of plasticity and trait means. This approach eventually came to be called the "polynomial approach" (Via et al. 1995) [or, somewhat quixotically, the "reaction norm" approach (Van Tienderen and Koelewijn 1994; de Jong 1995)]. Scheiner and Lyman (1991) suggested that selection acts on the coefficients of the relevant "polynomial"; Gavrilets and Scheiner (1993a) were the first to produce a formal model. In this approach, "the reaction norm is described by a polynomial function of the phenotypic values expressed by a genotype across a range of environments, and evolutionary models are based on the population means and genetic (co)variances of coefficients of the polynomial" (Via et al. 1995, p. 212). De Jong (1990b) extended standard quantitative genetics to construct purely additive (no dominance or epistasis) models in which allelic effects (the effects of the substitution of single alleles at a time) were linear functions of the environmental variables. In these models, the additive genetic (genic) variances of traits and the additive genetic covariances between them become quadratic functions of the environmental variables. The covariances can change sign over some range of environmental variables, in qualitative agreement with some experimental data. Subsequently, Van Tienderen and Koelewijn (1994), who extended these models to include epistasis, and de Jong (1995) showed that the character state and polynomial approaches are formally equivalent for discrete environments and continuous environments provided that the NoR is a Taylor-expandable function.

Gabriel and Lynch (1992), extending their earlier work on environmental tolerance (Lynch and Gabriel 1987), developed models for the evolution of Gaussian NoRs where the environment induced irreversible switches in development. Gavrilets and Scheiner (1993a) constructed models that allowed the NoR to assume any shape, and incorporated genetic constraints. Selection led populations to converge on the genotype with the maximum mean geometric fitness over all environments. Under temporal variation, it also forced convergence to a linear NoR. Even if selection acted on the trait itself in each population, the evolution of plasticity was partly independent of the evolution of the trait's mean (across all environments).

These formal issues were of marginal importance in the emerging controversy over the evolution of NoRs (and plasticity). Rather, the dispute was about whether plasticity was a direct target of selection—this question was usually framed as one of the existence of "plasticity genes"—or whether plasticity was a by-product of selection as Via and Lande (1985) had argued. Support for the former option was supposed to come from experimental data.

Scheiner and Lyman (1989)—extending the results of Scheiner and Goodnight (1984)—developed methods for measuring the heritability of plasticity. When these methods were applied to the dependence of thorax size on temperature in *Drosophila melanogaster*, an important result was that the heritability of plasticity was less than the heritability of the trait mean. This suggested that the plasticity of the trait was distinct from its mean. Scheiner and Lyman (1989) criticized Via and Lande's model for not allowing the evolution of plasticity independently of trait means and for not allowing genes for the "plasticity of a trait [to be] separate from those for the expression of that trait" (p. 105). However, they did not choose between the models of Via and Lande (1985) and Lynch and Gabriel (1987), although their remarks indicated a preference for the latter.

Their preference for plasticity loci became fully explicit in Scheiner and Lyman (1991). Selection on plasticity (for the same trait as before, using a family selection scheme) resulted in a response that was partially independent of selection on the trait mean. They tested three models of the genetic basis for plasticity: (1) an overdominance model—plasticity decreased as the number of homozygous loci increased; (2) a pleiotropy model—plasticity was a function of differential expression of the same allele in different environments [this model was a genetic instantiation of the quantitative genetic models of Via and Lande (1985)]; and (3) an epistasis model—"plasticity is due to genes that determine the magnitude of response of environmental effects which interact with genes that determine the average expression of the character" (p. 25). Their results generally supported the epistasis model although the support for it over the pleiotropy model was not unequivocal. They elaborated the epistasis model to distinguish between three types of loci: (1) "response/no response loci" that determined whether the genotype responded to an environmental cue at all; (2) "size loci" that controlled the mean response of the phenotype; and (3) "amount of response loci" (p. 45).

Additional support for the claim that the plasticity of a trait may evolve separately from the trait itself came from comparative studies of *Phlox* populations by Schlichting and Levin (1984, 1988, 1990; Schlichting 1989). Schlichting and Levin (1984) compared the plasticities of 18 traits of three species of annual *Phlox* (*P. cuspidata*, *P. drummondii*, and *P. roemeriana*) in six environmental treatments. Three hypotheses were tested: (1) a heterozygosity hypothesis—there was an inverse relation between plasticity and heterozygosity; (2) a relatedness hypothesis—plastic responses become less similar with phylogenetic distance; and (3) an ecological hypothesis—plastic responses were more similar for species that evolved in similar habitats. There was some support for all three hypotheses, underscoring the complexity of the etiology of plasticity. There was also a lot of divergence between the plasticities of different traits on the same species. Schlichting and Levin (1988) focused on 10 traits in five populations of *P. drummondii*. They generally found much more variation in trait means than in plasticities. More systematic work on natural populations of *P. drummondii* (Schlichting and Levin 1990) showed significant differences between patterns of divergence of plasticities and means of traits, lending more plausibility to the idea that plasticity evolved at least partially independently of trait means and was under independent genetic control.

Controversy and Consensus?

Via (1993a) disputed these inferences, especially the claim that there were specific "genes for plasticity," arguing that

> this view is (a) inconsistent with the action of either stabilizing or directional selection within environments, (b) misleading because of an ambiguity in the assertion that plasticity and trait means are independent, and (c) unnecessary for a description of reaction norm evolution because the proposed effects of separate genes for plasticity can either be produced by environment-specific gene expression or by allelic effects that vary across environments, both of which are already incorporated into current genetic models of reaction norm evolution. (p. 353)

More specifically, suppose that selection acted on plasticity itself. Then, Via argued, at least in some environments two genotypes with the same plasticity should have the same fitness (p. 358). However, if selection also acted on the trait itself, this was impossible. If selection acted on both plasticity and the trait, then these two forces would either be in opposition or in concert. The former option was incorporated in Van Tienderen's (1991) model. For the second option, Via found no "plausible biological rationale for selection for increased phenotypic plasticity . . . that is different [from] selection toward different trait values within different environments" (p. 360).

For Via, the claim that plasticity was independent of trait means involved a semantic equivocation: the trait mean could either be its mean in a single environment or its mean over all environments (the "grand mean"). Via argued that all the data collected by Scheiner, Schlichting, and their collaborators only showed independence between plasticity and the grand mean. But this was consistent with the quantitative genetic models that assumed selection for the trait mean within particular environments. Finally, positing plasticity genes was unnecessary: "To say that there are genes for plasticity that are different from genes for trait values must mean that such loci influence the degree of phenotypic change across environments per se, regardless of the values of the traits within environments" (p. 361). This was different from the view, which went back to Schmalhausen and was incorporated in Lande and Via's (1985) models, that "the same alleles that influence trait values create plasticity by virtue of allelic effects that depend on the environment" (p. 362). Via went on to argue that Scheiner and Lyman's (1991) experiments that provided support for the epistasis model actually involved selection on trait values. According to Via, Scheiner and Lyman's "amount of response" loci were simply "modifier loci with environment-specific expression" that were already incorporated in the quantitative genetic models discussed above.

Scheiner (1993b) responded by claiming that "Via attempts to define away 'plasticity genes' by calling them 'modifier loci with environment-specific expression.' . . . By any generally accepted meaning, both of those phrases describe the same phenomenon" (p. 371). Schlichting and Pigliucci (1993) defined "plasticity genes as regulatory loci that exert environmentally dependent control over structural gene expression and thus produce a plastic response" (p. 366). This definition excluded allelic sensitivity as a mechanism for plasticity but was more restrictive than Via's definition (quoted above). Schlichting and Pigliucci summarized a variety of evidence for the existence of such regulatory loci. They proposed a two-stage process for the evolution of NoRs: (1) selection exploits the available allelic variation to move the trait means towards the optimum as much as genetic and other constraints permitted, and (2) further adjustments of the NoR occur through a modification of the underlying genetic architecture, including the regulatory loci. Finally, they argued that genetic systems with regulatory loci would not produce the Gaussian (normal) phenotypic distributions that quantitative genetic models assumed, and that the necessarily epistatic interactions between regulatory and structural loci could not be incorporated into those models (which only allowed additive variation). Schlichting and Pigliucci (1995) further develop these arguments.

Via accepted Schlichting and Pigliucci's (1993) focus on regulatory genes but argued:

> The environment-specific control of phenotypic expression by regulatory genes is generally compatible with quantitative genetic models for three reasons: (1) quantitative genetic models are not meant to be mechanistic; they are blind to the precise genetic mechanisms that generate the patterns of genetic variance and covariance among characters, (2) regulatory gene systems do not necessarily cause violations of the normality assumptions of the quantitative genetic models, and (3) epistatic gene action between regulatory loci and the trait loci does not invalidate the reliance of the models on additive genetic (co)variance. (p. 374)

However (2) is only strictly valid if a large number of trait and regulatory loci are involved. Reason (3) is even more uncertain. Via could only argue that "epistatic gene action contributes heavily to additive genetic variance" (p. 376). She acknowledged that explicating the detailed genetics of NoR evolution was necessary and that this may well require different kinds of models than what quantitative genetics provided (p. 377).

The dispute had by now become a methodological one about two modeling strategies: (1) a theoretical top-down one that deployed the formidable analytic machinery of quantitative genetics without incorporating the messy contextual detail of the specific genetic basis of plasticity of any trait, and (2) a phenomenological bottom-up one that began with molecular detail. That the phenomenological strategy has promise will be emphasized in the next section. Whether the top-down strategy is still valuable remains more questionable: it depends on whether quantitative genetics is a good approximation for the locus-specific genetics of plasticity—in philosophical terms, whether it supervenes successfully over the molecular level.

Several of the participants in this dispute, after several days of lively discussion at the 1992 meeting on evolution at the University of California Berkeley, agreed to jointly publish "Adaptive Phenotypic Plasticity: Consensus and Controversy" (Via et al. 1995). After discussing a variety of empirical work that remained to be done, they noted the following theoretical problems:

- How do epistatic variance and epistatic gene action affect the long-term evolutionary response to selection on reaction norms? How robust are predictions from current models that are based on additive genetic variance?
- Can explicit genetic models . . . be formulated that will complement the genetically less-detailed quantitative genetic models, and can we learn enough about plasticity to adequately estimate the parameters of such models?
- Do we need models that specifically address the evolution of regulatory mechanisms?

- Can we improve the extent to which the character state and polynomial models can be interchanged for continuous environments? . . . When the approaches are not interchangeable, will it be useful to apply both approaches? (p. 217)

As should perhaps be expected of a consensus document, the top-down and bottom-up strategies receive equal time in this list of questions.

The Molecular Era

Starting in the 1940s and continuing to this day, biology—and particularly genetics—has been transformed by what is justly regarded as the molecular revolution. Molecular characterization has not only revealed the mechanisms of inheritance. It has also showed how genes, by encoding specific proteins, and by being subject to various subcellular regulatory regimes, play a critical role in phenogenesis. It is therefore not unreasonable to expect that elucidating the mechanisms by which plasticity is brought about may provide novel insights into the etiology of plasticity (and NoRs in general). In particular, molecular genetic data may resolve the controversy over the evolutionary etiology of NoRs that was discussed in the preceding section. Certainly, Schlichting and Pigliucci's (1993, 1995) claim that regulatory loci can account for patterns of plasticity can only be tested at the molecular level. The bottom-up strategy is likely to yield the type of evidence that can decide whether epistasis or pleiotropy is responsible for plasticity.

Surprisingly little is known about the molecular basis for plasticity and related phenomena such as canalization (see chapter 3). This reflects the fact that, despite decades of effort and some success, molecular developmental biology is yet to emerge from a prolonged infancy. In the case of plasticity, three factors have contributed to this situation.

First, by definition, plasticity involves sensitivity and variable response to environmental differences. Because scientific results—by any reasonable epistemological criteria—must be repeatable, experimental systems are typically designed to minimize variability. Although systematic historical exploration of these issues remains a project for the future, there is good reason to believe that variability was intentionally removed during the establishment of model organisms for research, including *Drosophila melanogaster*, *Saccharomyces cerevisiae*, and *Caenorhabditis elegans* (Kohler 1994). It is hardly surprising that the exploration of the molecular biology of any of these organisms has shed little light on plasticity.

Second, as part of the dominance of genetics in twentieth-century biology, strains of model organisms that became prevalent in laboratory systems were selected by screening at the genetic level. One very likely result of this practice is selection against genes that would confer variability of any kind, including plasticity, and the creation of lines with minimal underlying genetic variability. This is not only an inadvertent result of laboratory practice. The National Research Council explicitly recommended that model systems be selected to ensure "genetic uniformity of organisms," "generalizability of the results," and "ease of experimental manipulation" (NRC 1985, p. 73).

Third, most—but not all—of the model organisms that dominated biomedical research in the twentieth century at the molecular level were microorganisms (*Escherichia coli*, *Saccharomyces cerevisiae*, etc.) or animals (*Drosophila* species, *Caenorhabditis elegans*, etc.). Plasticity, on the other hand, is more characteristic of plants (Sultan 1992). Consequently, it is not particularly surprising that little molecular work on plasticity has been done.

Much of what is believed about plasticity at the molecular level remains allusive. Smith (1990) speculated that differential expression of individual members of multigene families, each under the control of its own regulatory element, provides the molecular basis for plasticity in plants. Each regulatory element is supposed to respond to a different environmental signal. Consequently, the same genotype would show variable but predictable phenotypic responses in different environments. For instance, barley (*Hordeum vulgare*) has three alcohol dehydrogenase loci, *Adh1*, *Adh2*, and *Adh3*. As in maize (*Zea mays*), *Adh1* is expressed in seed and pollen; *Adh1* and *Adh2* are induced in root and other tissues by anaerobiosis (Freeling and Bennett 1985; Bailey-Serres et al. 1988). *Adh3* is induced by extreme anoxia (Harberd and Edwards 1983; Hanson and Brown 1984). Smith (1990) argued that "[a]cquisition of different regulatory elements of the three loci presumably underlies the differential expression in response to environmental, tissue-specific and developmental signals" (p. 589). Regulatory elements of this sort are perfect candidates for Schlichting and Pigliucci's (1993, 1995) plasticity genes. However, it remains hard to see how Smith's scheme would incorporate the type of plasticity that consists of a graded response to a continuously changing environmental variable (as depicted by a continuous NoR). Nevertheless, Smith's model remains the only innovative use of molecular data to tackle the problem of plasticity.

Six years later, Pigliucci (1996a) endorsed the model, but little further progress had been made toward establishing the molecular bases for plasticity. However, in the late 1990s, Pigliucci and co-workers began work on plasticity in *Arabidopsis thaliana*, initially to study the evolution of plasticity in an explicit phylogenetic framework (see, e.g., Pigliucci and Schlichting 1995; Pigliucci et al. 1999), but eventually developing a model system to study light-sensitive plasticity genes at the molecular level (Callahan et al. 1999; Schmitt et al. 1999). Isogenic lines of *A. thaliana* (a wild type and seven photomorphogenic mutants) were exposed to different regimes of light intensity and water availability. Altered plasticity was found for sensitivity to both blue and red to far-red light. The results are consistent with models of coaction between these two distinct photoreceptor systems that have been characterized at the molecular level. The molecular mechanisms of plasticity remain to be elucidated, but this is a promising beginning.

At the molecular level, canalization has fared no better than plasticity. Wilkins (1997) implicates gene families again. Members of gene families are expected to have similar functions and thus be "paralogues." Each of the members is supposed to be a backup for any of the others and assume that function should the original member become deactivated in some way. This story is less than convincing as a general account for canalization that seems to require identical function of different "paralogues" in a wide variety of environments. Nevertheless, whether it be canalization or plasticity, explicating the details of molecular mechanisms seems to be the only promising strategy, not only for resolving the disputes discussed in the preceding section but also for progress in understanding the evolutionary etiology of plasticity and NoRs.

Acknowledgments For hospitality and support during the period when this chapter was written, I thank the Max-Planck-Institut für Wissenschaftsgeschichte in Berlin. For comments on an earlier draft of this chapter, I thank Thom DeWitt, Rees Kassen, and Sam Scheiner.

3

Genetics and Mechanics of Plasticity

JACK J. WINDIG
CAROLIEN G. F. DE KOVEL
GERDIEN DE JONG

All organisms function in variable environments. Each organism is plastic in at least some traits—typically in many traits—and almost all traits are plastic in at least some organisms. Plasticity is thus pervasive and exists in many different forms. The only thing that all plasticities have in common is that the phenotype is induced by the environment; in other words, the environment acts as a cue to form the phenotype. Two different processes, described by Smith-Gill (1983), may be involved in plasticity. *Phenotypic modulation* is the simple case. Here the environment directly influences the phenotype without active involvement of the organism, for example, when the environment simply forces a different phenotype by the laws of physics. In *developmental conversion*, the organism is actively involved in the production of the phenotype. Specific receptors may be present to perceive the state of the environment, and a complex "machinery" may produce dramatically different phenotypes. Although these two processes underlying phenotypic plasticity are fundamentally different, plastic phenotypes are, more often than not, shaped by both processes simultaneously. How the evolution of plasticity proceeds has been the subject of considerable debate (see chapter 2). In order to understand the evolution of plasticity, we need to understand the genetics of plasticity. Moreover, we need to identify what similarities and differences exist in the mechanisms of different plasticities, before we can generalize on the genetics of plasticity and its analysis.

Genetics

Given the enormous variety that exists in plastic responses, we expect the underlying genetic systems involved in regulating plasticity will also show enormous variation.

Understanding of the genetic system is further complicated by the many ways in which the genetics of plasticity have been analyzed. We discuss two questions that have dominated the debate on evolution and genetics of plasticity: (1) whether plasticity is a target or a by-product of natural selection, and (2) whether there are genes for plasticity *per se*. First, however, we briefly outline how the genetics of plasticity can be analyzed, and whether genetic variation in plasticity exists.

Analysis of the Genetics of Plasticity

Once it has been established that phenotypic plasticity is present for some trait, for example, by common garden experiments, the next question is whether there is an interaction between *genotype* and *environment* ($G{\times}E$ interaction), that is, whether all genotypes show the same response to environmental variation or not. If $G{\times}E$ interaction is present, selection for a specific relation between phenotype and environment is possible. The genotypic response to systematic environmental variation is called the *reaction norm* (Woltereck 1909). Split "family" designs are often used to get information on $G{\times}E$ interactions. Here individuals from "families" (consisting of full sibs, half sibs, clones, iso-female lines, or any other group of related individuals) are raised in different environments. By plotting the family mean trait values in each environment and connecting these by lines, a "bundle of reaction norms" (Van Noordwijk 1989) can be displayed (figure 3.1). These bundles are generally quite informative because they provide information on the difference in phenotype between environments and on the form of the reaction norms (linear, quadratic, or other; see also chapter 4). They also give some indication of whether genetic variation exists for trait values within environments, and on the form of the reaction norms themselves.

An analysis of variance (ANOVA) with "families" and environments as factors is a test of whether there is an overall effect of the environment and whether families differ on average. Significant family × environment interaction indicates that differences in family means across environments are not the same for all families. This can be due to genotypes having different reactions ($G{\times}E$ interaction). However, both bundles of reaction norms and ANOVAs should be interpreted with care because family means are influenced by residual (i.e., microenvironmental) variation. A difference between family means may thus be due to residual variance and not genetic differences. In a simulation study of a large half sib–full sib design (offspring of 100 males mated with two dams), the ANOVA indicated significant $G{\times}E$ interactions in 29.4% of the cases, despite parallel reaction norms (Windig 1997). Therefore, a significant interaction between family and environment cannot be taken as evidence of genetic variation for plasticity (Fry 1992).

A more sophisticated quantitative genetic analysis is needed to quantify the different contributions of additive, dominance, residual, and environmental effects. This information gives insight in the underlying genetic system and shows what part of the variation is available for selection. There are two main classes of analysis: one using genetic covariances across environments (Falconer 1952; Via and Lande 1985; Via 1987) and one based upon a function or polynomial approach (de Jong 1989, 1990a,b, 1995; Gavrilets and Scheiner 1993a,b), similar to sensitivity analysis often used in plant breeding (Falconer 1990). The first approach is based on the idea that one character in two environments can be seen as two separate characters between which a genetic covari-

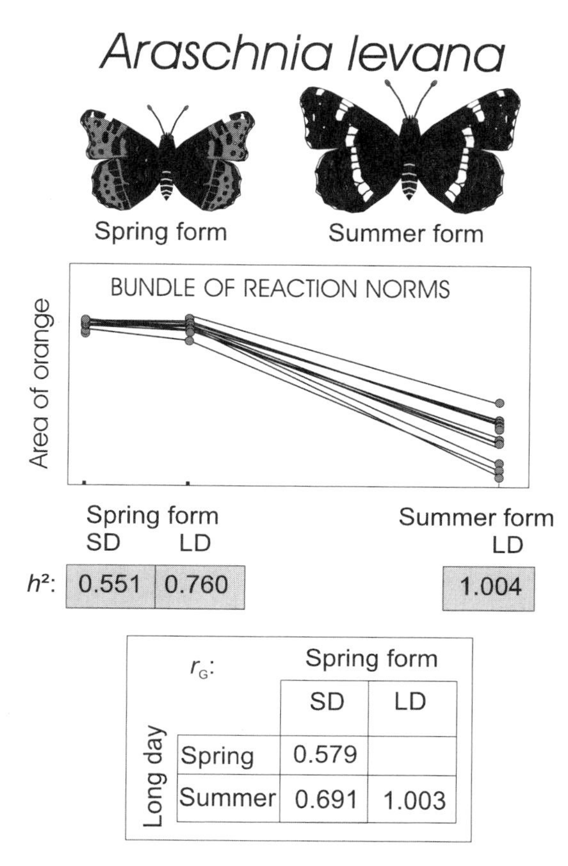

Figure 3.1. Example of a genetic analysis of plasticity. The trait measured is the relative amount of orange in wings of the males of the seasonally polyphenic map butterfly *Araschnia levana*. Full-sib families were split between long day (LD, 16 hr light) and short day (SD, 12 hr light). Spring forms were produced both in LD and SD; summer forms, only in LD. The bundle of reaction norms consists of lines connecting family means. Heritabilities are broad-sense heritabilities calculated with a resetricted maximum likelihood (REML) analysis from a full-sib design (dark orange indicates significant difference from 0, $P < 0.01$). Note the correspondence of the width of the bundle and the value of the h^2. r_G values are genetic correlations across environments. Note that the low amount of crossing of reaction norms results in relatively high r_G values (all correlations were not significantly different from 1 at the 5% level).

ance or genetic correlation can be calculated (Falconer 1952). These genetic correlations across environments (r_G) can be calculated by making use of the variance components from the family × environment ANOVA mentioned above or one of many other methods (reviewed in Windig 1997). The use of r_G values is known as the *character state approach*. An alternative, *function approach* is based on the idea that a plastic phenotype can be seen as a function of the environment. In this approach, one typically estimates genetic parameters for a function describing continuous reaction norms. One can, for example, estimate the heritability for the slope in case of linear reaction norms, as an estimate of the heritability of plasticity (Scheiner and Lyman 1989). Character state and function approaches are mathematically related (overview in de Jong 1999) and are simply different representations of the same variation (see figure 3.2 for a simulated example). Which approach to use is, biologically speaking, a formality of minor interest. The choice is best guided by the type of plasticity and the questions analyzed.

Genetic correlations across environments are more suited for cases when the environment or the phenotype is of a discrete nature (de Jong 1995; Via et al. 1995). A typical question that is best answered by a character approach is whether evolution can proceed independently in two environments. Examples are the performance of herbivores on different host plants (Carrière and Roitberg 1995; MacKenzie 1996),

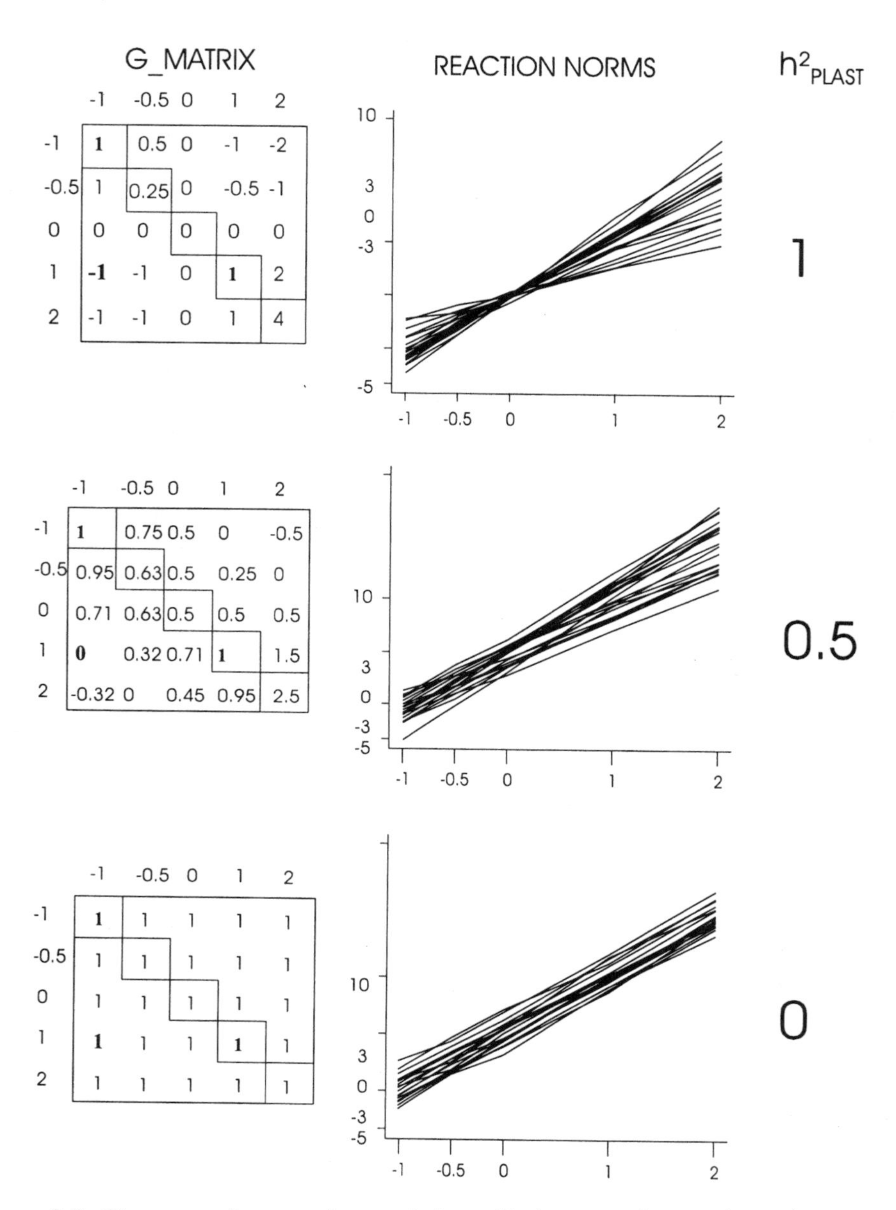

Figure 3.2. Three ways of representing populations with, from top to bottom, decreasing amounts of genetic variation in plasticity. (Left) Character approach: genetic correlations across environments. On the diagonal the additive variances are given, above the diagonal the covariances, below the genetic correlations across environments. (Middle) Bundles of reaction norms. (Right) Function approach: additive variance of slopes. We started with environment 1 and –1 with an additive variance of 1, and a genetic correlation as specified. G-matrices and additive variance of slopes are based on equations from Van Tienderen and Koelewijn (1994). Bundles of reaction norms are from a simulation study, with means of –3 and +3 in environments 1 and –1.

or the wing pattern of the European map butterfly, which occurs in two discrete forms (figure 3.1).

The function approach is better suited for cases where the environment changes on a continuous scale and where two individuals seldom experience exactly the same environment (e.g., Weis and Gorman 1990). A typical question that is best addressed by a function approach is whether a response to an environmental factor can change by natural selection if the relationship between environment and optimal phenotype changes. The relation between day length and season, for example, changes from higher to lower latitudes. Indeed, in the butterfly *Lasiommata petropolitana*, French and Swedish butterflies have different responses of development time to day length (Nylin et al. 1996; Gotthard 1998).

Often both approaches can be applied, each giving different insights. In the study of herbivore performance on host plants, r_G can give information on how easily separate phenotypes can evolve on different plants. A function approach might be applied when food with different concentrations of some compound (e.g., nitrogen or toxic alkaloids) is offered. Such an experiment might give insight as to the optimal concentrations of the compound and advance the understanding of the preference for different host plant species. It will also show whether genetic variation for host plant use is determined by different composition of host plants or by some other aspect.

Genetic Variation for Plasticity Is Prevalent

There are many examples of $G{\times}E$ interactions. Most studies that look for $G{\times}E$ interactions find them. A small selection of examples of $G{\times}E$ interaction includes the following: *Plantago lanceolata* from two populations showed $G{\times}E$ interactions in response to red:far-red light ratios in the leaf number and petiole length (Van Hinsberg 1996). Different genotypes of apomictically reproducing *Taraxacum officinale* in one population had their largest size and highest growth rates in different seasons (Vavrek et al. 1997). Sultan and Bazzaz (1993a,b) found $G{\times}E$ interactions within two *Polygonum persicaria* populations for a number of morphological traits with respect to soil moisture and light intensity; traits with an assumed high correlation with fitness, such as total fruit production, showed little variation in plasticity within the population. $G{\times}E$ interaction variance was found for populations of *Drosophila melanogaster* in experiments on the influence of larval temperature on size (body size: Noach et al., 1996; ovarian size: Delpuech et al., 1995). Not all *D. melanogaster* populations, however, show these $G{\times}E$ interactions (Noach et al. 1996). Interestingly, in these *Drosophila* studies the relationship between the correlation with fitness and the amount of $G{\times}E$ interaction was opposite to the *Polygonum* study mentioned above. $G{\times}E$ interaction variance within populations was significant for characters that are likely to correlate with fitness, such as ovarian size and body size. In contrast, characters that probably have no relationship with fitness, such as the distance between some specified chaetae, did not show significant $G{\times}E$ interaction variance.

$G{\times}E$ interaction is highly visible in situations where in one environment scarcely any genetic variation exists and in another environment genetic variation is high. Such a situation is found for traits relating to body size in the great tit, *Parus major*, where heritability is low in poor environments and high in rich environments (Van Noordwijk et al. 1988). In the butterfly *Lasiommata petropolitana*, genetic variation for larval development time is high at a day length of 16 hours compared with shorter day lengths (Gotthard 1998). In such situations, *G*×*E* interaction is caused by a change in genetic variance, whereas crossing reaction norms is another, and fundamentally different, cause of *G*×*E* interaction.

Studies looking for genetic variation for plasticity usually find significant amounts. Most research has been carried out using genetic correlations across environments (r_G values; Falconer 1952). Milk production in cows all over the world is a typical example. It has been evaluated using r_G values across countries: r_G with production in the United Kingdom varies from 0.82 (New Zealand) to 0.95 (Ireland). Generally, the closer two countries are geographically, the higher the r_G (Simm 1998). Studies on other organisms and traits invariably find that most r_G values are positive (e.g., Andersson and Shaw 1994; Rhen and Lang 1994; Carrière and Roitberg 1995; Guntrip and Sibly 1998), although negative r_G values do occur (e.g., Windig 1994a). Only a few of these studies (e.g., Shaw and Platenkamp 1993) test for a difference with the r_G of 1, which would indicate genetic variation for plasticity. Research using the function approach is much scarcer. Examples are gall size in the *Eurosta–Solidago* system (Weis and Gorman 1990), life history traits in *Daphnia* (Ebert et al. 1993), wing pattern in *Bicyclus* butterflies (Windig 1994b), and wing and thorax size in *Drosophila melanogaster* (Scheiner and Lyman 1989; De Moed et al. 1997a,b). Most prominent in this respect is the work of David and co-workers on various traits in *D. melanogaster* (see chapter 4).

Plasticity Genes: What Are They and Do They Exist?

Because genetic differences for responses to the environment are universal, genes that influence such responses must exist. There has been considerable debate on what kinds of genes induce differences in the response to environmental change and whether "genes for plasticity" exist (chapter 2). Bradshaw (1965) was the first to indicate that a response may have its own genetic control: "[S]uch differences (in phenotypic plasticity between species) are difficult to explain unless it is assumed that the plasticity of a character is independent of that character and under its own specific genetic control" (p. 119). Scheiner and Lyman (1989) made this view more explicit by talking about genes that influence the mean of a trait and genes that influence the response. Via (1993) does not agree with this view and states that plasticity genes ("loci [that] influence the degree of phenotypic change across environments per se, regardless of the values of the traits within environments," p. 361; figure 3.3A) do not exist. She argued that there is no evidence for such genes, which have to be different from genes with loci that are expressed in only a single environment, or loci that have variable allelic effects across environments. Scheiner (1993a,b) defines plasticity genes as "loci that determine the shape of the reaction norm function" (p. 371), Schlichting and Pigliucci (1993) restricted the definition of genes for plasticity to regulatory genes (figure 3.3B), and both argue that there is overwhelming evidence for such genes.

Genes that have variable expression across an environmental range can be modeled by a function that is a polynomial of the environmental values characterizing the environments (de Jong 1989; Gavrilets and Scheiner 1993a,b). The polynomial coefficients decide the way the reaction norm curves and its position on the environmental axis. In this function model for reaction norms, the coefficients are governed by loci (figure 3.3C). A specific case of polynomial functions are "tolerance curves," where genes influenc-

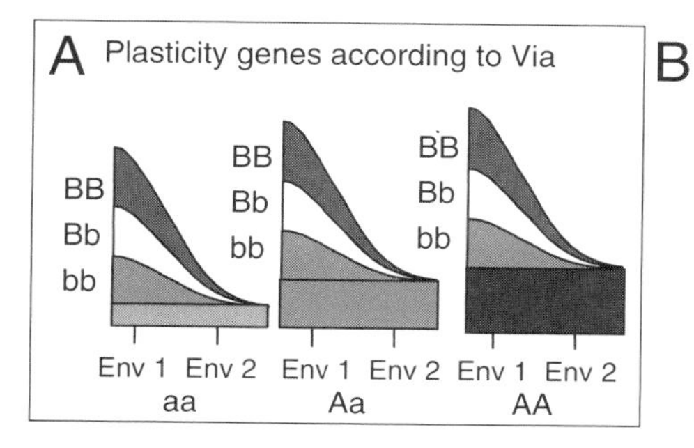

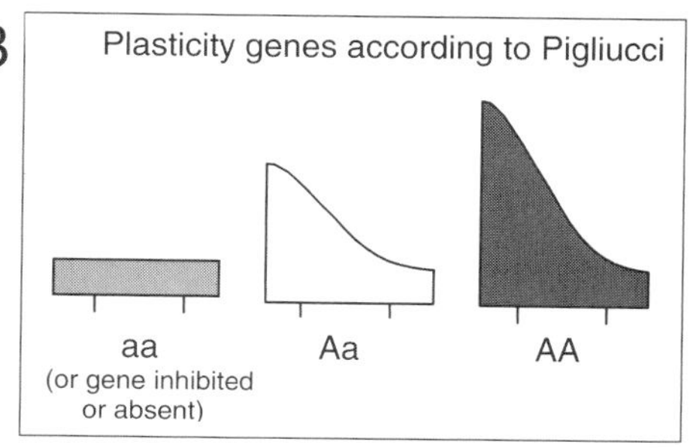

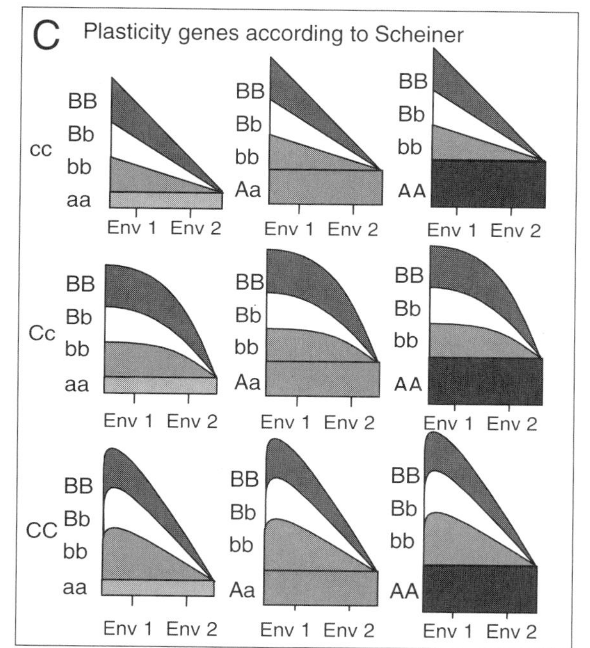

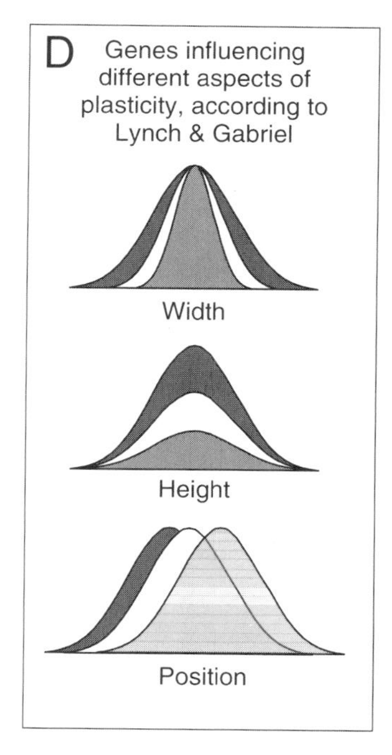

Figure 3.3. Four ideas on plasticity genes. (A) According to Via (1993a,b), plasticity genes have to change the response (slopes) independently of the means and do not exist. Gene A is not a plasticity gene, it influences the means. Gene B is a plasticity gene that influences the response irrespective of the mean. (B) According to Schlichting and Pigliucci (1993), plasticity genes are regulatory genes. Here, the phenotype does not change if the regulatory gene is absent, its product inhibited, or homozygous recessive (aa). If one or two dominant alleles are present the phenotype is enlarged in environment 1. (C) According to Scheiner (1993a,b), there are separate genes influencing the overall mean (gene A), the slope (gene B), and the curve (gene C). (D) Gabriel and Lynch (1992) modeled genes influencing the width, height, and position along the environmental axis of normal curves. Note that there is a large overlap in these ideas. For example, the plasticity gene in A is the same as the right half of the "height gene" in D.

ing plasticity are modeled as normal curves (Lynch and Gabriel 1987; Gabriel and Lynch 1992), which can have loci that vary the width, height, and/or position along the environmental axis (figure 3.3D). Some of these genes will be plasticity genes, as defined by Via (1993a,b); others probably will not, but it is difficult to draw the line. Via's (1993a,b) definition of plasticity genes is a rather unfortunate restriction, because many genes will influence both response and values of the traits within environments. This is obvious as soon as one considers plasticity of a continuous trait such as body size. Genes that appear to have alleles that are "expressed in only one environment or that have variable effects across environments" (Via 1993a, p.361), if examined in two environments, might have alleles with normal curves of expression differing in the position along the environmental axis (figure 3.4).

The difference between these definitions is small (figure 3.3). It is not fruitful to discuss what exactly a plasticity gene is and whether they exist or not. The study of the genetic control itself is the best way to answer questions about the role of different genes in the evolution of plasticity. It will probably reveal more variation than genes that simply fit into the two categories of plasticity and nonplasticity genes. In this respect, it will be interesting to connect the approach of David and co-workers of characteristic values (chapter 4) to the study of the underlying genes.

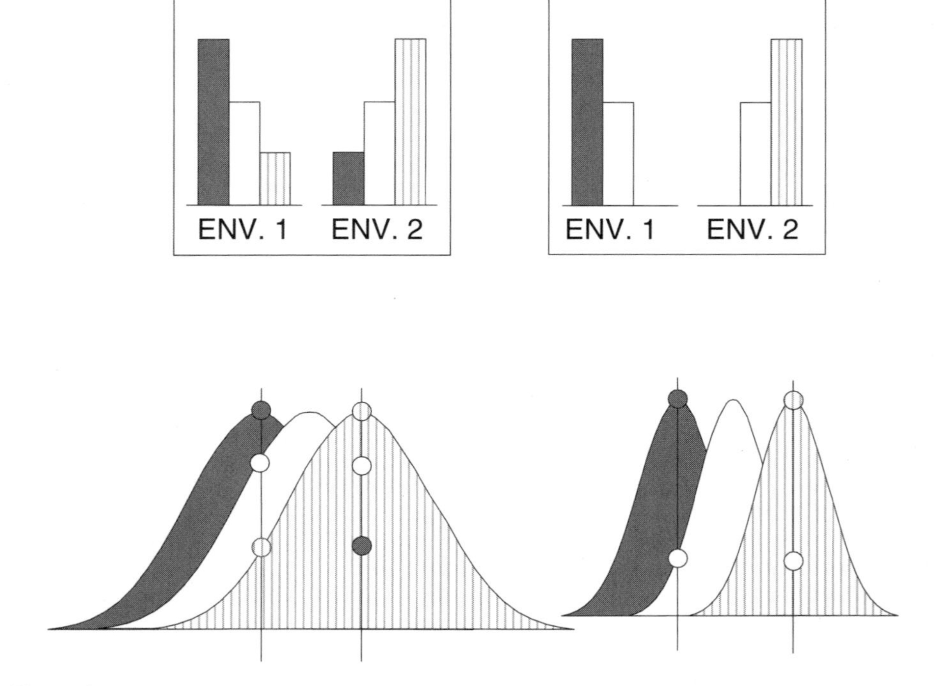

Figure 3.4. Genes may seem to have alleles that differ in environmental sensitivity (top left) or may seem to be expressed in a single environment (top right). In both cases, however, the observed alleles may be points on a continuous scale of genes that have alleles that differ in their point of optimal expression (bottom).

Plasticity By-product or Target of Selection?

Whether plasticity genes exist was part of the debate on whether plasticity is a by-product of selection or the target (chapter 2). Via (1993 a,b) argued that selection does not act on plasticity itself but on trait values within environments: "I have not read, nor have I been able to think of a plausible biological rationale for selection for increased plasticity . . . different from selection on trait values" (1993a, p. 360). We can think of two scenarios in which direct selection for plasticity will occur and where selection within environments is not sufficient.

First, if one factor influences the phenotype on a grand scale, the *response* to a second factor may be adaptive regardless of the level of the first factor. The darkening of the skin in humans in response to sunlight may be such an example. A dark skin is optimal in tropical climates; a pale skin, in colder climates. Yet, regardless of the color of the skin, darkening of it in response to sunlight seems always adaptive and occurs in all humans. Genes that code for such a response, regardless of skin color, are likely to spread much faster than genes that invoke a color change for only one specific skin color.

Second, if the optimal phenotype changes as a continuous function of the environment, and if the environmental factor can have any value within a certain range, selection is also more probable for plasticity than for trait means within environments. In such cases, individuals that are adapted to two or three environments may be at a disadvantage relative to individuals that possess a continuous response adapted to the whole range of possible environments. Even if an environmental value never occurred before, and therefore selection has never occurred in that environment, individuals that possess genotypes that express a phenotype according to a function that describes the optimal phenotype will produce the optimal phenotype.

An example of the latter may be animals where the optimal body size depends on the physics of temperature. Body size in *Drosophila melanogaster* is temperature sensitive. It is a phenotypically plastic character that readily responds to selection, and to selection on plasticity (Scheiner and Lyman 1989). Anyone working with *Drosophila* will recognize that it is more biological to describe any aspect of body size as a function of temperature than as a series of character states. Therefore, even if selection acts at specific temperatures, it is genetic variation in the function that is under selection. It is, however, probably better to try to understand the mechanics of plasticity, and how these are influenced by genes, before debating how evolution operates on plastic traits and what plasticity genes are.

Mechanics

It is clear that (genetic) variation in plasticity is generally present, but it is not clear what causes this variation—how many genes and what kinds of genes are involved, and whether the variation is generated in a particular stage of phenotype development or over all stages. Plasticity can be seen as a chain of steps going from a particular environmental condition to a particular phenotype. The chain starts with a cue indicating the environment. This cue may lead to another, secondary cue. The cue can be detected by a receptor that generates a signal, which in turn may lead to a second or even third signal. This signal may be transported and stored, after which it has to be read and translated into a process in which the phenotype is formed. In many examples of phenotypic plasticity, one or more of these steps are absent. In the case of pure phenotypic modulation, only the first step, environmental cue, and the last step, phenotype formation, are present.

A few examples of the steps involved in phenotypic plasticity are tabulated in table 3.1. When we were compiling this table, it became clear that there are few cases of phenotypic plasticity for which sufficient data are available on its mechanics. Table 3.1 probably also reflects our ignorance, because the mechanics of many more examples of phenotypic plasticity may be known. Steps in the chain have been published in isolation for many cases. In some cases, steps have been analyzed in different species, and we had to assume that the results are general for all species in the group displaying the plasticity. Often the details of the mechanics do not fit exactly into the scheme as presented here. This is a reflection of the enormous variation that exists in plasticity and, consequently, in its mechanics. Table 3.1 can be seen as an approximation highlighting the similarities in quite different processes but ignoring interesting details. Some of these details are described more comprehensively below.

The Environmental Cue

The environmental cue is the first, very essential part of the chain leading to the phenotype. Without a cue preceding the development of the phenotype, no plasticity can evolve (Bradshaw 1973; Shapiro 1976). The adaptive value of plasticity depends on how accurately the cue predicts the future environment. In the ground cricket *Allenobius socius*, populations can be found with different ovipositor lengths (Mousseau and Roff 1995). Crickets with longer ovipositors have a demonstrated selective advantage in colder environments, because they can oviposit their eggs deeper. Differences in ovipositor length are generally genetic because there are no reliable cues predicting the harshness of the coming winter. In bivoltine populations, however, plasticity is involved because day length and temperature are reliable cues predicting the coming season. In autumnlike conditions, ovipositors are formed that are longer than those formed in spring conditions.

If the change in phenotype occurs rapidly, the cue can be the same environmental variable to which the phenotype is adapted. Plant roots, for instance, proliferate locally in response to locally higher concentrations of nitrogen and phosphorus (e.g., Drew 1975; Zhang and Forde 1998). Juveniles of *Daphnia* develop into morphs with defensive neck spines when exposed to a water-soluble chemical, a kairomone, produced by larvae of the phantom midge *Chaoborus americanus* that predate on *Daphnia* (Parejko 1992).

Often, however, a substantial time elapses before the phenotype is formed. In the meantime, the environment may change. In these cases, the environmental cue may be different from the environmental variable to which the phenotype is adapted. Examples are tropical butterflies with wet- and dry-season forms. Although the amount of rain is the main factor determining the difference between the seasons, the different forms in the butterflies tend to be formed in response to changes in temperature (McLeod 1968; Brakefield and Larsen 1984). Temperature usually changes about a month before the end or start of the rains. Therefore, in the larval stage, some weeks before the adult is formed, temperature is a better predictor of the future environment than is humidity.

The interpretation of a cue may itself depend on the environment. In the butterfly *Lasiommata maera*, larvae growing under shorter days have higher growth rates in autumn but lower growth rates in spring. In autumn, short days indicate that the growing season is drawing to an end and that larvae have to "hurry" to reach a stage suitable for overwintering. In spring, short days indicate that summer, the season for adult flight, is still a long time away and fast growth is not needed (K. Gotthard, unpublished observations).

Sensitivity for the cue may vary depending on the age of the organism. In many animals, the sensitivity for a number of cues is limited to certain developmental stages (e.g., Kooi and Brakefield 1999). In the cabbage white butterfly *Pieris brassicae*, sensitivity for photoperiod is absent in the first larval instar and just before pupation and is strongest at the start of the last larval instar (Spieth 1995). In plants, the age of the module may affect its response. Inducibility of the defensive compound nicotine changes with leaf age in *Nicotiana* species (Baldwin 1999).

From Cue to Signal

The organism has to detect the cue in order to develop the appropriate phenotype. The detection can be done in a number of different ways. In some cases special receptors exist. For example, plants have special receptors to perceive the spectral quality of light: the phytochrome pigments. Two stable forms of these pigments exist: one with the highest light absorbance at 660 nm (the red form P_r) and one with the highest absorbance at 730 nm (the far-red form P_{fr}). Light absorption converts one form into the other, so the ratio of the two forms reflects the quality of the light spectrum. Light spectral quality contains important information, for example, the presence of neighboring plants, which indicates (future) competition. In reaction to shading, plants show a whole suite of morphological changes, often called the *shade avoidance syndrome* (Smith 1982; Schmitt and Wulff 1993). In addition to phytochrome other receptors detect light in the ultraviolet and blue region of the spectrum.

In tropical butterflies of the genus *Bicyclus*, which develop different morphs (dry-season forms and wet-season forms) in response to temperature, no receptor seems to be involved. The phenotype is better correlated with larval development time than with temperature itself (Brakefield and Reitsma 1991; Windig 1992). If development time is manipulated by other means than temperature, for example, food plant quality, the phenotype also changes (Kooi et al. 1996). The log-linear relationship between development time and phenotype is nicely explained by a model in which during each day of the larval period a fixed fraction of a hormone is synthesized (or broken down). In this model, the initial amount of the hormone determines the level of the reaction norm, and the fraction added each day determines the slope (Windig 1992). A similar mechanism is proposed for the determination of photoperiod in temperate insects. In these models, a hormone is produced daily in the dark phase of the day, and diapause is induced when this hormone exceeds a certain threshold (Truman 1971; Vaz Nunes et al. 1991a,b).

Animals can perceive the environmental conditions in many ways and respond to them. Greenfinches (*Carduelis chloris*), for example, stop foraging when they perceive a predator and, consequently have a reduced body mass at the end of the day (Lilliendahl 1997). In many butterflies, the background color of a pupation site is perceived by prepupae through ocelli on the larval head. Subsequently, pale (often green) pupae are formed on lightly colored backgrounds, and dark pupa on dark backgrounds. In these cases, the plasticity does not depend on special receptors adapted specifically for a plastic trait.

Table 3.1. Some examples of plasticity systems where information on its mechanics is present.

Trait	Primary cue	Secondary cue	Receptor	Signal	Transport/ translation	Machinery	Phenotype
Butterfly pupae	Background color	—	Photoreceptors in larval head	Neural signal	Neural peptide (PMRF)	?	
	Dark				Low	Melanin formed	Brown
	Light				High	No melanin formed	Green
Butterfly wing patterns	Temperature/day length/food quality	Larval development time	Hormone fraction added daily?	?	Ecdysteroid release	DOPA decarboxylase	Seasonal form
	high/long/good	Short			Early	Blocked, melanin formed	Wet season form/summer form
	Low/short/bad	Long			Late	Red pigments formed	Dry season form/ spring form
Sex in reptiles	Temperature	Size at particular stage		Aromatase	Conversion of androgen into estrogen		Sex
	Low	Small		High	High estrogen		Female (reversed in turtles)
	High	Large		Low	Low		Male
Diapause/ migration in insects	Day length, general conditions	?	?	Juvenile hormone (JH)	Degradation by JH-esterases		
	short days, unfavorable			Low	High	Wing muscles	Dormant or migrating
	Increasing or long days, favorable			High	Low	Ovaries	Developing or reproducing

Plant defenses	Pathogens	Cell wall fragments (glycopeptides)	a.o. membrane-linked ACC-synthase	Ethylene and lipoxygenase	Gene transcription	Phytoalexin production	Resistance
	Absent						Low
	Present						high
Plant shoot elongation	Shading, density of plants	Light red to far red ratio	Phytochrome pigments	?	Gibberelline?		
	High	Low	P_{fr}		Active form	Cell stretching and cell division	Long
	Low	High	P_r		Inactive form	Reduced cell stretching	Short
Plant stomatal opening	Blue light		H^+ pumps	Osmotic potential		Water	Stomata
	Present	Breakdown of starch to malic acid	K^+-flux inward	Up		Influx	Open
	Absent		K^+-flux outward	Down		Efflux	Closed
Body size in *Drosophila*	Temperature					Growth	
	High					Larger cell sizes,slightly more cells	Small size
	Low					Smaller cell sizes, slightly less cells	Large size

Pupal melanization reducing factor (PMRF)

In some cases a number of different environmental cues can set the same pathway in motion. For example, the stomatal opening in plants responds to the external factors light, temperature, and O_2 and CO_2 concentration (Fricker et al. 1990).

The Signal

When the phenotype is not induced directly by the environment, an internal signal has to be generated that designates which phenotype will be formed. These can be neural signals, but most known examples are hormones. Examples of hormones involved in plasticity in plants are auxins, gibberellins, cytokinins, and brassinosteroids, which are intermediates in response to light, temperature, drought, and probably gravity. In many insects, juvenile hormone is involved in the determination of the length of the larval stage, the induction of diapause, and the formation of wings and flight muscles. Neural peptides are involved in the determination of pupal color in lepidopterans. Sex steroids regulate sex determination in reptiles. Often, the signal is formed by not just one hormone but by a whole chain of messengers. In plants, mineral ions are often involved in information transport within the cells (Fricker et al. 1990; Ward and Schroeder 1997). Calcium is involved in gravitropic bending of roots (Sievers 1990), and in the transduction of the light signals that are perceived by phytochromes (Johnson 1989; Bowler 1997) and many other responses (Malhó 1999). Stomata opening results after a blue-light–induced breakdown of starch to malic acid has increased the H^+ content of the cytosol, which in turn leads to activity of (ATP-dependent?) proton pumps. Pumping out of the H^+ produces an influx of potassium ions, and this will eventually create the necessary higher osmotic pressure that leads to water influx and stomata opening (Schroeder 1990; Salisbury and Ross 1992).

Often an enzyme converts one hormone into another, thereby starting another developmental path. If that enzyme is blocked by an inhibitor, the default path is followed. In the snapping turtle (*Chelydra serpentina*), sex depends on temperature. Gender is determined by sex steroids: individuals with high levels of estrogens develop into females, and those with high levels of androgens (e.g., testosterone), into males. Estrogens are produced from androgenic precursors by aromatase (Rhen and Lang 1994). Aromatase production has been hypothesized to be temperature dependent (Pieau et al. 1994). Blockage of aromatase inhibits development of females at otherwise female-producing temperatures. *Plantago lanceolata* forms plants with few long, upright leaves in shade. Application of an inhibitor of the hormone gibberellin to plants growing under shaded conditions produces plants with phenotypes normally produced in sunny conditions (Van Hinsberg 1997). In lepidopterans, dark, melanized pupae are formed by default. A neural peptide called the pupal melanization reducing factor induces green pupae (Starnecker 1996).

From Signal to Phenotype Formation

In many cases a long time elapses between the generation of the signal and the start of the formation of the phenotype. The phenotype may be generated at another location in the organism than the signal. This suggests that information is transported and/or stored before it is used. It is not clear, in most cases, how information is stored. Storage of information must take place, for instance, during vernalization in plants. A certain period of low temperatures is required before higher temperatures and light conditions can induce flowering. This gives the impression that the plants store information on how much cold they have already received. A proposed mechanism is that certain flowering genes are demethylated during vernalization, because the fidelity of the maintenance methylase is related to temperature (Metzger 1996).

In the European map butterfly (*Araschnia levana*), the wing pattern is determined early in the pupal stage by the release of ecdysones. However, the wing pattern is formed at the end of the pupal stage. In the case of the spring form this is up to 6 months later. High or low temperatures at the end of the pupal stage may modify the wing pattern in the direction of the other form (Reinhardt 1969), but the basic pattern is fixed. Again, it is not exactly clear how this information is stored.

Transport of information in plants from the place where the cue is perceived to where it has an effect probably involves phytohormones. For instance, the day length that induces flowering is detected in the leaves and probably transmitted to the flower primordia through a phytohormone. In grafting experiments, induced leaves caused uninduced plants to flower (Chailakhyan 1968). A special case in information transport is the gaseous phytohormone ethylene. Ethylene synthesis is often enhanced by environmental stresses, but it also plays a role in normal development. The gas can permeate freely to all cells, yet tissue sensitivity is variable. Often ethylene increases enzyme production, but the kind of enzyme depends on the tissue (Salisbury and Ross 1992).

Formation of the Phenotype

At the end of the chain, the phenotype is formed. The signal is somehow read, and the machinery that forms the phenotype is set in motion. Because phenotype is a very general notion and may mean anything from color or size to the ability to use a certain substrate, the formation of the phenotype can involve almost any kind of mechanism. This includes, of course, gene transcription. Especially at this stage, phenotypic modulation can come into play. Phenotypic modulation, however, is not restricted to the last step. All biological processes are influenced by temperature, and it is therefore likely that, for example, the production of a hormone or the transport of it to another part of the organism will be influenced by temperature outside of the control of the organism itself. It is also not the case that continuously varying traits are only caused by phenotypic modulation and that discrete plasticity can only be caused by developmental conversion. For example, larval growth rate in butterflies varies continuously and is influenced by day length. This influence, however, is interpreted differently depending on the season, latitude, and altitude in which the larvae grow. On the other hand, variation in flower color in *Hydrangea macreophylla* is almost discrete (either pink or blue), yet it is determined by soil pH and the consequent availability of aluminum in the sepals (Takeda et al. 1985).

Developmental Conversion

Often phenotypic modulation and developmental conversion both act at the same time on the same phenotype. Genes, of course, use the mechanisms that are available from physical and chemical laws. However, the mechanics of the two processes are quite different. Whereas developmental conversion has evolved and is thus likely to be adaptive, phenotypic modulation can be useless or even harmful.

The wing pattern of the European map butterfly (*Araschnia levana*) is an example of developmental conversion. The phenotype is formed from the signal generated in the preceding steps: the timing of the release of 20-hydroxy-ecdysone (20E) determines whether spring forms or summer forms emerge (Koch and Bückmann 1987). In spring forms, 20E is released late in the pupal stage and predominantly red wings are formed. In summer forms, early release of 20E results in predominantly black wings. In another butterfly (*Precis coenia*), and probably also in *Araschnia levana,* dopa decarboxylase (DDC) is involved in the formation of the red pigments (Koch and Kaufmann 1995). 20E inhibits the activity of dopa decarboxylase. The earlier 20E is supplied, the darker the wing parts become, because of conversion of tyrosine into melanin. The developmental gene *Distal-less* determines where and in what form wing pattern elements are produced, both in *Precis coenia* (Caroll et al. 1994) and in *Bicyclus anynana* (Brakefield et al. 1996), and maybe in *Araschnia levana* as well. It is unknown how exactly 20E, dopa decarboxylase, and *Distal-less* interact, or how they are related to the form-inducing photoperiod in the larval stage.

Phenotypic Modulation

The influence temperature or food availability has on the size of organisms is an example of phenotypic modulation. Food level influences age and size at maturity in *Drosophila* (Gebhardt and Stearns 1988). Ectotherms generally decrease in size if raised at higher temperatures (Atkinson 1994). The influence of temperature on size in *Drosophila* is one of the most extensively studied systems. Plasticity depends on the species, the population, and the genotype, but the patterns of the plasticity are maintained across species (Morin et al. 1996; Moreteau et al. 1997; chapter 4).

The generality of the size decrease with increasing temperature over many groups of organisms points to a direct physical explanation: temperature may directly affect enzyme activity or metabolic rates. All metabolic rates are temperature sensitive. Temperature sensitivity of metabolic rates has a long-standing biophysical description in the Arrhenius equation (Hochachka and Somero 1984), which can be used to describe the temperature sensitivity of body size (Van der Have and de Jong 1996). Such direct effects of temperature on metabolic rates easily underlie size effects of temperature, not only in ectothermic animals but also in many plants.

Even in *Drosophila* the story is more complicated. Wing size is determined by both cell size and cell number: wing size, cell size, and cell number are all phenotypically plastic (Robertson 1959; James et al. 1995). Temperature and food both influence cell size and cell number in the wing. Perhaps temperature primarily influences cell size and food cell number (Robertson 1959). De Moed et al. (1997b), however, concluded that iso-female lines showed genetic differences for their combined trajectory of cell size and cell number under variation of both temperature and food levels. De Moed et al. (1997a) found that in *Drosophila melanogaster* genetic variation in plasticity of wing size originates in genetic variation of plasticity in cell number at low wing size plasticity, and originates in genetic variation in cell size at high wing size plasticity. At high wing size plasticity, plasticity in cell number is at its maximum. At low wing size plasticity, plasticity in cell size is at its minimum. The analysis revealed that genetic variation in phenotypic plasticity of wing size was intricately regulated, even though wing size plasticity might be described as a simple effect of temperature.

Mechanics and Genetics

It is clear that genetic variation for plasticity is prevalent. From our summary of the mechanics, it may also be clear that this variation might be generated at many different points in the chain leading from environmental cue to the phenotype. Selection pressures may also act on other aspects of plasticity than solely on the character value of the resulting phenotype. Strong selection on the speed of response may exist in the case of induced defenses. In many cases, the organism must not respond to a rapidly fluctuating cue. A short spell of heat in early spring, for instance, should not induce leaf unfolding in plants or termination of diapause in insects. Therefore, a threshold could develop for the duration or the strength of the cue (see also Gabriel 1998). One can also imagine selection for different types of receptors in different environments. Determination of day length may be different in shaded and open habitats.

To our knowledge, however, there is no example of phenotypic plasticity where the genetic variation in each step is documented and where the different contributions of the steps to the total genetic variation have been evaluated. Typically, there is no information at all, or only one gene is known. This is often a regulatory gene, which is known in the form of a mutation that influences some aspect of plasticity. Sometimes, significant additive variance is documented for one component in the formation of a phenotype, such as a hormone.

Single Genes

Single genes can influence plasticity; that is, mutations in single genes can affect the reaction norm significantly, even flatten it completely. An example can be found in commercially grown plant cultivars that are unusually stocky under virtually all light conditions. Many of these genotypes arise from mutations in gibberellin synthesis. This phenotype can be caused by mutations at a number of different loci. In *Zea mays*, at least four nonallelic recessive mutants are known that all produce a similar dwarf phenotype (King 1991). Extra long and slender forms sometimes originate from a mutation in the receptors for gibberellin that under all light conditions react as in full light. Many other single-gene mutants are known that produce a phenotype with a flat or abnormal reaction norm (e.g., Koorneef and Kendrick 1986; King 1991; Pigliucci 1996a). These mutants are mostly extremes in which a certain locus is not functional at all. They may appear during breeding but, more often, are induced by mutagens. Natural variation for specific genes influencing plasticity is much harder to find.

Polygenic Influences

An example of polygenic influence on plasticity is the activity of the enzyme juvenile hormone esterase in insects. This enzyme degrades the juvenile hormone and is involved in diapause induction and the formation of wings. In the cricket *Gryllus assimilis*, the heritability of enzyme activity was 0.26 (Zera and Zhang 1995). Aestival reproduction in *Chrysoperla carnea* provides an example where both single genes and polygenic influence is involved. Aestival reproduction can be influenced both by day length and by prey availability. Responsiveness to day length is determined by alleles at two unlinked loci both in western and eastern North America but is also under polygenic control in western North America. Responsiveness to prey is under polygenic control, and it is only expressed in western North American if the two loci determining day-length responsiveness are recessive (Tauber and Tauber 1992).

Pigliucci (1996b) suggests that plasticities may generally be under control of only a few principal genes. We suggest, however, that often many genes interact to produce a plastic response to the environment. In general, the finding of one gene that influences a trait to a large extent does not mean that that trait does not have a polygenic base (Roff 1997). The finding of a gene that determines the response to an environmental cue (e.g., a phytochrome mutant), and thus influences the plasticity to a large extent, does not exclude the possibility that many other genes, further down the path leading to the phenotype may also have an influence. In fact, as noted above, mutations at different loci can induce the same phenotype (e.g., Koorneef and Kendrick 1986).

Plasticity Genes?

Although it is not clear exactly how a "plasticity gene" should be defined, they probably exist. There is no doubt that regulatory genes exist that can switch on and off developmental paths, according to the environment in which the organism develops (i.e., plasticity genes *sensu* Schlichting and Pigliucci 1993). An example that would even fit Via's definition in the strictest sense might be found in the butterfly wing patterns of *Bicyclus*, or indeed generally for diapause induction in insects. Here models are proposed (see subsection above titled From Cue to Signal) in which a constant fraction is added daily in the larval stage to the amount of a hormone. At the end of the larval stage, the amount of hormone determines which phenotype is formed. In such a model, genes that influence the initial amount will influence the mean of the phenotype across environments. Genes that influence the fraction of hormone added each day will influence the strength of the response irrespective of the mean. Such a system allows for an accurate fine-tuning of the response to the environment. If the change from one season to the other becomes more gradual, a more gradual response can evolve by a change in gene frequencies that influence the fraction of the hormone synthesized. If a population is founded in an area where the change of season is similar to the area of origin, except that it occurs at a higher temperature, a response that occurs at a different temperature level may evolve by a change in the frequencies of those genes that influence the initial amount of the hormone. There may also be genes that influence a threshold for the hormone above which a particular phenotype is formed, which is likely in the European map butterfly *Araschnia levana* (Windig 1999). Such genes do not fit very well in the concept of genes that influence either the mean or the response, nor do they fit in the idea that there are only genes that are expressed in single environments or genes that have different sensitivities across environments.

Concluding Remarks

In the last 10–15 years of the twentieth century, a lot of information on the genetics and mechanics of plasticity has been gathered, but the details of how a plastic phenotype is formed remain obscure in most cases. The presence of genetic variation on the black box level is fairly well documented. Often, the steps between the environmental cue and the formation of the phenotype are unknown. In some systems (e.g., response of plants to light/shade, and response day length of butterfly wing patterns to temperature/day length), the picture is fairly complete. Information on these and other systems will probably accumulate in the immediate future. With the recent development of molecular techniques, much can be gained (Schlichting and Pigliucci 1998). Quantitative trait locus (QTL) analysis may bring us closer to the identification of the number and type of genes involved in plasticity (e.g., Gurganus et al. 1998). However, isolated consideration of components in the metabolic chain involved in the formation of a plastic phenotype is less interesting than is integrating knowledge available on different components. Our lack of understanding of how a plastic phenotype (or any phenotype, for that matter) is formed is, probably, a reflection of the lack of cooperation between physiologists and evolutionary biologists. There is, for example, an enormous amount of medical literature that deals with the effect of hormones on, for example, growth, fattening, fertility, and so forth, and it might be rewarding to examine these in the light of phenotypic plasticity.

Models currently developed in plasticity research provide information on how selection can lead to the evolution of plasticity, given the presence of genetic (co)variation for different aspects of plasticity. These aspects of plasticity can be formulated either in terms of polynomials (Gavrilets and Scheiner 1993a,b) or in terms of characteristic values (chapter 4). The models do not tell us the origin of the genetic variation, or what causes genetic covariation or the lack of (co)variation. Nor do they show how the mechanisms responsible for plasticity can have evolved. Information on mechanics integrated with information on genetic variation for components of plasticity may provide that information.

The structure outlined in this chapter, describing plasticity as a chain of steps leading from an environmental cue to a signal that can be transported and stored before it leads to the formation of the phenotype, may help to integrate different components of the mechanics of plasticity. Combining knowledge of this chain with genetics will provide interesting insights in the evolution of plasticity. Although many examples of cues resulting in phenotypes are known, the specific target of selection is not unambiguously clear. In many cases, therefore, the adaptiveness of the reaction norm is also unknown. For example, *Drosophila*, like many other insects, grows to a larger adult body size at lower temperatures (chapter 4). It is unknown whether the actual body size at maturity, or some aspect of larval development, is the trait under selection. A combination of knowledge of physiology and of genetic variation in physiology among different populations is necessary to identify the trait under selection and its adaptive significance.

Questions about the adaptive value of plasticity and its evolution need information on the machinery in order to be answered. The presence of specific mechanisms regulating the plastic response is an indication of the adaptive significance of that response. Information regarding genetic variation among populations for different steps in the chain enables us to evaluate the possibility of independent evolution of a plasticity. Genetic variation within and between populations shows alternatives and constraints. Whether plasticity can evolve more easily in some traits than in others begs identifying the constraints to plasticity. The best answer to this question involves the mechanics of plasticity, combined with genetics. Such an answer may provide predictions on how plasticity can evolve further, and even on the role of plasticity in speciation (West-Eberhard 1989; chapter 12). A body of detailed and integrated information on specific systems will lead to a clearer picture of the way plasticity evolves in nature.

4

Evolution of Reaction Norms

JEAN R. DAVID
PATRICIA GIBERT
BRIGITTE MORETEAU

Elsewhere in this book are examples of adaptive phenotypic plasticity. Adaptive plasticity implies that the present form of a trait's plasticity was shaped by natural selection (chapter 5). That plasticity might itself be shaped by natural selection has become appreciated only slowly since about the late 1940s to the mid 1970s, at which time rapid development of the field of plasticity research began (Stearns 1992; chapter 2). Empirical work on this topic is now common. Yet controversies abound and studies may have vastly different emphases. Our purpose here is to indicate what factors are important in the evolution of phenotypic plasticity, with a specific focus on reaction norms to illustrate some evolutionary patterns in a model system, using a general climatic factor, temperature.

Plasticity can be favored over fixed responses when environments vary predictably. The best plastic response depends on the types of environments that occur, their frequency, and the fitness consequences for various phenotypes in each environment. Because different populations experience alternative patterns of environmental variability, we should expect local differentiation to be favored, as long as genetic variation is present and gene flow is restricted. Thus, the specific patterns of plasticity that evolve not only depend upon local habitats but also on the nature of genetic variation for plasticity in relevant traits and population-level patterns of dispersal.

Some specific approaches to this field include descriptive surveys of reaction norms for given populations, population comparisons of reaction norms, genetic studies of plasticity, and functional ecological studies of trait performance in alternative environments. Obviously, the complexity of this topic requires good model systems for which many of these details can be adequately fleshed out. Such systems abound, but for present purposes we will illustrate several such empirical patterns with our work on *Drosophila*.

This group offers well-known advantages, such as a short generation time and the capacity for accurate control of experimental conditions. In this chapter, we describe recent observations concerning the evolution of reaction norms of morphometric traits in that group and consider the data as a general reference that might be called a *Drosophila* paradigm. First, however, we must provide some conceptual distinctions and discuss several technical issues before we can adequately examine the evolutionary processes that generate reaction norms.

Reaction Norms: A Four-Box Model

Patterns of phenotypic plasticity can be classified based on whether the phenotypic variability is continuous or discontinuous, and whether changes are reversible or irreversible (figure 4.1). We present this conceptual classification to emphasize the unity of reaction norms. Typically, the use of the term "reaction norm" is restricted to continuous variation. And most studies of reaction norm evolution have focused on irreversible plasticity. However, models of reaction norm evolution may apply equally to discontinuous variation. Although the overt phenotype is discontinuous, it may represent an underlying continuous environmental response with a threshold response.

Box I: Discontinuous, Reversible Reaction Norms

Discontinuous, reversible reaction norms encompass a diversity of inducible biological phenomena, such as many defense mechanisms (Tollrian and Harvell 1999). Another example is enzyme adaptation in bacteria. A species may respond to the presence of a potential substrate (e.g., lactose) in the medium by producing the corresponding metabolic enzymes. The substrate amount may vary continuously, but the bacteria exhibit only two phenotypes. Enzymatic adaptation is generally considered in terms of gene induction, not phenotypic plasticity. However, a reaction norm approach might be used by considering the reactivity of the bacteria to a gradient, defining a threshold for induction, and comparing it among populations or species.

	DISCONTINUOUS	CONTINUOUS
REVERSIBLE	I Enzymatic adaptation in Bacteria	II Physiological regulations acclimation
IRREVERSIBLE	III Aphid life history social castes	IV Size variation and resources

Figure 4.1. A four-box model of phenotypic variations that may be observed according to an environmental gradient. Reaction norms may be discontinuous or continuous, reversible or irreversible.

Box II: Continuous, Reversible Reaction Norms

Continuous, reversible reaction norms are exemplified by physiological reactions. For example, O_2 consumption in ectotherms increases with temperature. Comparisons of response curves have been made among species and populations and adaptations inferred (e.g., Precht et al. 1955). Quite often, the response curves change over time as a consequence of acclimation. The reaction norm approach might be useful for describing such changes.

Box III: Discontinuous, Irreversible Reaction Norms

Discontinuous, irreversible reaction norms are numerous, especially among plants (Bradshaw 1965) and ectotherm animals. Insects offer extraordinary examples of adaptive plasticity, such as the complex life cycle of aphids or the castes (workers, soldiers, reproductives) of social species. These examples imply extremely sophisticated regulation of development, some of which are known (e.g., hormones or specific chemicals). However, models are needed for understanding their origin and evolution.

Box IV: Irreversible, Continuous Reaction Norms

Irreversible, continuous reaction norms are the ones most commonly observed. They are also the most controversial with regard to adaptiveness. All organisms, faced with a shortage of resources, will react by a slower growth rate and eventually by a reduction in adult size (Stearns 1992). Plants are especially sensitive to their abiotic environment because they cannot escape and generally are highly plastic (chapter 9). Ectothermic animals also respond in a highly plastic fashion to variation in resources and temperature.

The Adaptive Significance of Reaction Norms

For boxes I, II, and III, the adaptive significance of phenotypic change is generally considered both as evidence for and as a consequence of a history of selection. Producing a specific enzyme when the substrate is available is obviously useful. Producing the same enzyme without the substrate will be costly and will be easily replaced by an inducible system, when available. Interpretations are far more difficult for box IV, despite its general occurrence. It is quite easy to understand, from a physiological point of view, that a shortage of resources will reduce both metabolic activity and growth rate. Why such a process will also reduce size at maturity is more difficult to explain and needs a more elaborate theoretical approach (Stearns 1989, 1992; Charnov 1993).

The difficulty of adaptive interpretations is best evidenced by considering temperature effects in ectotherms. In many insect species, body size decreases with increasing temperature, due to either genetic changes (e.g., latitudinal clines) or phenotypic plasticity (see review in Atkinson 1994). Despite numerous hypotheses, however, we lack any consensus concerning an adaptive interpretation (Atkinson and Sibly 1997). The problem is yet more complex if we consider that, at least in *Drosophila*, body size exhibits a significant decrease not only at high temperatures but also at very low temperatures (David et al. 1983). We deal with the *Drosophila* case in detail below.

Further complicating the attempt to determine the adaptive significance of plasticity is deciding on the correct null hypothesis. Most past studies have assumed that the nonadaptive state is no plasticity. However, many traits show passive plasticity. For example, temperature is able to modify any biological process, either by acting directly on cell metabolism and cell division, or by modifying physiological regulation and differentiation. What is likely to be strongly selected and canalized is development, resulting in a progressive differentiation of cells, organs, body parts, and functions. The development–temperature interaction cannot be avoided and is likely to have a strong passive component. For example, pigmentation variations in *Drosophila* are best interpreted as an interaction between developmental constraints on successive segments and natural selection favoring darker phenotypes in a cold environment (Gibert et al. 2000).

Methods for Describing Reaction Norms

A lack of plausible interpretations is not the only reason for neglecting the analysis of reaction norms. The other reason is a lack of a convenient, descriptive model. When reaction norms are investigated over a broad environmental gradient, they generally turn out to be nonlinear (figure 4.2), whereas most theoretical analyses have considered, for simplicity, only two environments, that is, linear norms. A consensus has emerged recently (Scheiner 1993a; Gavrilets and Scheiner 1993a,b; Via et al. 1995) that polynomials might be a general, all-purpose, descriptive model.

The practical implementation of this method in *Drosophila* has revealed, however, several difficulties in the use of a polynomial model as first proposed. The first hurdle is the choice of the degree of the polynomial. We (Moreteau et al. 1997; David et al. 1997; Gibert et al. 1998a) suggested that the best method for determining polynomial degrees will often be to calculate an empirical derivative. Empirical derivatives are determined by calculating the slope of the reaction norm for each successive pair of environmental states (figure 4.3). If replicate genotypes or family groups are available, an analysis of variance of the slope values can be used to determine if they differ among environments, that is, whether the reaction norm is nonlinear. If it is, a graphical analysis is used to determine its overall shape. If the empirical derivative is linear, then the reaction norm is a quadratic curve. A more or less bell-shaped derivative implies a cubic norm.

A more critical problem is that the polynomial coefficients of the reaction norm are often very variable and highly correlated (Delpuech et al. 1995; David et al. 1997). The coefficients of variation among iso-female lines (a measure of genetic variation) sometimes surpass 50%. This variability increases with the order of the polynomial, complicating the analysis of quadratic or cubic norms. For example Delpuech et al. (1995), in a study of ovariole number, could not discriminate the polynomial coefficients of very different populations. In addition, the coefficients of the polynomial are not genetically independent. Rather, they are always highly correlated, with absolute values generally > 0.9 (David et al. 1997). This problem is inherent in the mathematics of polynomial coefficients. Finally, position along the environmental gradient of the intercept is arbitrary, creating problems for biological interpretations of the coefficients.

The solution to these problems is the use of characteristic values (figure 4.4; Delpuech et al. 1995; David et al. 1997; Gibert et al. 1998a). A linear reaction norm has two char-

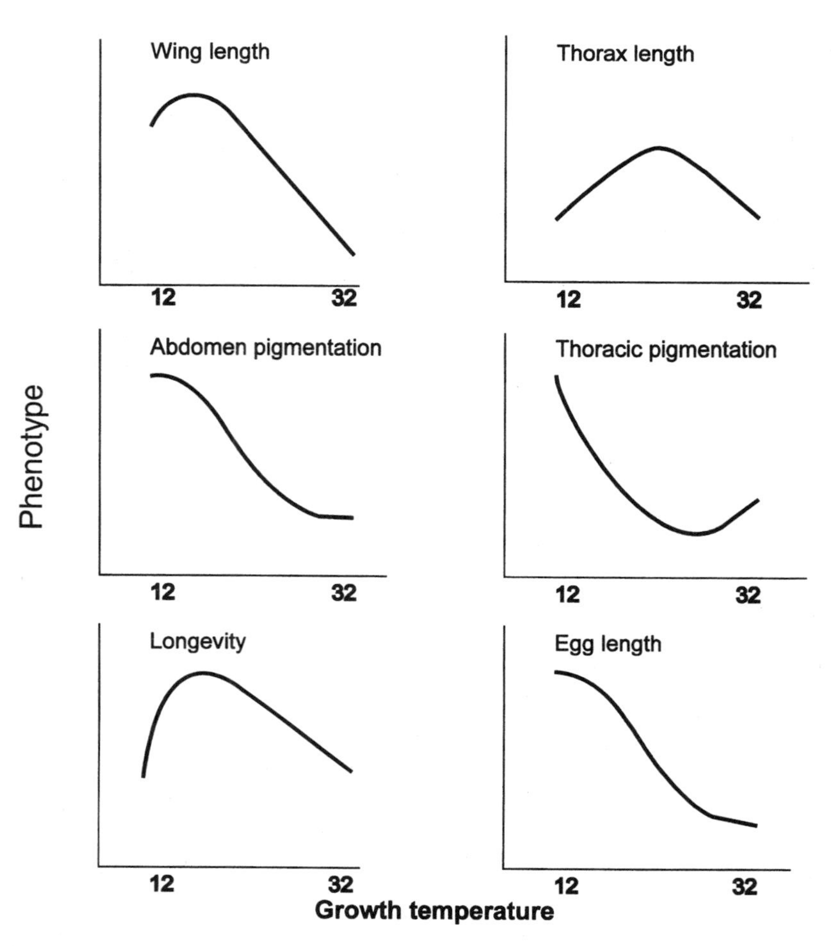

Figure 4.2. Illustration of the diversity of reaction norms observed in *Drosophila* according to growth temperature. Reaction norms correspond to box IV of figure 4.1 and are nonlinear.

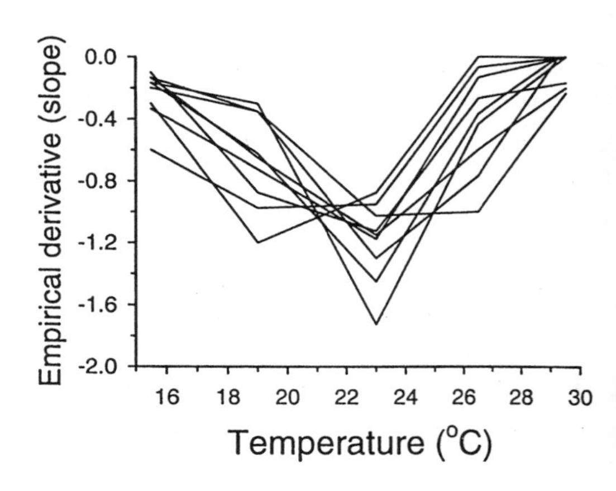

Figure 4.3. An example of empirical derivatives.

acteristic values, the slope (g_1) and the average response (P_a) at the midpoint of the environmental range (E_a), in effect standardizing the intercept to the environmental midpoint. A quadratic reaction norm has three values, (1) the environment of the minimum/maximum (E_m), (2) the phenotypic value at E_m (P_m), and (3) the curvature (g_2). Again, parameters are standardized to an environmental midpoint defined by the minimum/maximum of the curve. Two of these characteristic values (E_m and curvature) concern plasticity, whereas P_m is a trait value. For a sigmoid reaction norm, the choice of possible models increases. Based on observed reaction norms (see, e.g., figures 4.5, 4.8), we have sometimes focused on the logistic (Gibert et al. 1998a). It has four values, the slope at the inflection point (S), the environment at the inflection point (E_i), the phenotypic value at the minimum asymptote (P_m), and the phenotypic value at the maximum asymptote (P_M). Again, the curve is standardized to the environment of the inflection point.

The analysis of nonlinear reaction norms is always a compromise between mathematical adequacy and biological interpretation. For example, a sigmoidal reaction norm could be replaced by a linear reaction norm, if the asymptotes are outside the natural environmental range. The evolution of the linear reaction norm might be much easier to interpret. In another instance, an ostensibly quadratic reaction norm for wing length was better fit by a cubic polynomial function (David et al. 1994; Morin et al. 1996; Karan et al. 1999a), especially for estimating the temperature of the maximum. In this case, the other possible characteristic values, such as the inflection point, could be ignored because they were well outside the thermal range.

Selection on Reaction Norms

Characteristic values are naturally interpretable with regard to targets of selection. For linear reaction norms, selection on the average response (P_a) is selection to either raise or lower the entire reaction norm or, equivalently, to push it right or left (mathematically these two operations are the same). Selection on the slope (g_1) is selection to increase or decrease phenotypic responses in extreme environments relative to the median environment.

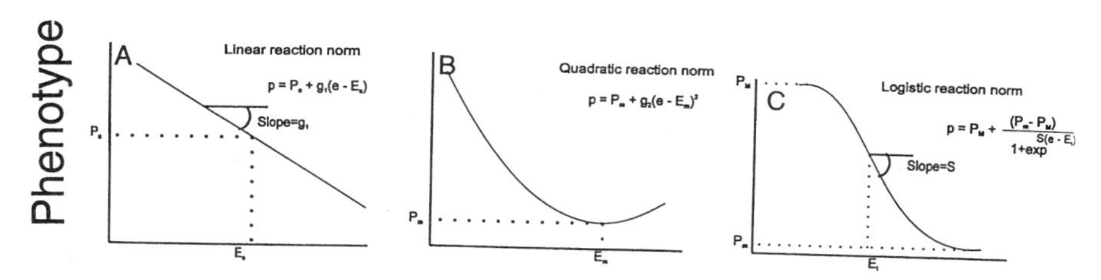

Figure 4.4. Characteristic values of reaction norms. (A) First-order (linear): P_a is the average value of the trait, E_a is the average value of the environment, and g_1 is the slope. (B) Second-order (quadratic): P_m is the minimum/maximum phenotypic value, E_m is the environment at the minimum/maximum, and g_2 is the curvature of the reaction norm. (C) Third-order (logistic): E_i is the environment at the inflection point, P_m and P_M are the phenotypic values at the minimum and maximum asymptotes, and S is the slope at the inflection point. [From Gibert et al. (1998a).]

For quadratic reaction norms, selection on the phenotypic value at the minimum/maximum (P_m) is selection to alter the range of phenotypic responses. For example, if the reaction norm is bowed downward (negative curvature), then an increase in the maximum corresponds to an absolute increase in the expression of the trait. Selection on the environment of the minimum/maximum (E_m) is selection to displace the curve to the right or left, which changes the direction of the response in the region around the minimum/maximum. Selection on the curvature of the reaction norm (g_2) is, again, selection to increase or decrease phenotypic responses in extreme environments relative to the median environment.

For logistic reaction norms, selection on the phenotypic value at the minimum or maximum (P_m, P_M) again is selection to alter the range of phenotypic responses. Selection on the location of the inflection point (E_i) is similar to selection on the average response of a linear reaction norm (P_a), pushing the reaction norm to the right/left or up/down. These are no longer equivalent changes, however, because of the asymptotic form of the curve. Finally, selection on the slope at the inflection point (S) is selection to change the phenotypic response between gradual and abrupt.

Temperature-Induced Reaction Norms in Drosophila

For *Drosophila*, temperature is a major environmental factor, explaining the geographic distribution of species, and various latitudinal clines, especially body size (David et al. 1983). In *D. melanogaster*, complete development from egg to adult is possible from 12°C to 32°C. Over this temperature range, all traits are plastic but the shapes of their reaction norms are highly variable (figure 4.2). Such diversity is difficult to interpret in adaptive terms.

Some of these curves were described more than 60 years ago (Riedel 1935), but their interpretation remained elusive. It was easier to consider them as natural curiosities, passive and contingent responses, as described above. The similar shapes observed, for example, for abdominal pigmentation and egg length (figure 4.2), are just due to chance. For these two traits, the physiological and genetic mechanisms that are modified by temperature are likely to be completely different. From this kind of reasoning, we should expect more similar norms if we compare traits related either by function or development. We illustrate this problem by considering some examples.

In *D. melanogaster*, genetically determined size is known to increase with latitude, producing bigger flies in colder environments (Capy et al. 1993, 1994). Similar clines are observed in various traits, such as wing and thorax length, body weight, and ovariole number. All of these traits exhibit concave reaction norms with a maximum within the developmental thermal range. A usual interpretation is that a maximum corresponds to an optimum (Stearns 1992; Charnov 1993; Roff 1997). So, if ambient temperature is really the selective factor, in a given population different characters should have similar or at least convergent maxima.

Experimental data clearly show that such is not the case. Maximum values are observed at 22.2°C for ovariole number (Delpuech et al. 1995) and at 19.5°C for sternopleural bristle number (Moreteau et al. 2003). Wing and thorax length arise from the development of a single imaginal disk, but their maxima occur at different temperatures: 19.1°C for thorax length and 15.6°C for wing length of females (David et al. 1994;

Morin et al. 1996). Indeed, when parts of the wing are looked at in detail, they do not react in the same way with maximum values ranging from 8.7°C to 18.8°C (Moreteau et al. 1998). Nor are the maxima for males and females the same: 17.9°C versus 19.3°C for thorax length and 14.8°C versus 15.5°C for wing length in a temperate population (Karan et al. 1999b). It is generally assumed that body weight is a better estimator of size than are wing and thorax length. A detailed analysis (Karan et al. 1998) revealed that these three traits had significantly different reaction norms.

Consider the pigmentation of the last three abdomen segments in females. These segments are determined by a single gene, *Abdominal B* (Fristrom and Fristrom 1993). The shapes of the reaction norms are clearly different for different segments (figure 4.5). Yet, we can make a single adaptive interpretation for the reaction norm: greater black pigmentation in colder environments is thermally superior (Gibert et al. 1996). Darker phenotypes do better in colder environments because they absorb more irradiance and, thus, increase body temperature. In hot environments, where thermal cooling is critical, the opposite argument holds. Although this interpretation holds for all segments, the different shapes of the reaction norms are not explainable in an adaptive way and might be the consequence of developmental constraints that specify the fate of each segment (Gibert et al. 1998a, 2000). Under ideal circumstances, a definitive demonstration of adaptation would also include field measurements of fitness. Laboratory measures in a uniform environment are limited because behaviors (e.g., habitat selection) may mean that the temperatures actually experienced by the organism deviate substantially from ambient conditions.

Genetic Variability and Correlations

For reaction norms to evolve, there must be genetic variation for their components. Numerous studies have now demonstrated the existence of genetic variation for plasticity in natural populations (Scheiner 1993a). This problem is extensively treated in chapter 3, so here we deal only with issues specific to reaction norm evolution.

Less well documented is genetic variation for parameters of reaction norm shape. In part, this lack exists because most studies are done in only two environments, so at best, genetic variation for the slope of the reaction norm can be estimated. Few studies have measured reaction norms across many environments of a single environmental gradi-

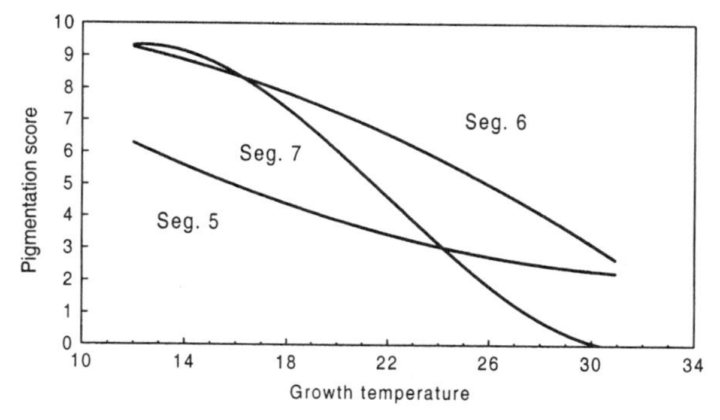

Figure 4.5. Variation of phenotypic plasticity and of the shape of reaction norms in the last three abdomen segments of female *D. melanogaster*. Segment 7 is described by a cubic norm, segment 6 by a linear norm, and segment 5 by a quadratic norm.

ent. Such studies are needed to accurately estimate quadratic, cubic, or higher-order curves. In our laboratory we have demonstrated genetic variation for reaction norm shape, including the characteristic values of quadratic, cubic, and sigmoidal reaction norms (David et al. 1994; Delpuech et al. 1995; Gibert et al. 1998b; Morin et al. 1996; Moreteau et al. 1997).

For example, we estimated the amount of genetic variation for three measures of thorax length: the trait in each environment, the maximum value (MV) of the trait, and the temperature of the maximum value (TMV) (Karan et al. 1999b, 2000). The latter two measures are characteristic values of a quadratic reaction norm. We can compare genetic variance between traits using the genetic coefficient of variation (CV_G), sometimes called evolvability (Houle 1992). For the trait in each environment, CV_G at intermediate temperatures was about 2.3% (range, 2.2–2.4). For MV, the CV_G was much less, 1.3% and 1.5% in males and females, respectively. For TMV, the CV_G was higher, 3.6% and 3.9% in males and females, respectively Thus, the genetic variability of TMV (i.e., plasticity) was relatively higher than that of the trait itself. Such a surprising conclusion suggests that natural selection in a novel environment might be more efficient in changing overall plasticity than in changing the trait value itself. More extensive investigations are needed to check the overall validity of this hypothesis.

Genetic correlations among traits can shape the trajectory of evolution. The situation becomes more complex for reaction norms because we are now faced with an additional form of correlation, that between the same trait expressed in different environments (Falconer 1952, 1990; Via and Lande 1985, 1987; Windig 1997). Or, considered from another perspective, the characteristic values of the reaction norm may be genetically correlated.

When examining genetic correlations for the same trait measured in different environments, we might expect the correlation to decrease as more distant environments are compared. This pattern was found for body pigmentation in *D. melanogaster* and *D. simulans*. (David et al. 1990; P. Gibert, unpublished observations). However, the situation was more complicated for thorax and wing length in *D. melanogaster* (Karan et al. 2000). Although the pattern held for females, for males the correlations increased again for the most different environments. This difference in correlation structure implies either that selection may be acting differently on male and female reaction norms or that the genetic bases of the reaction norm differ between the sexes. Given that the characteristic values correlate across the sexes, evolution in one may be constrained by evolution in the other.

The characteristics values of a norm (e.g., the coordinates of a maximum) are genetically determined traits. If so, patterns of cross-environmental correlations can be expressed as genetic correlations among characteristic values. For example, in our study of *D. melanogaster* MVs of wing and thorax were positively correlated, the TMVs were positively correlated, but MV and TMV were not correlated for either trait (Karan et al. 1999b). A lack of correlation between MV and TMV was also observed for ovariole number (Delpuech et al. 1995) and sternopleural bristle number (Moreteau et al. 2003). Quite surprisingly, a significant positive correlation has been found for sternopleural bristles between MV and curvature (g_2) (Moreteau et al. 2003).

A different issue is raised by considering body pigmentation. Gibert et al. (1998c, 2000) found that the genetic correlation decreased as more distant body segments were compared, showing that the genetic control of pigmentation differs among segments.

These results imply that genetic correlation among these traits will vary among environments. The generality of these results is currently unresolved. This issue is central to evolutionary studies because constancy of the genetic variance/covariance matrix is assumed in most studies.

The issue of the effects of genetic correlation on reaction norm evolution is barely touched. For example, are correlations influenced by the shape of reaction norms? Although the character-state and polynomial approaches are complementary, this linkage has not been explored. Do these patterns of genetic correlations hold among different environmental gradients? What about genetic correlations of characteristic values for the same trait across different gradients? Can the evolution of reaction norms inform more general issues of the constancy of genetic variance/covariance matrices? These are totally unexplored issues.

Differentiation among Geographic Populations

Some of the best evidence for the adaptive significance of a trait comes from comparisons among populations and species (Endler 1986). Are trait differences those one would expect if the populations were adapting to local conditions? Are parallel differences found in different species? We have used *Drosophila* extensively to address such questions.

Here we summarize work done in our laboratory comparing reaction norms of populations from different climates. First are studies that compared populations of *D. melanogaster* found in France, a temperate climate, with ones from northern India, which experiences a mild winter but a very hot and arid summer, and ones from equatorial Africa. The French and Indian populations were compared for body pigmentation (Gibert et al. 1998b). Reaction norms differed in their shape resulting in a maximal divergence at high temperatures, which mainly discriminates French and Indian climates (figure 4.6A). In contrast, a comparison of ovariole number in African and French populations (Delpuech et al. 1995) evidenced a major divergence in trait mean value but an overall parallelism of the response curves (figure 4.6B). Only a very slight difference was found: maximum ovariole number, and presumably egg production, occurred at 22.2°C in French flies and at 22.7°C in African ones. A similar conclusion was recently obtained by comparing sternopleural bristle numbers in French and African populations (Moreteau et al. 2003).

A series of studies were conducted that demonstrate evidence of parallel adaptation in two sibling species (*D. melanogaster* and *D. simulans*) reviewed in Gibert et al. 2003). Body pigmentation was compared between two French localities (Bordeaux and Villeurbanne) differing in their cold-season temperature. The reaction norms differed among localities, but only at low temperatures: plasticity was greater in the locality experiencing maximum seasonal fluctuations (figure 4.7A; Gibert et al. 1996; David et al. 1997).

In a comparison of Caribbean and French populations of the two species (Morin et al. 1999), reaction norms of wing and thorax length differed in the shapes of the reaction norms (figure 4.7B). More precisely, the position of the MV was always shifted to the right (higher temperatures) for the populations living in the warmer climate. Ample evidence exists for assuming that, in both species, tropical populations are ancestral and temperate populations are derived (David and Capy 1988). In other words, during their

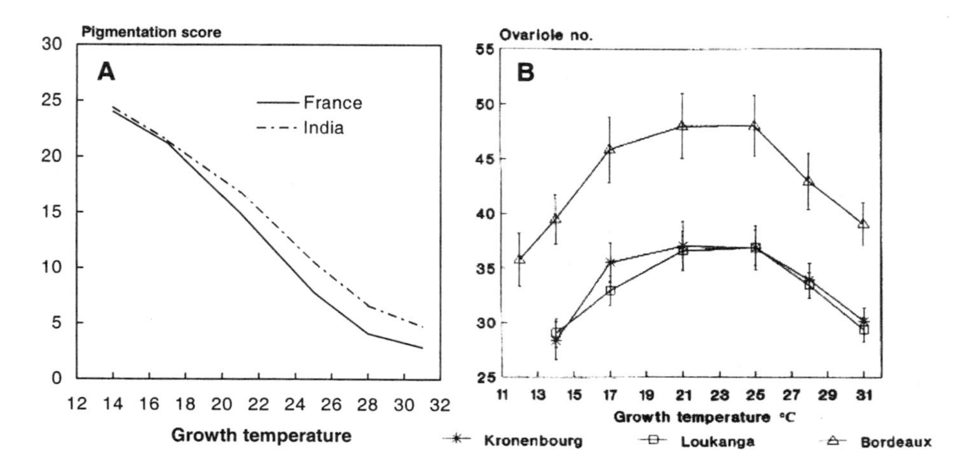

Figure 4.6. Reaction norms of populations of *D. melanogaster* living under different climates. (A) Abdominal pigmentation in populations from France and India. Divergence occurs at high temperatures that mainly distinguish the two localities. (B) Ovariole number in populations from France and Africa (Kronenbourg, Loukanga). Only slight differences are found in the position of the maximum.

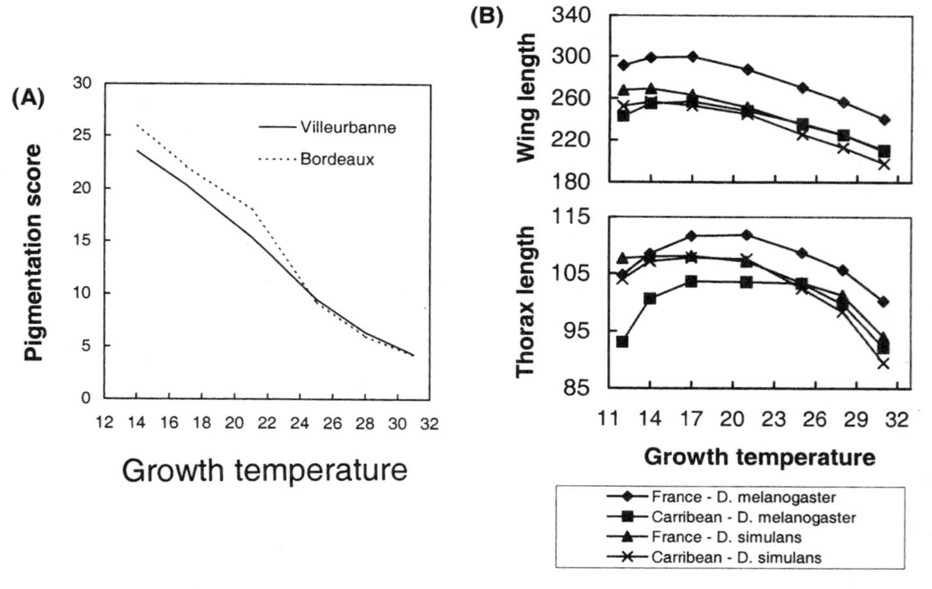

Figure 4.7. Reaction norms of two sibling species (*Drosophila melanogaster* and *Drosophila simulans*) living sympatrically. (A) Abdominal pigmentation in populations from two French locations (Bordeaux and Villeurbanne) differing by their cold season temperature. Values of both species were averaged in each locality. Reaction norms diverge at low temperatures, which discriminate the localities. (B) Wing and thorax length in Caribbean and French populations.

adaptation to colder climates, the two species underwent a modification of their reactivity to temperature, and a displacement of the maxima toward colder temperatures. This is a likely adaptive reaction if we assume that the reaction norm maximum is more or less akin to an optimum.

The ratio of wing and thorax length is also a character with a reaction norm of its own (figure 4.8; David et al. 1994; Morin et al. 1996). The wing:thorax ratio is the inverse of wing loading, which is related to flight capacity (Pétavy et al. 1997). At higher temperatures, plasticity results in increased wing loading and imposes a higher wing beat frequency (Pétavy et al. 1997). As first argued by Stalker (1980), flying in a colder environment is likely to select for a lower wing loading. Thus, wing loading and wing:thorax ratio might be direct targets of natural selection. As predicted, we found a smaller ratio in the tropical populations of both species.

Differentiation among Species

We analyzed reaction norms of body size and ovariole number in several species with different developmental thermal ranges and geographic distributions. The Indian *D. ananassae* is cold sensitive; it can be raised between 16°C and 32°C (Morin et al. 1996). By contrast, the European *D. subobscura* is cold tolerant and heat sensitive and can be raised between 6°C and 26°C (Moreteau et al. 1997). The cosmopolitan *D. melanogaster* has an intermediate range (12–32°C).

For the three traits investigated, important differences were found in the position of the TMVs, with the highest values from the tropical *D. ananassae* and the lowest values in the temperate *D. subobscura* (figure 4.9). For example, the TMV of ovariole number was 13.25 ± 0.25°C in *D. subobscura*, 22.17 ± 0.14 in *D. melanogaster* and 24.61 ± 0.25 in *D. ananassae*. Differences persisted among traits: lowest values were always found for wing length, and highest ones for ovariole number. Similar trends were also described more recently in a tropical drosophilid, *Zaprionus indianus* (Karan et al. 1999a).

Significant differences were also observed between species for the norms of wing:thorax ratio, but the shape differences are difficult to interpret. The major trend is a lower ratio in warm-living species, as already found among geographic populations of *D. melanogaster* and *D. simulans* (figure 4.7).

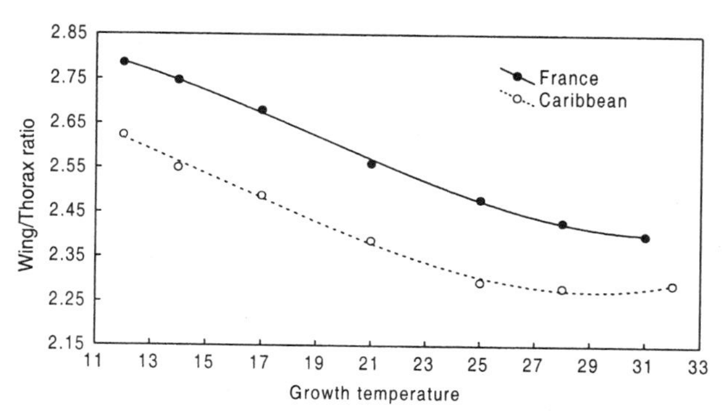

Figure 4.8. Wing:thorax ratio in Caribbean and French populations of *D. melanogaster*. Living in a warmer environment (Caribbean) results in a decrease of the ratio and in a translation of the inflection point to the left. [After Morin et al. (1999).]

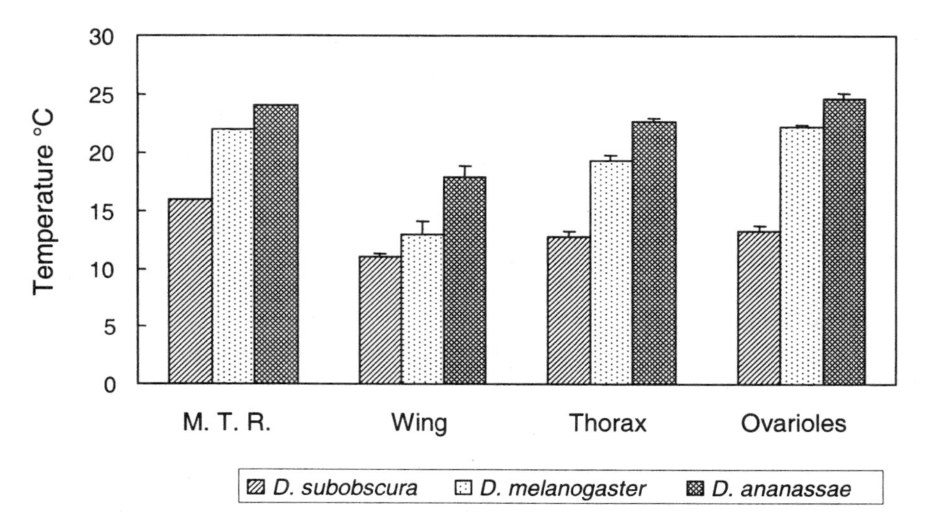

Figure 4.9. Comparison of the thermal adaptation of three species of *Drosophila* (*D. ananassae*, *D. subobscura*, *D. melanogaster*) (after Moreteau et al. 1997).

A comparison between *D. melanogaster* and *D. simulans* provides, however, slightly discordant conclusions (Morin et al. 1996, 1999). Based on wing:thorax ratio, *D. simulans*, with its lower ratio, would be classified as more warm adapted than *D. melanogaster*. On the other hand, the TMVs of wing and thorax length are lower in *D. simulans* than in *D. melanogaster*, suggesting cold adaptation (Morin et al. 1996, 1999; Moreteau et al. 1998). More comparisons among species are needed before we can determine the general pattern of reaction norm evolution in *Drosophila*.

Conclusions

The evolution of reaction norm shape is a largely untouched area of research. Investigations in this area face two kinds of challenges. First, what are the genetic bases of shape variation? What is the nature of plasticity genes? Can we connect characteristic values of reaction norms with genetic elements? A major, still controversial issue is whether plasticity results from gene regulation *per se* or allelic sensitivity (Via 1993a,b; Scheiner 1993a,b; Schlichting and Pigliucci 1993; chapter 3). One promising approach is the use of quantitative trait locus (QTL) mapping that is just starting to be used in plasticity studies. We predict that plasticity QTLs will be easy to demonstrate for some traits, but will not exist for others. Considering the complexity of developmental pathways and the diversity of their genetic determinism, we cannot expect any simple and general answer to the occurrence of plasticity genes.

Second, how can we interpret the diversity of reaction norms of different traits in the same organism (e.g., figures 4.2, 4.5)? To what extent are these differences due to developmental constraints or natural selection? All biological processes result from a permanent interplay between internal constraints and natural selection. This problem is especially significant when investigating reaction norms. The *Drosophila* model, including intraspecific geographic variations and a diversity of differently adapted species, has been a useful testing ground for obtaining conceptual and technical progress, although it has often generated as many questions as answers.

The reaction norm approach is not restricted to the analysis of mean values of morphometric traits. Indeed, it can be applied to physiological properties that may be more directly related to fitness, like starvation and desiccation tolerance (Da Lage et al. 1989; Karan and David 2000), locomotor performance (Gibert et al. 2001), chill coma temperature (Gibert and Huey 2001), and male sterility thresholds (Chakir et al. 2002). Polynomial adjustments can also be used for the analysis of variance parameters, calculating, for example, the temperature of minimum or maximum variability (Karan et al. 1999a; Imasheva et al. 2000; Moreteau et al. 2003).

Plants are strongly dependent on environmental fluctuations, have low ability to escape such variations, and are thus expected to be highly plastic (chapter 9). Plants also offer the possibility of easily replicating the same genotype across environments. Even among sexually reproducing species, many are autogamous and highly tolerant of inbreeding, so reaction norms can be compared among almost homozygous families. Finally, many different traits, more or less closely related to fitness, can be measured simultaneously on the same individual, and their variation can be monitored during growth and development. Maybe this wealth of advantages has been too much for plant biologists (Schlichting and Pigliucci 1998). Experimental data concerning phenotypic plasticity in plants are numerous, but few examine more than two environments or responses to an environmental gradient. Still fewer address questions concerning the shapes of responses curves. We hope that through the data presented here *Drosophila* might now be considered as a paradigm for stimulating similar investigations on other kinds of organisms.

5

Evolutionary Importance and Pattern of Phenotypic Plasticity

Insights Gained from Development

W. ANTHONY FRANKINO
RUDOLF A. RAFF

Development and Phenotypic Plasticity

Phenotypic plasticity in morphological traits is fundamentally a developmental phenomenon; developmental pathways are expressed differently in response to specific environmental factors to produce continuously varying traits or discrete, alternative phenotypes. Although the study of phenotypic plasticity currently enjoys a renaissance in evolutionary biology, it is generally considered a substantially less interesting topic in developmental biology (Raff 1996) for two reasons. First, the goals of developmental biology are largely concerned with determining how proximate mechanisms regulate and integrate trait ontogeny and how alterations of these mechanisms produce changes at a macroevolutionary scale; developmental biology seldom considers how proximate mechanisms generate the phenotypic variation among individuals that is important in microevolution. Second, phenotypic variation among individuals, especially environmentally induced variation, is usually viewed as noise in developmental studies—something to be minimized when the research goal is to understand the proximate mechanisms regulating ontogeny. Consequently, the organisms chosen as models of development share several characteristics and are raised in ways that may generate bias, suggesting widespread conservation of developmental mechanisms across taxa and typically canalized development that generates invariant phenotypes within taxa (Bolker 1995; Grbi and Strand 1998). Such perceived absence of developmental (phenotypic) plasticity in the laboratories of developmental biologists, coupled with a lack of focus on variation among individuals, may have contributed substantially to the lack of information on, or even interest in, the proximate bases of plasticity.

The paucity of information regarding the proximate basis of observed phenotypic variation, however, is not necessarily viewed as important by evolutionary biologists. This is because the statistical techniques typically used to study adaptive evolution are thought to incorporate, albeit indirectly, the important effects of development on phenotype production and evolution. For example, the developmental mechanisms that influence the distribution of traits or trait suites are thought to be captured in the variance/covariance matrix (Cheverud 1984) used to predict the multivariate response of a population to selection. Strong correlations among traits reflect developmental integration, which ties the trait complexes together evolutionarily (Lande 1979; Lande and Arnold 1983).

Such indirect, statistical approaches to studying what is fundamentally a developmental phenomenon have generated problems for the study of phenotypic evolution, and the evolution of phenotypic plasticity in particular. One well-known problem is the controversy over the genetic basis for phenotypic plasticity (chapter 3; see also Scheiner 1993b; Schlichting and Pigliucci 1993; 1993a,b, 1994; Via et al. 1995; Pigliucci 1996a). Although other problems introduced by a purely statistical approach (discussed below) are more subtle, they may be important, if not perhaps more important. In the next section, we illustrate how such methods that consider developmental processes only indirectly have the potential to mislead investigators into underestimating the magnitude, and consequently the evolutionary importance, of phenotypic plasticity.

Development and Reaction Norms

To examine the effects of the developmental environment on the distribution of phenotypes in a population, reaction norms are typically generated for the focal trait (figure 5.1; see also figure 3.1). These measurements of the cross-environmental phenotypic variance are thought to reflect the developmental mechanisms producing trait variance and covariance. Strong correlations are interpreted as resulting from expression of the same developmental processes across environments; looser correlations result from different expression of developmental processes across environments.

Predictions of the evolutionary response of a population to selection that are based on estimations of the variance and cross-environmental covariance in phenotypes (as manifested in reaction norms) may be inaccurate for at least two reasons, both of which may be addressed by considering the developmental basis of the plastic response. First, interpreting the cross-environmental covariation in trait values is difficult without knowing the developmental basis of the correlation. Strong cross-environmental correlations in trait values are thought to constrain the evolution of a plastic responses, whereas weak correlations are less problematic for the independent evolution of trait values (Via and Lande 1985; Via 1987). However, the degree to which genetic correlations represent real constraints on the independent evolution of traits is determined by exactly how the traits are developmentally integrated (Wolf et al. 2001).

In many insects, for example, negative correlations in the size or shape of some traits can result from interactions among developing tissues at specific points in ontogeny (Klingenberg and Nijhout 1998; Nijhout and Emlen 1998; Stern and Emlen 1999). Even if the measured negative correlation is strong, the impact of such trade-offs on phenotypic evolution will depend on the proximate basis of preferential resource allocation,

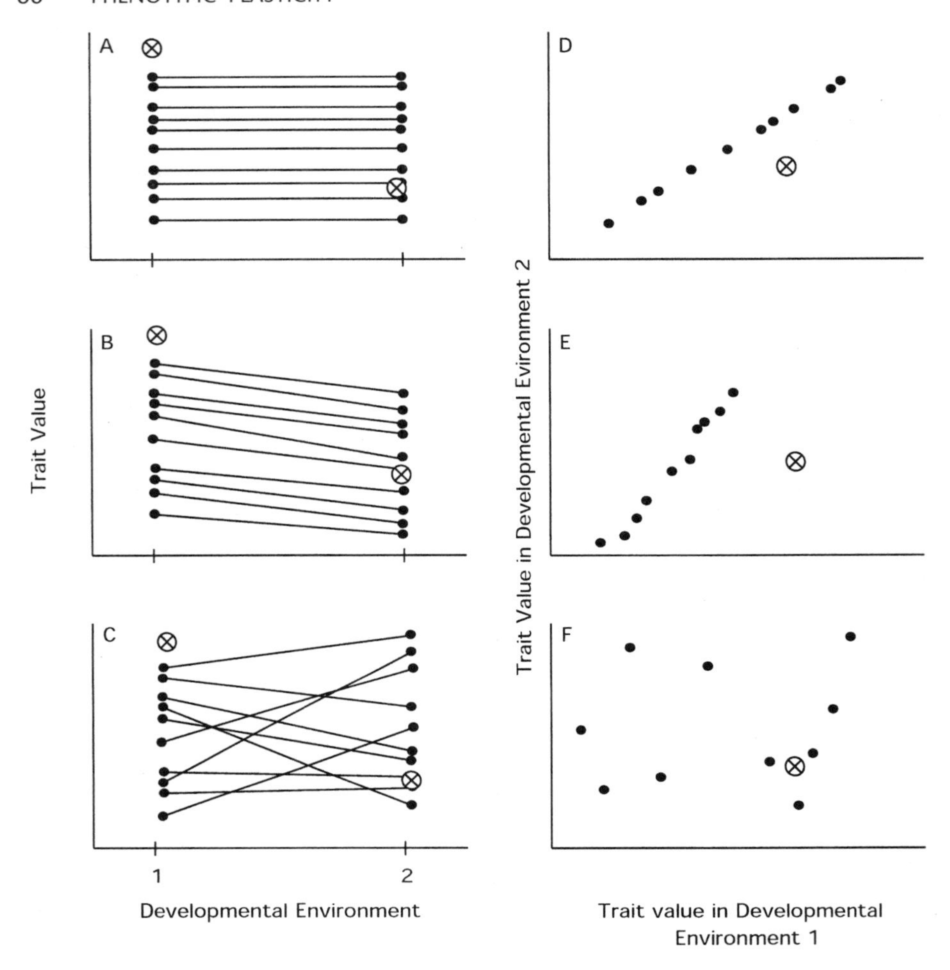

Figure 5.1. Representative linear reaction norms and cross-environmental trait correlations. Points indicate phenotypic values for hypothetical genotypes (or some surrogate such as sibship mean phenotype) in two developmental environments. Slopes in A–C represent levels of apparent phenotypic plasticity for a trait across environments. The phenotypic optimum in each environment is shown by the circle with the X. In D and E, this joint optimum is not achievable because the tightly correlated trait values prohibit that trait value combination, although it is easily achievable in E. [Reprinted with permission from Via (1987).]

resource competition, or other form of communication among developing traits (Nijhout and Emlen 1998; Stern and Emlen 1999; Wolf et al. 2001). The same arguments apply to the evolution of phenotypic plasticity in some trait; tight cross-environmental correlations may or may not represent constraints on the evolution of a plastic response, depending on the proximate basis of trait development in each environment.

Second, statistical estimation of the relationship between allelic and phenotypic variation may be inaccurate when based on the variance reflected in the slopes of reaction norms. The degree of phenotypic plasticity in some trait is often measured as the magnitude of the slope of a reaction norm (see chapter 3). However, some traits are environmentally canalized, where developmental regulatory action counters or resists the influences of the developmental environment through ontogeny.

For example, reaction norms with a zero slope are interpreted as the developmental environment not affecting the final trait value or no developmental response to the environment. However, the production of an invariant trait across developmental environments (figure 5.1A) may represent environmental canalization (Wagner et al. 1997), meaning that the mechanisms regulating development may counter environmental effects, reducing the impact of variation in the developmental environment on the final value of the trait. In such a case, significant effects of the environment on development are present; however, adaptive compensatory mechanisms act to resist the environmental perturbation. Here, a flat reaction norm represents adaptive (developmental) phenotypic plasticity; in the absence of developmental compensation, a different and less fit phenotype would be produced.

Countergradient variation, where genetically based developmental compensation resists environmentally induced phenotypic variation (reviewed in Conover and Schultz 1995), is one example of such compensation. Completely different, alternative ontogenies can also be employed to resist or compensate for effects of the developmental environment on terminal phenotype. For example, under stressful conditions some individuals may adopt alternative juvenile phenotypes that may differ dramatically in form and behavior and yet still converge on the same adult phenotype (e.g., Collins 1979; Pfennig 1992). In cases of countergradient variation and these environmentally induced polymorphisms (polyphenisms), environmental effects on trait development would manifest in the absence of such adaptive developmental compensation, transforming the reaction norm in figure 5.1A, for example, into that of figure 5.1, B and C. Hence, care must be taken when making inferences about the evolution of phenotypic plasticity based solely on reaction norms; flat reaction norms may sometimes conceal extremely interesting adaptive developmental plasticity underlying trait production. It is therefore critical to identify when developmental compensation alters the slope of a reaction norm, because failure to do so risks misinterpreting the effects of the environment on trait expression or, worst still, misunderstanding instances of cryptic adaptive plasticity in a focal trait.

Together, these two points illustrate that measuring the terminal phenotypes produced across a range of environments may be initially informative, but more information is needed to understand the evolutionary potential of an observed plastic response. Below, we describe a research program designed to determine the role of proximate developmental mechanisms in shaping the evolution of plastic responses.

The Research Program

The research program has three basic steps. First, putative developmental regulators believed to underlie a plastic response must be identified. Second, a series of experiments are performed to establish the hypothesized role of the focal developmental mechanism in producing the plastic response. Third, similar manipulations of development are applied to other lineages to determine if the same developmental mechanisms regulate similar plastic responses in those lineages and to detect a developmental bias or predisposition in lineages to produce given forms of phenotypic plasticity.

Step 1. Identification of the Hypothesized Proximate Basis of a Plastic Response

The first step in dissecting the proximate basis of a plastic response requires at least some knowledge of developmental regulation of the traits in question. Such understanding can narrow the field of inquiry and facilitate the identification of critical windows or sensitive periods during ontogeny, where developmental decisions or evolutionarily important trade-offs are likely to occur. Also, considerations of development can aid in the identification of a plastic response not reflected in a flat reaction norm, that is, the occurrence of countergradient variation or other form of resistance to environmental effects on development that may not be apparent from the examination of reaction norms. Once hypotheses about specific proximate mechanisms are made, one can establish the role of the putative developmental basis of a plastic response. Ideally, experiments can be designed such that some estimate of the evolutionarily important variation in the relevant developmental mechanisms can be made as well. There is a growing body of knowledge that can provide the required kinds of detailed information (Gilbert 2000; Stern 2000). In particular, investigators may have to consider the developmental modules, or constituent parts, that comprise the complex mosaic trait that is of evolutionary interest. Most morphological traits that are of evolutionary interest are composed of developmental modules; considering these modules separately may make the proximate regulation of trait ontogeny more developmentally tractable (Stern 2000) and more informative in a quantitative genetic setting (Wolf et al. 2001).

Step 2. Investigating the Developmental Basis of the Plastic Response

Just as there are well-established criteria for identifying a trait as an adaptation (Gould and Lewontin 1979), there are strict criteria for establishing the causal role of some process in development (Gilbert 2000). To meaningfully connect development to evolution requires a melding of both fields of inquiry. Failure to do so risks misdirecting entire research programs because it can lead to the description of false evolutionary patterns and a flawed understanding of how development shapes evolutionary change (Bolker 1995; Zera 1999; Zera and Huang 1999).

We recommend that investigators collaborate across evolutionary and developmental biology, bringing together their respective views and expertise on developmental regulation, experimental tools, phenotypic variation, environmental variation, and the complexities of selection and population structure. Through collaboration, developmentally tractable traits that exhibit evolutionarily interesting phenotypic plasticity can be identified and the proximate mechanisms regulating plasticity can be dissected. In the remainder of this section, we briefly review some of the most promising techniques for investigating the proximate basis of plastic responses in animals at a few different levels of biological organization. We focus on techniques not usually employed by classically trained evolutionary biologists; more standard approaches (e.g., the use of bioassays or more precise techniques to quantify levels of hormone among groups of interest) are probably familiar and are easily gleaned from reviews or textbooks (e.g., Nijhout 1994; Rose 1999; Gilbert 2000).

Physiology

Hormones coordinate the development of traits within an organism, regulate the production of continuous and discrete phenotypic variation, and create ontogenetic windows or sensitive periods during which plastic responses occur (see recent review in Nijhout 1999b). Establishing the role of a given endocrine mechanism in development is challenging, because it requires a variety of *in vitro* and *in vivo* tests. These include (but are not restricted to) documenting the endocrine profiles of individuals expressing typical and plastic ontogenies, and manipulating the circulating hormone titer through a variety of techniques, including elimination of endogenous sources of the hormone and replacement with endogenous hormone (removal and phenotype rescue). The specifics of such tests will differ with the biological system under study, and any single test may not provide sufficient information to unequivocally establish the role of a hormones in development (for an overview, see Gilbert 2000). For example, although differences in hormone titer may not be apparent between individuals expressing typical and alternative trait ontogenies, differences in the amount of physiologically active hormone may be achieved through alteration of hormone-binding proteins, receptor densities, and so forth. Each of these possibilities must be examined in detail to clearly establish the role (or nonrole) of a given hormone in trait development.

One approach typically not used in studies of phenotypic plasticity is cell or tissue culture. In culture studies, targets (cells or organs) are raised in cocktails that differ in chemical composition. By comparing the ontogeny of the targets in media, the physiological signals that affect trait ontogeny can be isolated. For example, cultures of amphibian tails and limbs have identified the role of different hormones and genes in organ regulation at amphibian metamorphosis (reviewed in Tata 1993; Rose 1999) and in the physiological basis of differences in development rate among amphibians (Buchholtz and Hayes 2002).

Embryology

Embryonic manipulations involving the destruction of incipient traits or the transfer of structures (e.g., limb buds, neural crest cells, imaginal disks, etc.) among individuals to create chimeras offer a powerful set of techniques for the dissection of signal–target interactions. These manipulations can be used to identify key factors that initiate, terminate, or regulate a plastic response to the environment.

Laser ablation and microsurgical approaches can be used to identify the role of a structure in the expression of other traits. Such experiments can determine, for example, if a group of cells act as a morphogen source or sink, and the ontogenetic timing of the release or uptake of the morphogenic factor. Somewhat surprisingly, major organs can be removed, such as the brain of an insect (e.g., Endo and Kamata 1985) or thyroid gland of a larval amphibians (e.g., Kanki and Wakahara 1999), or more subtle manipulations can be applied to smaller putative signaling centers (e.g., Nijhout 1980; Nijhout and Emlen 1998; Dichtel et al. 2001; Patel et al. 2002). Such experiments are particularly powerful when coupled with exogenous replacement of the putative signal to "rescue" the normal phenotype.

Transplant experiments involve removing structures from individuals possessing one phenotype and placing the structure into, or into contact with, a recipient differing in phenotype. Donors and recipients can be from different species, selected lines, seasonal morphs, instars, and so on. Transplant experiments can be of two general types. In the first kind of transplant experiment, the focus is on the effect of transplanted tissue on the development of the host phenotype. In the second, the focus is on what the host phenotype does to the transplanted tissue. Again, these types of manipulations can be dramatic, for example, involving the transplantation of brain tissue into the abdomen of a host or the fusing of two host types to identify a morphogen source that induces an alternative seasonal morph (e.g., Endo and Funatsu 1985; Endo and Kamata 1985). Alternatively, such manipulations can be more subtle. For example, Monterio et al. (1997) transferred signaling centers among individual *Bicyclus anynana* butterflies to determine the degree to which the response to artificial selection on wing pattern was due to changes in the signal or in the response cells. These techniques were also used to investigate how the artificial selection affected the expression of alternative seasonal wing pattern morphs in this butterfly (Brakefield et al. 1996). Gilbert (2000) has a thorough review of transplant experiments, and how they can be applied to address evolutionary questions.

Molecular Genetics

There are a number of molecular genetic techniques that can be used to investigate the proximate basis of phenotypic plasticity. Some model genetic systems exhibit evolutionarily interesting forms of plasticity and consequently provide an enormously powerful resource for understanding the relationship between proximate mechanism and the expression of alternative phenotypes. One of the genetic model systems, the nematode *Caenorhabditis elegans*, develops continuously from larva to adult or can employ an alternative life history and diapause as a dauer (Cassida and Russel 1975; Golden and Riddle 1984). The developmental switch is determined by physiological signals, especially the threat of starvation, during a particular point in larval ontogeny. The current model for this regulatory system involves signaling from the transforming growth factor (TGF) [2] pathway and from an insulin receptor signaling pathway, impinging *daf-12*, a gene that acts as a life history switch in conjunction with heterochronic genes that regulate life staging (Antebi et al. 1998; Gerisch et al. 2001). This research is being extended to homologous processes in other nematodes. Because the genetics are available, a mechanistic view of phenotype switching works is thus possible in *C. elegans* beyond that currently possible even for temperature-dependent sex determination in reptiles (Crews 1994, 1996; Godwin and Crews 1997). A number of technological advances developed for use in such model systems also hold promise for nonmodel systems.

Injection of novel oligonucleotide reagents (Summerton 1999) or RNA inhibition (Brown et al. 1999; Fire et al. 1998; Kennerdell and Carthew 1998; Paddison et al. 2002) may allow investigators to perform genetic manipulations in at least some nongenetic model systems. These reagents inhibit the action of a target gene, mimicking the effect of a knockout mutation. If injected early in ontogeny, this manipulation can prevent expression of the target gene in the entire organism; if it is applied later, mosaics can be created where some cells or tissues express the gene whereas others do not. These manipulations can be used to investigate the genetic basis of plastic responses and could be particularly powerful if coupled with other manipulations of development, such as the removal and phenotype rescue approach discussed above.

Genomic approaches offer new ways of investigating evolutionary changes in genetic architecture among closely related species or populations of nonmodel organisms (Peichel et al. 2001). The growing number of gene chips available for model systems also can be used to study the proximate mechanisms underlying plastic responses in nonmodel systems (Gibson 2002). Extractions from a given tissue or a whole organism are passed over a chip that contains an array of genes from the model organism. Levels of gene expression in the sample are indicated by the degree of illumination of each receptor in the array. Comparing the illumination pattern generated by individuals expressing typical and alternative phenotypes reveals correlations between specific gene products and the production of the plastic response. Although the treatment of the data generated by such chip tests is an active area of research in itself, the technology offers promise because many genes of interest (e.g., hormone receptors) are conserved phylogenetically (Hacia et al. 1999; Rast et al. 2000). Hence, gene chips developed for a model system may be used to investigate nonmodel systems.

These approaches, in combination with the use of more traditional methods drawing on the resources developed for model systems such as *in situ* hybridization or antibody staining, can be combined with traditional gene mapping (candidate gene approaches or genomewide searches using techniques such as subtractive hybridization or linkage map building) to relate quantitative variation in plastic responses to some genes associated with their production. A theoretical link between candidate gene approaches and microevolution is now being forged (Haag and True 2001).

Investigators may wish to focus attention on the *cis*-regulatory regions that control gene expression. Evolution of *cis*-regulatory regions are responsible for quantitative differences between species (Stern 1998; Sucena and Stern 2000) and between sexes within species (Kopp et al. 2000). The evolution of the *cis*-regulatory region allows for quantitative, independent changes in gene expression across tissues or times in ontogeny (Stern 2000). Consequently, *cis*-regulatory regions probably regulate expression of plastic responses at some level in both continuously varying traits and alternative, discrete traits.

Step 3. Investigation of Other Lineages

Once the developmental basis of the plastic response in a focal lineage has been identified, at least in part, other lineages can be investigated to address how proximate mechanisms may have shaped the evolution of the plastic response. Comparisons can be made among lineages (e.g., taxa, selected lines, mutants) that differ in the threshold for induction of a plastic response or the degree to which trait plasticity is exhibited. Such investigations can indicate what parts of a developmental cascade are involved in the microevolution of the plastic response. However, using the comparative method in combination with experimental manipulations also has the potential to reveal the role of development in the generation of phylogenetic pattern of adaptive phenotypic plasticity.

The degree of phenotypic plasticity exhibited by any trait will be related to how trait ontogeny is regulated. It follows that some aspects of developmental regulation may facilitate the evolution of phenotypic plasticity in some traits or periods of ontogeny and restrict it in other traits or times in the life cycle. We can identify latent potential in the existing developmental system that may facilitate the evolution of plasticity by comparing the response of individuals from several lineages to environmental and developmental manipulations. If done in a broad phylogenetic context, such comparisons may reveal a correspondence between mode of developmental regulation, action of some gene, and so forth, and the specific form of plasticity under investigation. Eventually, these comparisons may explain phylogenetic pattern in phenotypic plasticity and hint at the general evolutionary importance of a given plastic response.

One obvious way these topics can be addressed is to apply steps 1 and 2 above to lineages that exhibit independently derived, convergent plasticities. If the developmental basis of the plastic response is the same across lineages, then there may be a narrow range of developmental routes that can be taken to produce the plastic response, or it may be that repeated co-option of the same proximate mechanisms is responsible for the convergent plasticities. This approach has revealed convergence in the proximate basis of plasticity in pupal color across families of butterflies (Starnecker and Hazel 1999). Conversely, a variety of developmental processes underlying convergent plasticities would indicate that development may play little role in shaping or constraining the evolution of the plastic response.

A second way to investigate phylogenetic pattern in plastic responses is to perform manipulations similar to those described in step 2 to individuals from lineages that do not naturally exhibit the plastic response. Potential biases in the evolution of a plastic response introduced by the proximate mechanisms regulating development would be indicated when individuals from a lineage normally canalized for the focal trait exhibits plasticity in response to the manipulation. If the manipulation induces some degree of plasticity in the focal trait, this suggests that there is at least the potential to produce the plastic response in the outgroup via the same developmental means as in the focal lineage.

The degree of similarity between the naturally occurring plasticity in the focal lineage, the experimentally induced plasticity of manipulated individuals in the focal lineage, and the experimentally induced plasticity in the outgroup suggests how easily that form of trait plasticity might evolve, and perhaps how easily it occurred in the ancestor of the focal lineage. A high degree of similarity in response to the manipulation between individuals from the outgroup and the focal lineage indicates that the evolution of the plastic response was probably relatively easy and could be repeated among other members of the clade. Disparity in the response indicates that the evolution of the plastic response may represent more of a developmental, and consequently and evolutionary, challenge.

Using environmental manipulations, this approach has been used investigate the evolution of seasonal polyphenic wing color patterns in *Pieris* and the closely related *Tatochila* butterflies. Some species in these groups are polyphenic, producing melanized (cool season) adults during short day conditions and adults lacking melanization (immaculates) under long day conditions. Other species in these genera are monomorphic, expressing either the melanized or immaculate morph. Selection imposed by the thermal environment often conveys higher fitness to the melanized form in the cool season and the immaculate form during the dry season in *Pieris* species (reviewed in Kingsolver and Huey 1998), and this is likely to be the case with other, less studied *Pieris* and *Tatochila* species. In some lineages that do not exhibit the melanized phenotype, cold shock induces a phenotype superficially similar to the cold-season morph (reviewed in Shapiro 1976, 1980; Nijhout 1991). Differences between the typical and induced patterns indicate that the alternative morphs are not homologous (Shapiro 1980). It would be very interesting to know the degree to which the proximate basis of morph determination is shared between the typically expressed seasonal morphs and the induced, novel alternative forms, however, because this would suggest how easily the polyphenism can evolve in these speciose and widely distributed butterflies.

In the remaining portion of the chapter, we explore how determining the phylogenetic distribution and proximate basis of trait plasticity in single and composite traits can be particularly informative for investigations into how development shapes the evolution of adaptive plasticity. In so doing, we describe what is known about the proximate basis of adaptive plasticity in some amphibian and insect systems and illustrate the kind of integrative research approach we advocate.

Evolutionary Pattern in Phenotypic Plasticity

We have argued that examination of the phylogenetic distribution of adaptive trait plasticity can help to illuminate the role of development in shaping plastic responses and in creating phylogenetic pattern in plasticity. However, it can be a daunting undertaking to compare the proximate basis of trait plasticity across lineages, in large part because of the difficulty in identifying the "trait" of interest for a given plastic response. One way to approach this problem is to consider traits differently to address different research questions. For example, comparisons can be made for both morphologically and ecologically defined traits. Comparisons of plasticity among homologous morphological traits will indicate the role of development in producing a range of phenotypes from a single structure. Comparing ecologically defined traits (i.e., mosaic traits that may be composed of different, nonhomologous elements across lineages) can elucidate how development affects the evolution of plasticity in similarly adaptive phenotypes assembled from different constituent parts. Such comparisons can also suggest certain types of ontogenies or particular aspects of developmental regulation that may facilitate the evolution of adaptive developmental plasticity in particular traits. In the subsections that follow, we examine each of these topics in detail and provide empirical examples to illustrate our points.

Evolutionary Pattern in the Proximate Basis of Adaptive Plasticity

A first step toward understanding the evolutionary importance of phenotypic plasticity is to document the phylogenetic distribution of plastic responses. Knowing the ancestor–descendent relationship for plastic responses is central to the formulation and testing of hypotheses addressing the adaptive evolution of plasticity (Doughty 1995; Gotthard and Nylin 1995). However, identifying the point of origin of a plastic response may also hint at the general evolutionary impact of plasticity in the focal trait. Clusters of homologous plastic responses may suggest that trait plasticity could have facilitated adaptive radiation (i.e., plasticity in the focal trait represents a key innovation *sensu* Liem 1990). Widespread homoplasies (convergences or independent evolutionary origins), in particular, plastic responses, might indicate that the response is relatively easy to evolve when favored ecologically.

Once the phylogenetic pattern of plastic responses is known, comparing the proximate basis underlying trait plasticity could prove enlightening. For example, paedomorphosis, a life history strategy in which salamanders retain the aquatic, larval morphology into reproductive maturity has evolved independently several times among, and even within, the main urodele clades (reviewed in Shaffer and Voss 1996). Although there are numerous hypotheses addressing the conditions favoring the evolution of obligate and facultative paedomorphosis (Sprules 1974; Whiteman 1994), all hypotheses share the basic tenet that reproduction as a paedomorph confers higher fitness than reproduction as a terrestrial adult. When larval habitats can change in quality over the course of an individual's lifetime or when the relative success of the reproductive strategy is context dependent, then phenotypically plastic development, facultative paedomorphosis, should be favored (Whiteman 1994).

The suite of morphological traits that compose the paedomorph is the same across lineages. Yet the evolutionary loss of metamorphosis (i.e., the trait of larval reproductive mode) has occurred through modification of different proximate mechanisms among lineages. Investigations of paedomorphosis focusing on several species of *Ambystoma* suggest details about the genetic basis of alternative phenotype production. Amplified fragment length polymorphisms studies suggest that the switch to paedomorphosis in *A. mexicanum* involves one gene of major effect, and possibly numerous genes of smaller effect (Voss and Shaffer 1997). Differences among populations in the polygenic basis of facultative paedomorphosis in *A. talpoideum* are suggested by common garden experiments (Harris et al. 1990; Semlitsch and Gibbons 1986), as are genetically based differences in sensitivity to environmental conditions that induce metamorphosis in facultatively paedomorphic lineages (Semlitsch et al. 1990).

Mechanistically, paedomorphosis could result from any one of numerous changes that prevent metamorphosis (Rose 1996; Shaffer and Voss 1996). Timing and coordination of developmental events in amphibians are regulated by interactions of thyroid hormones (TH) with other hormones and response tissues (reviewed in White and Nicoll 1981; Galton 1992a; Shi 1994; Kaltenbach 1996; Rose 1999). As reviewed by Shaffer and Voss (1996), the independent evolution of paedomorphic life histories resulted from a variety of changes in proximate developmental mechanisms, ranging from a reduction of thyroid activity in some lineages to a decrease of tissue sensitivity to TH in other lineages. There are several candidate mechanisms that could lead to reduced tissue sensitivity to TH, including changes in the binding affinity or number of TH receptors, and alterations of enzyme systems that regulate or otherwise effect exposure of specific tissues to circulating TH levels (Galton 1985, 1992b; Rose 1996). Hence, the paedomorph phenotype is composed of changes in the same suite of morphological traits, but the ontogeny of these traits has been altered through modification of different mechanisms across lineages. This suggests that it is relatively easy for salamanders to truncate or delay somatic development when such ontogenies are favored.

Such variation in the proximate basis of the paedomorphic ontogeny stands in stark contrast to the apparent constancy in the basis of adaptive acceleration of development rate exhibited by many larval amphibians. Amphibians from lineages (populations, species) that typically experience changing environments during the larval stage can alter ontogeny in response to stresses in the developmental environment, whereas lineages that usually develop in more constant environments often lack such plasticity (Denver 1997b; Duellman and Trueb 1986). Declines in larval habitat quality can be signaled by decreases in per capita food availability or water volume; increases in larval density, water temperature, or concentration of chemicals; and the presence of chemical cues associated with predators or injured larvae (reviewed in Newman 1992; Denver 1997b). The proximate basis of the adaptive acceleration of larval ontogeny has been examined extensively for the developmental response to habitat desiccation, particularly in spadefoot toad larvae that exhibit both exceptionally high growth rates and exceedingly short larval periods (Buchholz and Hayes 2000, 2002; Morey and Reznick 2000). In response to declining water levels, production of corticotropin-releasing hormone (CRH) in the brain precociously activates the thyroid and interrenal axes, which in turn coordinate accelerated larval ontogeny by raising TH and corticosterone (B) levels (Denver 1997a, 1998a). TH regulates larval differentiation, pacing the approach toward metamorphosis by controlling patterns of gene expression in virtually all tissues (Kikuyama et al. 1993; Rose 1999). Corticosterone (B) seems to synergize the effects of TH, although the mechanism by which this occurs is not clear (reviewed in Denver 1997b; Kikuyama et al. 1993; Rose 1999). The conservation of the neuroendocrine response to habitat desiccation in amphibians is interesting because, from a developmental perspective, there are several other possible mechanisms that could be used to achieve the same effect (reviewed in Denver 1997b).

Elevation of CRH levels is part of a generalized, phylogenetically ancient and conserved response to stress in vertebrates (Denver 1997a). It would be interesting to know if the same basic endocrine mechanisms underlie the production of the environmentally induced, adaptive phenotypes exhibited by many larval amphibians. Trophic polymorphisms in tadpoles (e.g., Pfennig 1992) and salamander larvae (e.g., Collins and Holomuzki 1984; Loeb et al. 1994; Walls and Blaustein 1995) are produced under stressful larval habitats. Predators also induce developmental acceleration in many amphibian larvae (e.g., Skelly 1992; Warkentin 1995) as well as polyphenisms in larval fin morphology (McCollum and Van Buskirk 1996). In each of these cases, as well as in the environmentally induced metamorphosis in paedomorphic salamanders described above, alteration of typical ontogeny occurs when larvae experience stressful conditions. It seems likely that, at least on some level, the generalized stress response produced by increases in CRH levels will be involved in these cases of adaptive plasticity.

In summary, adaptive plasticity can have the same or different developmental basis across lineages. Facultative paedomorphs are morphologically similar among lineages, yet somatic development is interrupted differently to achieve reproduction in the aquatic environment. The independently evolved, morphologically divergent, stress-induced responses to declines in the larval habit quality may share involvement of the thyroid and interrenal axes, as mediated by CRH. Contrasting the developmental basis of these cases of adaptive plasticity raises the issue of how proximate developmental mechanisms can generate phylogenetic pattern in phenotypic plasticity. The variety of proximate mechanisms involved across lineages of facultatively paedomorphic salamanders is not surprising because this life history strategy represents an inhibition of metamorphosis or truncation of typical somatic ontogeny; there are many ways to interrupt typical development. The repeated co-option of the generalized CRH-mediated stress response may suggest that evolutionarily plastic novelties are constrained in how they can develop. It also suggests that there is something about the proximate basis of amphibian development that facilitates the evolution of adaptive plasticity in this group, and that this has allowed amphibians to exploit a variety of habitats. Interestingly, two genera of spadefoot toads (*Scaphiopus* and *Spea*) develop rapidly because of high levels of circulating TH and greater tissue sensitivity to this hormone (D. R. Buchholz, in preparation), suggesting that there is variation in the developmental cascade downstream and perhaps even semi-independent of CRH on which selection has acted. Moreover, the rapid development conferred by these changes may have facilitated diversification of the clade in xeric environments (Buchholz and Hayes 2002).

Developmental Windows and Morphological Plasticity

Adaptive flexibility in ontogeny or in the development of a trait may be characteristic of some lineages and wholly absent in others. The phylogenetic breadth and variety of traits exhibiting adaptive plasticity suggest that the problematic absence of detectable, reliable predictors of the future selective environment, not patterns of selection, limit the evolution of adaptive plasticity (Moran 1992). For such cues to be effective, they must be detected before or at a point in ontogeny when development can be altered to produce the most appropriate phenotype for the future selective environment. This necessary temporal match between the detection of cues in the developmental environment and ontogenetic points where adaptive adjustments in development can be made has the potential to produce developmental constraints on, and phylogenetic pattern in, the evolution of adaptive plasticity.

For example, the stress-induced modulation of the thyroid and interrenal axes of amphibian larvae discussed above can take place exceptionally quickly, allowing adaptive plasticity in development rate through much of larval ontogeny. Changes in the circulating levels of TH are detectable in tadpoles of *Scaphiopus hammondii* within hours of a decrease in the water level (Denver 1997a). These changes in TH titer rapidly alter patterns of organ-specific gene expression (Denver 1998b), which manifest as changes in larval ontogeny as quickly as 3 days (Denver 1998a). Furthermore, these ontogenetic responses are proportional to the relative change in water level and are reversible—the developmental response closely tracks environmental changes in real time (Denver et al. 1998).

However, there are limits to plasticity in developmental rate in amphibians. Adaptive acceleration of ontogeny in amphibians involving precocious activation of the thyroid and interrenal axes requires that these axes are developed and that tissues are competent to respond to changes in signal. Absence of a fully developed thyroid–interrenal axis may explain why amphibians do not alter development rate in response to changes in the environment early in larval ontogeny. Later in ontogeny, a developmental commitment is made to metamorphose, after which larvae are again unable to respond adaptively to changes in the larval environment. These two developmental events, the maturation of the thyroid and interrenal axes and the commitment to begin the endocrine–genetic cascade toward metamorphosis, bracket a window of ontogeny during which adaptive plastic responses are possible. Such ontogenetic windows of environmental sensitivity during which adaptive changes in development can occur are commonly observed in amphibian ontogenies (e.g., Hensley 1993; Denver et al. 1998; Morey and Reznick 2000).

The continuous development of amphibians contrasts with the ontogenies of insects where saltatory morphological change is a necessary consequence of moving stepwise through discrete instars. Because insects can only realize morphological changes at a molt, alteration in the development of a larval or adult structure necessarily precedes the appearance of the structure, sometimes by great periods of time. For example, male

Othophagus taurus beetles facultatively produce horns, depending on their projected adult size. Size at pupation in *O. taurus* is determined by the amount and quality of dung with which each larvae has been provisioned (Moczek 1998), similar to other *Othophagus* beetles (Emlen 1994, 1997b), and only large males develop horns, which they use to secure matings (Moczek 1998; Emlen and Nijhout 1999; Moczek and Emlen 2000; see chapter 9 for a review of the ecology this system). As larvae deplete the finite amount of dung with which they have been provisioned, male phenotype (long or short horned) is determined and the developmental commitment to pupate is made (Emlen and Nijhout 1999, 2001). Male morph determination is probably affected by the general mechanisms that regulate development in insects. Below, we discuss some of these mechanisms generally and then explore specifically how they influence the evolution of morphological plasticity in horn beetle morphology.

In insects, differentiation and the molting cycle are controlled primarily by the interaction of target tissues with two hormones, juvenile hormone (JH) and ecdysone (see reviews in Nijhout 1994, 1999b). JH is produced by the corpora allata and is generally thought to exist for only a short time in circulation. Because it has a short half-life in the hemolymph, reductions in JH production can lead to swift declines in circulating JH levels. Low titers of circulating JH induce the production of prothoracicotropic hormone (PTTH), which in turn initiates the synthesis and release of ecdysone, a steroid hormone that begins the cascade of developmental events leading to a molt. If the circulating levels of JH are above some threshold when ecdysone interacts with target cells, then the tissues retain their larval commitment and a larval molt ensues. If JH is below this threshold, then cell fates change and differentiation occurs; this is when changes in phenotype or shifts to different forms of instars (e.g., pupa) can occur. In hemimetabolous insects (e.g., true bugs, crickets) such a molt leads to an adult, whereas in holometabolous insects (e.g., butterflies, beetles) a pupa or adult is produced. This interaction of JH and ecdysone with target cells during specific points in ontogeny generates critical periods or developmental windows during which cell fates are determined (Nijhout 1994, 1999a,b).

As is the case with most adult structures in holometabolous insects, the horns of horned beetles develop from imaginal disks (Emlen and Nijhout 1999). Early in larval ontogeny, epidermal fields are sequestered into pockets in the body. These pockets become the imaginal disks, which grow slowly and semi-independently from the rest of the insect until late in the final larval instar, when their cell populations grow exponentially and differentiate. Eventually, the structures in disks develop cuticle and expand to form nearly all the external adult morphological structures (e.g., eyes, wings, antennae) (Nijhout 1994).

The current developmental model that explains phenotype development in *O. taurus* postulates two critical windows late in juvenile ontogeny (Emlen and Nijhout 2001). The first involves the setting of horn disk sensitivity to JH and the second involves JH-mediated horn disk growth. Late in larval ontogeny, male morph is determined by projected adult size. As the limited food resource approaches depletion, male larvae that fall beneath a critical threshold mass and all female larvae (i.e., larvae that are destined to become hornless adults) experience a transient a spike in ecdysone production. Male larvae that maintain a mass above the threshold (i.e., larvae destined to produce horned adults) experience no such spike. Just before the prepupal stage, JH levels climb, producing horns on large males and hornless small males. The role of JH in morph determination during this second critical window has been confirmed by producing small horned males via topical application of a JH mimic (methoprene) at this time (Nijhout and Emlen 1998; Emlen and Nijhout 1999, 2001).

Although JH has been implicated as being involved in the facultative development of horns and the critical window of JH action has been defined, the mechanism(s) through which JH affects male phenotype has not been established. As was the case with altered TH action in amphibians, there are many (nonexclusive) JH-related mechanisms that underlie the differences between horn morphs (Emlen and Nijhout 1999, 2001). First, JH levels may be the same across all male sizes, but imaginal disk sensitivity to JH may differ between morphs. The transient spike in ecdysone that occurs in presumptive hornless males could reprogram cells in the horn disks, decreasing disk sensitivity to JH. This hypothesis is supported by changes in JH sensitivity that result from topical treatment with methoprene during the first critical period; treated males experience an increase in the size threshold for horn induction (Emlen and Nijhout 2001). If this hypothesis is correct, it means that the methoprene treatment during the second critical period induces horn production by greatly exceeding the higher threshold required for JH response in small males. Second, rates of JH production or elimination may differ between morphs such that large males have higher JH levels than small males during the critical period. Indeed, levels of JH synthesis scale linearly with the size of the corpora allata, which scales with body size in other insects (Emlen and Nijhout 2001). This suggests that larger (horned) males may have higher JH levels during the second critical period. Third, JH may be produced for a longer period of time in horned males, increasing the period of population growth for the cells that comprise the horn disks in these individuals. Finally, juvenile hormone esterase (JHE), an enzyme involved in JH degradation, could have greater titers or activity in small males. Differences in JHE are involved in polyphenism expression in other systems (discussed below).

Determining which of these, or other, mechanisms regulate horn ontogeny in this and other species of beetles exhibiting facultative horn production could greatly inform our understanding of the evolution of switch points. It would be also be interesting to know where in the developmental cascade response to artificial selection and natural selection on plasticity in horn morphology (e.g., Emlen 1996) have occurred. Furthermore, how genetic correlations shape or limit the evolution of phenotypic plasticity (and morphological traits in general) can be explored with this and other insect systems where individuals facultatively express alternative phenotypes in which the relative sizes of morphological traits differ among morphs.

There is a trade-off, represented by a negative genetic correlation, between horn size and adult eye size (Nijhout and Emlen 1998). These structures develop from imaginal disks sharing close proximity to one another, and manipulation experiments indicate that altering the size of one disk causes an inverse change in the size of the structures produced by adjacent, but not distant disks (Nijhout and Emlen 1998). Although the proximate basis of these interactions is not known, the evolutionary potential of the response will be determined largely by the mechanistic basis of the trade-off (Wolf et al. 2001). For example, resource competition or chemically mediated epigenetic effects among disks could have different effects on evolutionary outcomes (Nijhout and Emlen 1998). Also, slight changes in the sensitivity of one set of disks to regulatory hormones (i.e., shifts in thresholds or critical periods) may alter the relative timing among disks of the initiation or termination of the exponential growth phase. Small temporal changes of this type are likely to enhance the growth of one set of disks at the expense of adjacent disks, whereas larger temporal changes may effectively dissociate the ontogeny of the disks, possibly eliminating the trade-off altogether. Because disk growth is restricted to a small window of time late in juvenile ontogeny (Emlen and Nijhout 1999, 2001), such large temporal dissociation among structures may not be possible in this system. The restrictions imposed by development on the possibilities for altering this trade-off raise the issue of how the proximate mechanisms that regulate and integrate trait ontogeny can produce phylogenetic pattern in morphology or plasticity in morphological traits.

Comparative work indicates that the negative relationship between the size of horns and traits that develop in proximity holds across the *Onthophagus* phylogeny (Emlen 2001), suggesting that the interactions among developing disks produce phylogenic patterns in phenotypes. It appears likely that *Onthophagus* beetles have broken the developmental constraint imposed by disk–disk interactions by moving the horns to regions where they interfere the least with the development of other ecologically relevant traits (Emlen 2001). Nocturnal species of beetle are more likely than diurnal species to have horns near their eyes. Presumably this is because the reduction in eye size caused by large horns has little impact on relative fitness in nocturnal species. In these cases of facultative horn production, integrating an understanding of the proximate basis of trait development with the ecology of the morphologies helps to explain the phylogenetic pattern in phenotypes and plasticity of the traits (Emlen 2001; Emlen and Nijhout 2000).

A well-characterized facultative polymorphism in insects for understanding proximate endocrine mechanisms, life history trade-offs, and quantitative genetic parameters involves a set of dispersal polymorphisms in crickets (reviewed in Zera and Denno 1997; Zera 1999). Dispersal polymorphisms, where some individuals facultatively develop wings and flight musculature, occur in a variety of insects, differ in the environments that induce them, and differ in the postdispersal maintenance of flight muscles. Furthermore, because the sensitive period for wing induction differs widely across taxa, ranging from embryonic development through each juvenile instar (Zera and Denno 1997), wing polymorphisms represent another promising experimental model for investigating how the relationships among suites of traits are affected by the proximate mechanisms that create ontogenetic windows during which plastic responses can occur.

Crickets are hemimetabolous insects, and although they lack the imaginal disks discussed above, the wings of crickets develop largely during the last instar. As reviewed by Zera and Denno (1997), the juvenile hormone–wing morph hypothesis (Southwood 1961; Wigglesworth 1961) posits that JH levels above some threshold late in juvenile development act to inhibit growth and development of the wings and associated flight muscles, producing a short-winged (nondispersing) adult. In the cricket *Gryllus rubens*, wing development can be inhibited by experimental increases in JH titer late in the final larval instar (reviewed in Zera and Denno 1997; Zera 1999). Furthermore, JHE activity is greater and associated with increased levels of JH degradation in juveniles that develop into long-winged crickets compared with those that become short-winged morphs. These data are consistent with the juvenile hormone–wing morph hypothesis, but Zera (Zera and Denno 1997; Zera 1999) notes that these data are not definitive because the difference in JH levels between developing dispersing and nondispersing morphs is small enough to call into question its functional significance. Interestingly, ecdysone titers rise earlier and to a greater degree in the final juvenile instar of immatures destined to develop long wings—suggesting a role for ecdysone in morph determination similar to that suggested for the transient pulse of ecdysone in horned morphs of *O. taurus*.

As a description of the endocrine basis of development in this dramatic morphological, behavioral, and life-historical polymorphism, these investigations are valuable. However, these studies are made more informative because they, and those of other investigators (e.g., Fairbairn and Yadlowski 1997; Roff et al. 1997), have estimated several quantitative genetic parameters of the endocrine system and even performed elaborate selection experiments on components of the endocrine system (reviewed in Zera 1999; Zera and Huang 1999). Wings, flight muscles, and ovarian tissues trade off, as do other insect traits that develop in close proximity to one another (e.g., Klingenberg and Nijhout 1998; Nijhout and Emlen 1998). However, because development of these traits is dissociated temporally, the observed negative relationships among them probably results from division of a fixed energy budget among competing sinks (Zera and Denno 1997), not from interactions among traits during ontogeny.

Thus far we have discussed only discrete facultative polymorphisms in insects. Insects also illustrate continuous plastic responses to the environment. Diapause, where insects undergo a virtual developmental arrest, is a particularly dramatic developmental response to the environment present in nearly all insect orders, and is often facultative (Denlinger 1985; Nijhout 1994). Among species, diapause occurs in every stage of development, and it is usually associated with a response to poor or harsh environmental conditions, such as periods of low food availability or water scarcity (Nijhout 1999b). The genetic architecture underlying facultative diapause has been thoroughly described for populations of the pitcher-plant mosquito, *Wyeomyia smithii* (reviewed in Bradshaw and Holzapfel 2000). Across its geographic range, *W. smithii* differs in the photoperiod required to initiate and maintain larval diapause, and the threshold for diapause induction is polygenic (Bradshaw and Lounibos 1977; Hard et al. 1992, 1993). A cline in photoperiodic response exists such that mosquitoes from the more northern, derived populations differ from the southern, ancestral populations in the minimum critical photoperiod for diapause induction by 10 standard deviations (Bradshaw and Holzapfel 2000).

Extensive studies of the proximate basis of larval diapause have been on the final instar larvae of just a few lepidopterans (reviewed in Nijhout 1999b). Diapause is achieved developmentally by continued JH production, which inhibits PTTH production, resulting in developmental arrest. Developmentally, diapause is broken when environmental cues trigger the cessation of JH production, which in turn initiates PTTH production and the resumption of ontogeny. This pattern suggests that the mechanisms that regulate the timing of the typical molting cycle have been co-opted to achieve larval diapause (Nijhout 1999b), as seems to be the case for wing polyphenisms discussed above. It would be interesting to know the relationships between the well-characterized genetic differentiation among populations in critical photoperiod discussed above for mosquitoes and the mechanisms believed to underlie diapause regulation. Of particular interest would be to know the degree to which different aspects of the endocrine system are involved in the evolution of the switch point for diapause induction or resumption of development.

Summary and Conclusions

Phenotypic plasticity is a developmental phenomenon, and as such, the likelihood of the evolution and maintenance of adaptive plastic responses will depend not only on the ecological patterns of selection favoring plasticity on some trait but also on broad aspects of the regulation of ontogeny and the proximate mechanisms underlying trait development and phenotypic integration. We advocate the inclusion of developmental mechanism to deepen our understanding of the forces that shape the evolution of plastic responses.

Knowing how proximate mechanisms regulate trait ontogeny and integrate it with that of other traits will help elucidate how developmental processes might bias evolutionary outcomes. Such bias can be in the form of constraints; development defines the ontogenetic windows during which environmental signals must be perceived and processed, thereby limiting the possibilities for timely plastic responses. Development also determines the limitations on the degree of trait plasticity that can be exhibited. Bias can also be positive, facilitating the expression of plasticity in some traits, perhaps promoting diversification of lineages that regulate ontogeny of traits or life histories in particular ways. Understanding the proximate basis of trait ontogeny and integration also allows a more accurate interpretation of the evolutionary impact of the genetic correlations among traits believed to affect the evolution of plastic responses.

To more fully understand the evolutionary importance of phenotypic plasticity, we advocate an examination of the proximate basis of adaptive plasticity in traits and trait suites across taxa. Documenting phylogenetic pattern in the proximate basis of adaptive plasticity can suggest hypotheses regarding the relative impact of development on adaptive evolution, which can then be tested through direct manipulations of development. We propose a research program that crosses the boundaries of many subdisciplines in biology. It requires that investigators classically trained in evolutionary and developmental biology consider phenotypic variation and phenotype development in new ways. By fostering collaboration across disciplines, the proposed research program has the potential to increase our understanding not only of phenotypic plasticity but also of phenotypic evolution in general. In particular, species that exhibit discrete polyphenisms—where the differences between phenotypes are so great that alternative morphs have occasionally been misdescribed as separate species—offer powerful tools for investigating how proximate mechanisms affect the generation and evolution of morphological diversity.

Acknowledgments We thank the many people who have discussed the issues presented in this chapter. In particular, members of the 1998–2000 Brodie, Lively, and Delph labs at Indiana University, and D. Emlen, C. P. Grill, C. Lively, R. Repasky, S. Via, and A. J. Zera provided valuable insights and comments on earlier drafts of the manuscript. W.A.F. was supported during the writing of this chapter by fellowships from the Research and University Graduate School and the Center for the Integrative Study of Animal Behavior, Indiana University. This work was also supported in part by a National Science Foundation Dissertation Improvement Grant and a Bioinformatics Postdoctoral Fellowship (2001) awarded to W.A.F. Any opinions, findings, and conclusions or recommendations expressed in this publication are those of the authors and do not necessarily reflect the views of the National Science Foundation.

6

Modeling the Evolution of Phenotypic Plasticity

DAVID BERRIGAN
SAMUEL M. SCHEINER

Besides prediction, a theory should offer plausible explanation for what we know.

Levins (1968)

[A] good mathematical model provides two types of insight: it explains what we know and it predicts what we do not know.

Harvey and Purvis (1999)

Mathematical models have long played a critical role in evolutionary biology. Models allow us to systematically characterize complex interactions and to explore the consequences of assumptions about genotypes, phenotypes, and the environment. Models of phenotypic plasticity have shaped research and raised a host of compelling questions for ecologists, physiologists, and geneticists. Most formal models of plasticity evolution are only about 15 years old (e.g., Orzack 1985; Via and Lande 1985; Gavrilets 1986; Stearns and Koella 1986), although the concepts are much older (chapter 2). To some extent models have outrun data, as we discuss in this chapter. Yet substantial gaps still exist among plasticity models.

Decades ago, Levins (1968) pointed out that models attempt to maximize three things: generality, precision, and realism. However, only two can ever be maximized within a single model. Until now, models of plasticity evolution have concentrated on the first two factors, generality and precision. Few models have attempted to maximize realism, especially with regard to the melding of genetics with traits such as morphology and life history. One goal of this chapter is to point out potential avenues toward this melding.

This chapter is organized around a series of questions about phenotypic plasticity (table 6.1). These questions fall into three main categories: (1) When will natural selection favor plasticity over fixed strategies? (2) What form will plasticity assume? (3) What are the evolutionary dynamics of phenotypic plasticity? Answers to these questions provide a framework for understanding the ecology and evolutionary biology of phenotypic plasticity.

Modeling Approaches

Before addressing the questions posed above, we briefly review the strengths and weaknesses of three approaches to modeling plasticity: optimality, quantitative genetic, and gametic (table 6.2) [for more details, see Scheiner (1993a)].

Optimality Models

In its pure form, the optimality approach ignores the complications of genetics and presumes that selection can act on variation resulting in the fixation of the optimal strategy or combination of strategies. The optimal strategy maximizes fitness, subject to an identified set of constraints (Parker and Maynard Smith 1990). Many optimality models that predict the shapes of reaction norms assume that one or more traits are fixed by the environment. An optimal reaction norm is then calculated for another trait or combination of traits. One well-developed and much debated example involves reaction norms for age and size at maturity when growth rates are imposed by the environment (e.g., Stearns and Koella 1986; Kawecki and Stearns 1993; Berrigan and Koella 1994; Day and Taylor 1997). We return to this topic when we discuss what form plasticity will take. The optimality approach is also related to the polynomial model used in quantitative genetics (see below), although specific links have yet to be explored. Although optimality models are the most realistic with regard to specific traits, they are the least realistic with regard to genetics.

The great strength of optimality models is their power to illustrate the quantitative influence of assumptions about phenotype and environment on the outcome of evolution. The outcome predicted by such a model defines a state of maximal fitness. The great weaknesses of the approach are that it does not generate predictions about the dynamics of evolutionary change and it ignores potential constraints associated with the underlying genetic architecture.

Table 6.1. Major questions addressed by models for the evolution of phenotypic plasticity.

Question	Issues
1. When should a plastic strategy be favored over a fixed strategy?	Temporal versus spatial heterogeneity, reliability of cues, uniformity of selection, costs of plasticity
2. What will the plasticity be like?	Discrete versus continuous plasticity, shapes of reaction norms, time scale of plastic responses
3. How fast will plasticity evolve?	Genetic basis of plasticity, strength of selection, population structure, migration characteristics

Table 6.2. Models of the evolution of phenotypic plasticity classified by modeling approach, environmental discreteness, and mode of heterogeneity (spatial vs. temporal).

	Environmental Gradient	
Modeling Approach	Discrete	Continuous
Optimality		
Spatial	**Lively (1986a)**	Stearns and Koella (1986)
	Fagen (1987)	**Kozlowski and Wiegert (1986, 1987)**
	Houston and McNamara (1992)	Werner (1986)
	Moran (1992)	Perrin and Rubin (1990)
	Padilla and Adolph (1996)	Sibly (1995)
	Sultan and Spencer (2002)	**Adler and Karban (1994)**
		Berrigan and Koella (1994)
Temporal	**Moran (1992)**	Caswell (1983)
	Leon (1993)	Clark and Harvell (1992)
	Jablonka et al. (1995)	Hutchings and Myers (1994)
	Padilla and Adolph (1996)	Gilchrist (1995)
	Sultan and Spencer (2002)	Abrams et al. (1996a)
Quantitative genetic		
Spatial	Via and Lande (1985, 1987)	Lynch and Gabriel (1987);
	Van Tienderen (1991, 1997)	Gillespie and Turelli (1989)
	de Jong and Behera (2002)	García-Dorado (1990)
		Gabriel and Lynch (1992)
		Gomulkiewicz and Kirkpatrick (1992)
		Zhivotovsky and Gavrilets (1992)
		Whitlock (1996)
		de Jong and Gavrilets (2000)
		Tufto (2000)
Temporal	Sasaki and de Jong (1999)	Gavrilets (1986, 1988)
		Lynch and Gabriel (1987)
		Gabriel and Lynch (1992)
		Gomulkiewicz and Kirkpatrick (1992)
		Gavrilets and Scheiner (1993a)
		de Jong and Gavrilets (2000)
		Tufto (2000)
Gametic		
Spatial	Levins (1963)	de Jong (1989, 1990a, 1999)
	Castillo-Chaves et al. (1988)	Zhivotovsky et al. (1996)
		Scheiner (1998)
Temporal	Levins (1963)	None
	Orzack (1985)	

Models that include costs are indicated in bold. The lists are relatively complete for genetic models but are a small and arbitrary subsample of the numerous optimality models that can be interpreted as models for the evolution of the shapes of reaction norms. [Modified from Scheiner (1993a).]

Quantitative Genetic Models

Quantitative genetic models incorporate genetic variance–covariance matrices into dynamic predictions about reaction norms in alternative environments. For phenotypic plasticity, there are two basic approaches to quantitative genetic modeling (figure 6.1). The first approach treats plasticity as a set of character states in a series of environments, with plasticity being defined by the cross-environment covariance (Falconer 1960). The second approach treats plasticity as a reaction norm defined as a polynomial function across an environmental gradient (Scheiner and Lyman 1989; de Jong 1990a,b; Gavrilets and Scheiner 1993a,b; Tufto 2000). Character state models (e.g., Via and Lande 1985, 1987; Gomulkiewicz and Kirkpatrick 1992) examine the evolution of reaction norms using within-environment means and genetic variances and covariances. In the polynomial approach, the reaction norm is described by an equation and the evolution of the reaction norm is modeled using genetic variances and covariances of the equation parameters (e.g., Gavrilets and Scheiner 1993a,b). A recent advance in this approach is the use of characteristic values to define the reaction norm (chapter 4; Gibert et al. 1998a), although these have yet to be applied in any model.

There was a brief but vigorous debate over the meaning and merits of these two approaches in the late 1980s and early 1990s. The debate was resolved by Van Tienderen and Koelewijn (1994) and de Jong (1995) for discrete environments. They showed that the approaches are mathematically equivalent for discrete environments as long as the reaction norm can be described as a differentiable function of the environment. The two approaches lend themselves to characterizing different kinds of plasticity. Character state models are most useful for analyzing discrete character states and environments such as predator induced changes in defense or seasonal polyphenisms. The polynomial approach is more appropriate for analyzing continuous characters and environmental variables such as temperature or amount of food (Gomulkiewicz and Kirkpatrick 1992; Via et al. 1995; Roff 1997). Phenotypic plasticity involving discrete morphs may result from continuously varying cues, however, so the nature of the plasticity might not determine the most informative modeling approach (Gomulkiewicz and Kirkpatrick 1992; Van Tienderen and Koelewijn 1994). Currently, no rigorous study has examined the relative efficacy of the different approaches to analyzing reaction norms of discrete morphs over continuous environments.

One notable feature of table 6.2 is the lack of quantitative genetic models for phenotypic plasticity across environmental gradients that show discrete variation on a tempo-

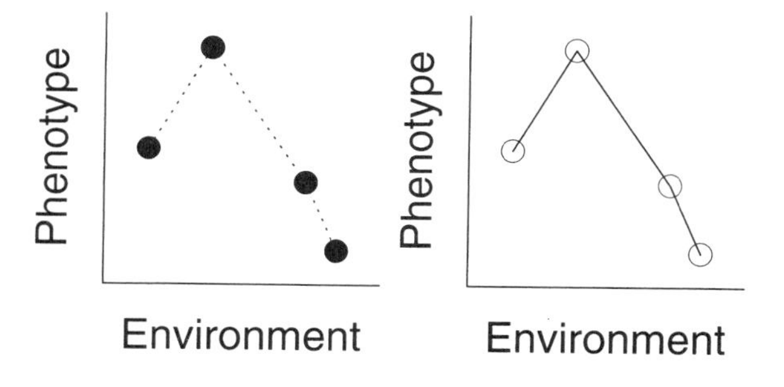

Figure 6.1. In the analysis of phenotypic plasticity, the character state approach (left) focuses on traits in each environment, whereas the polynomial approach (right) focuses on the reaction norm. [After de Jong (1995).]

ral scale. Perhaps this is because the environment seldom (if ever) changes in discrete steps in time. Current models that use the polynomial approach could be easily modified for temporal variation. On the other hand, it may be fruitful to consider the evolution of complex life cycles in this context. Once an organism evolves a life cycle with two or more components, it may then experience transitions from one habitat to another very different one, for example, the transition from aquatic to terrestrial of many insects and amphibians. It would be interesting to explore the consequences of quantitative genetic correlations across life stages for the evolution of phenotypic plasticity within a life stage. Such correlations are known (e.g., Loeschcke and Krebs 1996; Krebs et al. 1998), but we are unaware of theoretical work explicitly addressing this issue.

The most appropriate model will also depend on the focus of selection. The character state model is most appropriate if the focus of selection is on individuals within demes. In contrast, the polynomial approach is most appropriate if the focus of selection is on genomes across demes because selection directly on reaction norms requires some sort of group selection in a structured environment (Scheiner 1998). Until now the two approaches have been treated separately. We speculate that a hybrid approach might be fruitfully applied to the investigation of the effects of population structure on plasticity evolution. Much work remains to be done in this area, because there are likely to be subtle effects of source–sink population structure both on the dynamics of reaction norm evolution and on the form of the optimal reaction norm (Houston and McNamara 1992; Kawecki and Stearns 1993; de Jong and Behara 2002).

The strength of the quantitative genetic approach is that it can make dynamic predictions in addition to predictions about the optimal reaction norm. Its weaknesses are associated with the added complexity of incorporating genetics; namely, it is much more difficult to determine the potential importance of biotic and abiotic forces for the evolution of optimal strategies. Furthermore, various assumptions of quantitative genetic models are the subject of active debate (Barton and Turelli 1989; Roff 1997; Lynch and Walsh 1997). Although some assumptions are reasonable to a first approximation (e.g., many traits appear to be controlled by a reasonably large number of loci), the assumption of purely additive effects remains problematic (Crnokrak and Roff 1995). Even more doubtful is the assumption that genetic variances and covariances remain constant during evolution. This assumption limits the utility of this approach to making short-term evolutionary predictions (Pigliucci 1996b). To date, quantitative genetic models have been largely general, focusing on broad characteristics of plasticity and its evolution, and providing less insight into specific environmental factors influencing the form of plasticity.

Gametic Models

In gametic models, individual loci are followed, the simplest version being a one-locus, two-allele model (Haldane 1924). They have the advantage of most easily incorporating details such as dominance, epistasis, and linkage. A major limitation of such models is relating the evolution of individual loci to the evolution of traits such as morphology and life history. It is possible to create analytical models with large numbers of loci (Zhivotovsky et al. 1996). However, this can only be done by making simplifying assumptions such as weak selection, weak linkage, no epistasis, and random mating. An alternative approach is to use simulation models, especially as computing power becomes more and more accessible (e.g., Scheiner 1998). In studies of plasticity evolution, gametic models are the least developed (table 6.2). They have led, however, to some important insights (see below). An important next step will be extending these models to real traits in situations where we know the genetic mechanisms of plasticity (chapter 3). Simulation models will almost certainly be required to analyze the interaction of multiple demes and multiple loci.

When Will Plasticity Be Favored over Fixed Strategies?

Natural selection will favor plasticity over fixed strategies when the mean fitness of individuals with the plastic strategy exceeds that of individuals with the fixed strategy. Very generally this requires (1) environmental heterogeneity, (2) reliable cues, (3) benefits that outweigh the costs of plasticity, and (4) a genetic basis to plasticity. We now discuss the contributions of models to understanding the consequences of these factors for the evolution of plasticity.

Environmental Heterogeneity

Environmental heterogeneity is a necessary, but not sufficient, condition for plasticity to be favored. A uniform environment will always favor a fixed phenotype even if plasticity is not costly. Beyond this trivial insight, however, it is not at all obvious when, how much, and what form of heterogeneity is necessary for plasticity to be favored.

We can recognize two broad classes of heterogeneity, spatial and temporal. Temporal heterogeneity can be further subdivided into within-generation and between-generation. Spatial heterogeneity is similar in effect to between-generation heterogeneity in that a single organism experiences only one environment during its lifetime. Finally, for within-generation temporal variation, traits may be fixed at a single time during development or may be labile, continuing to vary over an organism's lifetime. A variety of modeling approaches have examined all of these patterns of variation, although quantitative genetic and gametic models have mostly focused on between-generation or spatial variation.

Optimality models have focused on several factors that limit the evolution of plasticity, notably unpredictability (e.g., Cohen 1968; Kaplan and Cooper 1984), costs of plasticity (e.g., Moran 1992; Padilla and Adolph 1996), and the timing of environmental fluctuations (e.g., Levins 1968; Jablonka ad Szathmary 1995). Unpredictability, costs of plasticity, and environmental fluctuations that occur more rapidly than the organism can change its phenotype, and all reduce the likelihood that plasticity will evolve. These models, and others listed in table 6.2, generally focus on either spatial or temporal variation in the environment. One notable exception is the work of Moran (1992).

Moran explicitly contrasts the consequences of spatial and temporal variation. She considers the evolution of a developmental switch that results in discrete differences between individuals. In her model there are two environments and two phenotypes. In the absence of costs of plasticity, if there is a trade-off between performance in one environment versus the other, then spatial heterogeneity always favors phenotypic plasticity. Costs of plasticity and unpredictability can result in selection favoring fixed strategies (figure 6.2). With temporal variation, similar caveats apply but plasticity is favored

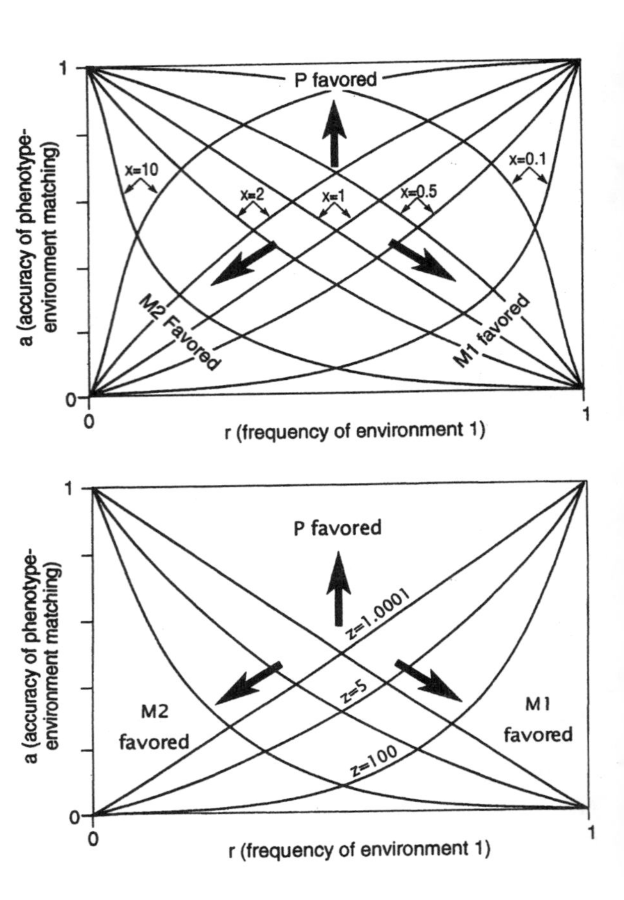

Figure 6.2. Consequences of environmental frequency, ability of the organism to match the environment (a), and proportional fitness advantage of the more fit phenotype (x or z) for the evolution of plastic (P) versus fixed (M1 and M2) strategies. With both spatial heterogeneity (A) and temporal heterogeneity (B), plasticity is favored when the phenotype reliably matches the environment and when the proportional fitness advantage is high. Overall, plasticity is more favored under temporal heterogeneity than under spatial heterogeneity. [Reprinted with permission from Moran (1992).]

under a broader range of conditions. Moran points out that this is consistent with the relative frequency of seasonal polyphenism compared with polyphenisms associated with spatial heterogeneity. However, note that plasticity for continuous traits is also very common under spatial heterogeneity, as in the case of age and size at maturity in insects. Moran does not consider changes in the timing of environmental fluctuations. Jablonka et al. (1995) showed that details of the timing of environmental change can result in either selection for plasticity, or selection for carryover effects, that is, changes in phenotypes that persist for several generations.

Spatial heterogeneity and between-generation temporal heterogeneity are explicitly compared in one gametic model with two discrete environments (Levins 1963). Levins concludes that, given sufficiently large differences among environments, plasticity will be favored in both cases. However, spatial variation will select for discrete phenotypic states, whereas temporal variation will select for continuous phenotypic change. All three types of heterogeneity are compared in three quantitative genetic models with continuous environmental variation (Lynch and Gabriel 1987; Gomulkiewicz and Kirkpatrick 1992; Gabriel and Lynch 1992). Lynch and Gabriel (1987) conclude that for nonlabile traits, spatial variation is likely to result in alternative fixed strategies, whereas both within-generation and among-generation variation will result in plasticity. Finally, Gabriel and Lynch (1992) considered the case where the mean is fixed and evolution can alter only the phenotypic variance. In this instance, both spatial and between-generation temporal variation favor plasticity, whereas within-generation variation favors a fixed strategy. Given these disparities in model conclusions, more work is needed to determine when and how different patterns of environmental heterogeneity affect plasticity evolution.

With few exceptions, models concerning temporal heterogeneity are highly abstract, and environmental fluctuations are assumed to be either periodic or stochastic. However, some characteristic features of the environment that are likely to influence fitness are easy to measure. For example, temperature fluctuates predictably on daily and seasonal cycles but can be quite unpredictable on intermediate (e.g., weekly) cycles (Kingsolver and Huey 1998). More realism (*sensu* Levins) could be added to models for the evolution of plasticity if they explicitly incorporated information about environmental variables such as temperature, rainfall, or salinity and used this information to compare life history differences along environmental gradients. Because most environmental factors vary both in time and space, more work is needed to integrate both types of variation into single models.

The Reliability of Cues

Cue reliability is related to patterns of within-generation environmental heterogeneity, the lag between the time a trait becomes fixed during development and when selection occurs. All models agree that an unreliable cue will result in a fixed strategy. The evolutionary outcome is quite sensitive to cue reliability and the frequency distribution of environments (Orzack 1985; Clark and Harvell 1992; Moran 1992; de Jong 1999; Saski and de Jong 1999; Tufto 2000; Sultan and Spencer 2002). A key factor is developmental delay, phenotypic constraints on how fast an organism can respond to a change in the environment (Levins 1968; Clark and Harvell 1992; Leon 1993; Padilla and Adolph 1996). More empirical work on such developmental constraints is necessary (chapter 5).

Migration

Nearly all models of plasticity evolution ignore the effects of population structure. Temporal variation is modeled within a single population, or spatial variation is modeled with only two populations and panmixia each generation. Metapopulation structure can be added to such models with multiple demes (populations) in which only a portion of the population migrate each generation. Migration interacts with spatial heterogeneity to create variation in the environment experienced by a lineage over time, converting spatial variation into temporal variation. Because temporal variation is more favorable for plasticity, we would predict that increased migration rates would favor plasticity. Three genetic models confirm this prediction (Scheiner 1998; de Jong 1999; Tufto 2000). On the other hand, migration might disfavor plasticity if cues are not reliable. An optimality model (Sultan and Spencer 2002), which updates that of Moran (1992), includes both spatial and temporal heterogeneity, as well as incorporating effects of migration and cue reliability. It reiterates that plasticity is favored when environmental heterogeneity and cue reliability are high. It also emphasizes that environmental heterogeneity, migration, and cue reliability interact such that greater migration rates or temporal heterogeneity favors plasticity only when cues are reliable. Similarly in a genetic model, de Jong and Behera (2002) show that adult migration after the phenotype has been determined can lead to the evolution of nonoptimal reaction norms. In their model, source–sink dynamics can also result in a compromise reaction norm weighted toward the phenotype favored in the source population.

Costs of Plasticity

Environmental heterogeneity is apparent on a variety of spatial and temporal scales, yet not all organisms change all elements of their phenotypes to match this heterogeneity. One possible reason for incomplete matching is costs of plasticity (Bradshaw 1965; DeWitt 1998; Scheiner and Berrigan 1998). Costs can manifest themselves in a variety of ways (DeWitt et al. 1998). Several models explicitly address the consequences of various combinations of three of these costs: maintenance, production, and information (table 6.2). Four models use an optimality approach to examine developmental polyphenism or reversible plasticity in two environments. Lively (1986a) analyzed the effects of an aggregate cost of plasticity on the evolution of alternative developmental polymorphisms. The influence of costs depends on the reliability of the cue. When the cue is reliable, intermediate cost levels result in selection for developmental plasticity. Curiously, both very low costs and very high costs increase the circumstances under which fixed strategies are favored. Moran (1992) analyzes a somewhat more abstract example of an optimality model for a developmental polyphenism (figure 6.2). A recent expansion of this model reaches a similar conclusion (Sultan and Spencer 2002). Here costs generally reduce the range of circumstances in which plasticity is favored. Padilla and Adolph (1996) consider constitutive (maintenance) and production costs in a model for reversible responses to the environment with temporal heterogeneity. They show that costs of reversible plasticity influence the sensitivity in the time required to change phenotypes. This model seems relevant to a variety of biotic and abiotic factors that result in plastic changes in morphological or physiological traits. Information costs have been analyzed in models of foraging behavior (Stephens and Krebs 1986), but we know of no examples explicitly addressing them in the context of plasticity occurring over longer time scales.

Models of inducible defenses in plants and animals, a form of plasticity, have also shown that costs of defense influence the evolution of inducible resistance. As might be expected, the higher the cost, the more stringent the conditions for the evolution of plastic responses to predators or herbivores (Edelstein-Keshet and Rausher 1989; Riessen 1992; Adler and Karban 1994).

Van Tienderen (1991, 1997) reports the influences of costs of plasticity in a quantitative genetic framework. He analyzes an aggregate measure of costs that is linearly related to the magnitude of plasticity. Costs reduce the slope of the reaction norm. This is a satisfying result, but we wonder if it depends on the simple relationship assumed between the slope of the reaction norm and the fitness costs of plasticity.

Models demonstrate that costs can substantially alter the evolution of plasticity. But, how prevalent are costs? If they are rare or of small effect, then we need not consider them further. Conversely, if they are common or of large effect, then we should work to incorporate costs into a wider variety of models. To date only a few studies have demonstrated significant costs of plasticity (e.g., Krebs and Feder 1998). Other studies have either failed to find costs or the costs were very small (DeWitt 1998; Scheiner and Berrigan 1998; Krebs and Feder 1998; Winn 1999; Donohue et al. 2000, Relyea 2002).

So far, costs have appeared only where the actions of one or a few genes are involved. Where entire life histories are altered, costs appear to be spread sufficiently across the entire genome as to be negligible. We call for more empirical work on the costs of plasticity and for models explicitly and systematically contrasting consequences of costs (and the lack thereof) that manifest themselves in different ways. Analyses of immune responses could be a fruitful area. For example, a recent report concerning *Drosophila* suggests that successful immune response to parasitoids results in decreased resistance to abiotic stress (Hoang 2001), and *Drosophila* are known to show genetic variation in their degree of resistance to parasitoid attack. Given our current observations on the manifestation of costs, efforts should also be directed to incorporating costs into gametic models.

Genetics of Plasticity

As discussed in chapter 3, the genetic basis of plasticity has been a substantial concern for some time. We are now accumulating increasing evidence that trait values and plasticity result from a complex interaction of many genes. These complex interactions can influence the outcome of evolution. Current models are just beginning to explore these effects, and these issues are the realm of gametic models, because they deal with details of genetic architecture.

At a minimum, the phenotype of plastic traits may be determined by two types of loci, those sensitive to the environment—plasticity loci—and those insensitive to the environment—mean loci (Bradshaw 1965; Gavrilets 1986; Scheiner and Lyman 1989; de Jong 1989; Via et al. 1995). Models of the evolution of plasticity can be constructed with just plasticity loci or with both plasticity and mean loci. Scheiner and Lyman (1989) labeled these two types of models pleiotropic and epistatic, respectively. In a *pleiotropy model*, the only loci are those whose allelic expression differ across environments. An *epistatic model* has a second class of loci whose expression is independent of the environment. This second type of model is epistatic in the sense that the final phenotype is not a simple additive function of the two sets of loci but, rather, depends on an environmental interaction across locus types. Pleiotropy models can be considered a special case of epistasis models in which there is no genetic variation at the environment-independent loci. In both types of models, all gene action within environments is strictly additive. Most gametic models have only considered the existence of plasticity loci (e.g., Levins 1963; Orzack 1985; de Jong 1989).

The one model that compared the two types of genetic architectures is Scheiner (1998). He used a gametic simulation model to demonstrate that, in a structured population, the two types of genetic architectures lead to different evolutionary outcomes. A pleiotropy model always maximizes phenotypic plasticity. In contrast, an epistatic model typically results in a mixed outcome. De Jong (1999) found the same result using an algebraic model, although she attributed the result to the unpredictability of selection due to adult migration and population structuring. Although local fitness is maximized, global fitness is a compromise between plasticity and a fixed strategy. Tufto (2000) developed a similar model and reached similar conclusions. Scheiner (1998) speculated that with multiple loci affecting both the mean and plasticity, populations can get trapped on local fitness peaks. Further work looking at the effects of starting allele frequency is necessary to prove this assertion.

In the above discussion, the terms "pleiotropy" and "epistasis" are being used outside their typical context (Phillips 1998). Alleles are epistatic across environments (genotype–environment interaction), whereas they are strictly additive within environments. Nonadditive within-environment effects have not yet been explored. All models to date assume strict additivity. Clearly, this is an area in need of exploration.

What Specific Form Will Plasticity Assume?

Plasticity can occur for continuous or discrete traits. Predictions about plasticity in continuous traits involve comparisons of the shapes of reaction norms. Predictions about discrete traits involve probabilities of alternative phenotypes conditional on the environment. Discrete traits can be treated in a quantitative genetics framework by postulating an expression threshold and an underlying set of multiple loci responsible for the threshold (Roff 1997).

Optimality models, tailored to particular systems, are a rich source of predictions about the shapes of reaction norms and the agents of selection. Specific examples of models predicting the shapes of reaction norms are common in the literature on age and size at maturity (e.g., Stearns and Koella 1986; Satou 1991; Berrigan and Koella 1994; Day and Taylor 1997). Many other examples can be found in the literature on life history evolution and behavioral ecology (e.g., Charnov 1982; Stephens and Krebs 1986; Stearns 1992; Roff 1992). Overall, any factor influencing the optimal strategy in a single environment will also influence the shape of the optimal reaction norm. Additionally, important recent work has shown that spatial heterogeneity can influence the position of the optimal reaction norm, independent of the direct effects of the environment on performance. These influences manifest only when the intrinsic rate of increase is the appropriate fitness measure. Therefore, it is critically important to integrate studies of demography with experimental studies of phenotypic plasticity in the field.

Houston and McNamara (1992) and Kawecki and Stearns (1993) are responsible for the insight that source–sink population structure can influence reaction norm shape. If net reproductive rate is the appropriate fitness measure, then selection will favor the life history strategy that maximizes lifetime offspring production in each habitat. In contrast, when fitness is determined by the intrinsic rate of increase (r (0), then the optimal reaction norm is influenced by the probability of settlement in a particular patch (Kawecki and Stearns 1993). For example, clutch size in an iteroparous organism is predicted to be sensitive to juvenile survival in a model that ignores population structure, but insensitive to it when population structure is explicitly incorporated (figure 6.3). Predictions about the effects of population structure are testable. For example, many insects oviposit in discrete hosts, and these hosts may differ in quality. It would be very interesting to extend a model for optimal clutch size to include population structure and determine if such structure was indeed influencing the reaction norm of clutch size among hosts.

When population structure does not influence fitness (and therefore the shape of optimal reaction norms), predictions about phenotypic plasticity can be generated simply by examining any optimality model and exploring its predictions for a range of

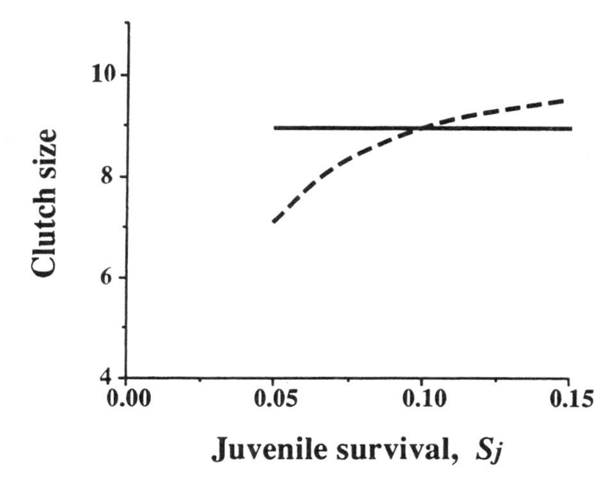

Figure 6.3. Reaction norms for clutch size when juvenile survival varies. The dashed line gives the predictions from a model that ignores populations structure (Stearns and Koella 1986), and the solid line is from a model that explicitly incorporates it. [From Kawecki and Stearns (1993).]

parameters (Roff 1992). Thus, the entire array of optimality analyses can be interpreted as a set of predictions about reaction norms. However, inspired by the insights of Houston, McNamara, Kawecki, and Stearns, much new theoretical work is being devoted to the relationship between fitness, population structure, and reaction norm evolution (e.g., Day and Rowe 2002; de Jong and Behara 2002). These recent results make it clear that age- and stage-specific differences in migration behavior and the physiological constraints associated with transitions between developmental stages can both act to strongly influence the evolution of reaction norms. It remains to be determined how strong these influences are in practice, and of course, all these predictions assume that the organism can detect environmental cues that predict environmental change, and alter its phenotype appropriately and cheaply.

Currently, quantitative genetic and gametic models have relatively little to say about the shapes of reaction norms, primarily because most models either only consider two environments or only consider linear reaction norms. Four models have explicitly explored shape evolution. One quantitative genetic model found that temporal variation always results in linear reaction norms (Gavrilets and Scheiner 1993a). Another quantitative genetic model showed how hard and soft selection results in different reaction norm shapes and that genetic correlations can constrain the reaction norm far from the optimum (Gomulkiewicz and Kirkpatrick 1992). Two gametic models also found that genetic architecture can overwhelm ecology, resulting in curved reaction norms even when a linear reaction norm was optimal (figure 6.4; Scheiner 1998; de Jong 1999). These results suggest that genetic, physiological, or developmental constraints may be more important than environmental effects in determining reaction norm shape. Other models should be extended to multiple environments that include spatially explicit patterns of environmental heterogeneity. Another promising, but as yet unexplored, direction of attack is the influence of selection on the characteristic values of reaction norms (chapter 4). More work is required in this area.

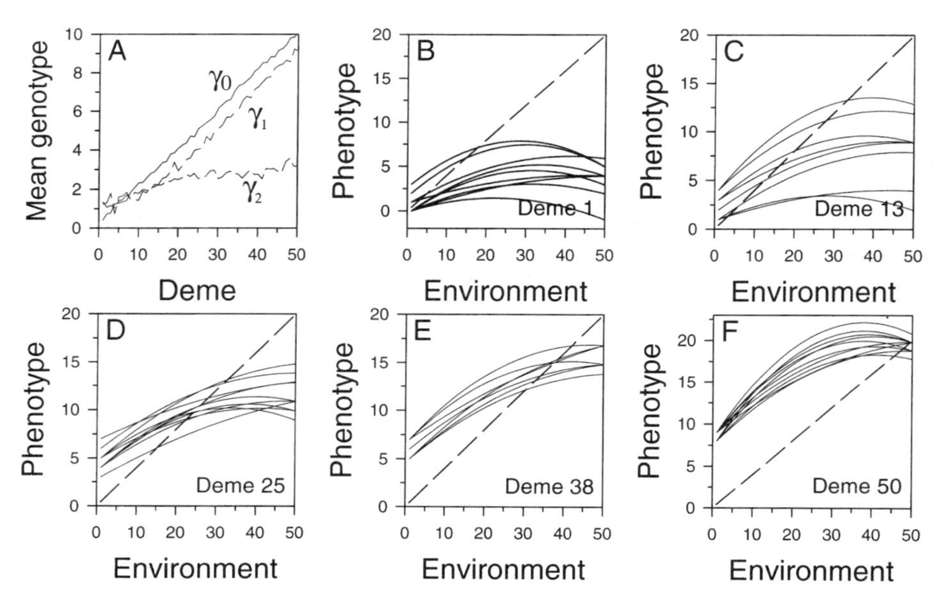

Figure 6.4. Genotypic and phenotypic values for a gametic model with a linear selection function and loci that affect are environmentally insensitive (γ_0), environmentally sensitive with a linear effect (γ_1), and environmentally sensitive with a quadratic effect (γ_2). The environment consists of a stepping-stone array of 50 demes. The migration rate was 10% with a maximal migration distance of 5 demes. (A) Mean genetic values for γ_0, γ_1, and γ_2. (B–F) Reaction norms for ten randomly chosen individuals from each of five demes along the cline (solid lines). The optimal phenotype is indicated by the dashed lines. [From Scheiner (1998).]

What Are the Evolutionary Dynamics of Phenotypic Plasticity?

Given that selection favors the evolution of plasticity, it is important to consider how, if, and when selection will result in fixation of the optimal strategy in a population. Costs of plasticity, the genetic basis of plasticity, the strength of selection, and the demography of natural populations are all likely to play a role in the dynamics of plasticity evolution.

Few models have explicitly examined evolutionary dynamics. The vast majority are concerned only with the evolutionary equilibrium or end point (e.g., Orzack 1985; Gabriel and Lynch 1992, García-Dorado 1990). Other models provide equations for dynamics but do not explicitly explore trajectories (e.g., Gavrilets and Scheiner 1993a; Zhivotovsky et al. 1996). The most extensive examination of dynamics is contained in one class of quantitative genetic models (Via and Lande 1985, 1987; Gomulkiewicz and Kirkpatrick 1992; Van Tienderen 1991, 1997). Those models show that the structure of the genetic variance–covariance matrix can substantially alter the direction of approach and time to equilibrium. The population may spend substantial time away from the evolutionary optimum and then suddenly shift to the optimal phenotype. Dynamics differ under hard and soft selection, with the approach to equilibrium usually being more gradual under soft selection (figure 6.5).

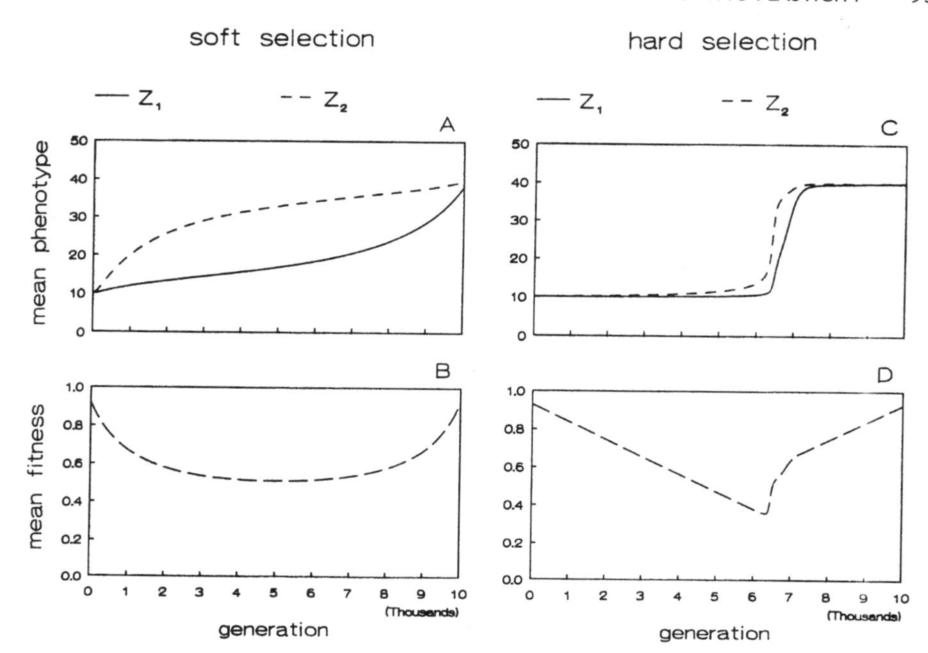

Figure 6.5. Evolution over 10,000 generations during which the frequency of one habitat gradually declines from 1 to 0. Change under soft selection in mean phenotype (A) and mean fitness (B) in each environment (Z_1 and Z_2). Plasticity is the difference between Z_1 and Z_2. Change under hard selection in mean phenotype (C) and mean fitness (D). [Reprinted with permission from Van Tienderen (1991).]

Empirical studies of the evolutionary dynamics of phenotypic plasticity are also rare. Plasticity can evolve rapidly under artificial selection (Scheiner 2002); Roff (1997, p. 198) lists 11 examples. But we know little about the dynamics of plasticity evolution more generally. Quantitative genetic models predict that plasticity will evolve rather slowly because heritabilities of plasticity are often low (Scheiner 1993a). In addition, selection on plasticity is based on an allele's global fitness, which can be much weaker than are opposing local fitness effects (Zhivotovsky et al. 1996; Scheiner 1998). These predictions are borne out by three quasi-natural selection experiments. Such experiments are laboratory mimics of natural selection. A series of experiments involving both spatial and temporal variation and 350–950 generations of *Chlamydomonas reinhardtii* obtained a significant response (Bell and Reboud 1997; Reboud and Bell 1997). In contrast, an experiment involving temporal variation and 10 parthenogenetic generations of *Daphnia pulex* failed to get a change in plasticity (Scheiner and Yampolsky 1998). On the other hand, a recent study of the evolution of phenotypic plasticity in resistance to algal toxicity in a lake population of *Daphnia galeata* suggested a change in the reaction norm in about 15 years (Hairston et al. 2001). An obvious place to look for additional examples would be cases where trait means have evolved rapidly in natural populations. Huey et al. (2000) describe the rapid evolution of a geographic size cline in *Drosophila subobscura* introduced to North America, for example. Studies of phenotypic plasticity in such introduced species could help characterize the evolutionary dynamics (or statics) of plasticity.

More theoretical work on the evolutionary dynamics of phenotypic plasticity could help guide empirical work. For example, most current comparative studies of phenotypic plasticity examine differences among closely related populations (e.g., Dudley and Schmitt 1995; Rodd et al. 1997), although more distant comparisons have been done (e.g., Barker and Krebs 1995; Gibert et al. 1996). The implication of the quasi-natural selection experiments is that adaptive plasticity differences will most likely be found in distantly related populations, or sibling species. We need models to tell us what phylogenetic level will provide the most fruitful comparisons.

Conclusions

Although theory has made numerous predictions, and there are many experiments characterizing plasticity, the link between the two often appears tenuous. However, two kinds

Table 6.3. Areas for further development.

- Use of characteristic values in quantitative genetic models (chapter 4; Gibert et al. 1998a)
- Quantitative genetic models for phenotypic plasticity across environmental gradients that show discrete temporal variation.
- Consequences of genetic correlations across life stages for the evolution of phenotypic plasticity.
- Combinations of character state and polynomial quantitative genetic models to explore consequences of population structure.
- More empirical and theoretical work on source–sink population structure and its consequences for plasticity (de Jong and Behera 2002).
- Gametic simulation models, particularly the incorporation of pleiotropy, epistasis, and nonadditive within-environment genetic effects, and the incorporation of empirical information on the genetics of plasticity (Scheiner 1998).
- More work on effects of different patterns of spatial and temporal variation, especially in quantitative genetic models, with the aim of resolving differences in predictions among current models (Lynch and Gabriel 1987; Gomulkiewicz and Kirkpatrick 1992; Gabriel and Lynch 1992).
- Explicit incorporation of data concerning natural environmental variability into simulation models for the evolution of phenotypic plasticity (Kingsolver and Huey 1998).
- Experimental and theoretical analyses of the costs of plasticity and their influence on the evolution of plasticity, particularly those involving different kinds of costs (DeWitt et al. 1998). Such empirical guidance is needed before we begin the process of reassessing models in the light of costs. Whether costs are large, small, or nonexistent will potentially greatly alter model conclusions and tell us whether such modeling efforts are even warranted.
- Empirical work on demography in populations differing in the amount and form of plasticity.
- Models of the dynamics of plasticity evolution (Via and Lande 1985, 1987; Van Tienderen 1991, 1997).
- Further work on the comparative biology of plasticity, both empirical and theoretical.
- Integration of biomechanical analyses with optimality and genetic approaches to modeling the evolution of plasticity. Biomechanical models can clearly delineate the performance consequences of alternative phenotypes and give clues as to how to model the costs and benefits of plasticity (e.g., Grunbaum 1998; Van der Have and de Jong 1996; Woods 1999; chapters 6–8).
- Experimental work devoted to identifying nonadaptive versus adaptive plasticity (Huey and Berrigan 1996; chapters 9 and 10). This information will tell us which types of traits are likely to respond to selection on plasticity and what types of nonadaptive constraints we need to include in models.

of experimental results are consistent with the ubiquity of plasticity for various traits. First, when plasticity is present, it is cheap, as testified to in a variety of morphological and behavioral studies. Second, when environmental cues are predictable, there is plasticity. Plastic responses to temperature, photoperiod, predator kairomones, and herbivory are examples of good cues for biotic or abiotic factors influencing fitness. Thus models for the evolution of plasticity explain some of what we know. But do they predict what we do not know? We answer in the affirmative, notably with regard to predictions involve costs of plasticity, effects of time lags and uncertainty on the evolution of plasticity, and effects of spatial and temporal heterogeneity. We challenge ecologists and experimentalists to identify natural systems that allow robust comparative and experimental tests of such hypotheses.

The development of models of plasticity evolution is continuing. Throughout this chapter we have noted areas of model building that require further work. In addition, we need better empirical information to guide model development and parameter choice. We summarize these suggestions in table 6.3. We especially encourage more work on gametic models, particularly in structured populations. It seems likely that this approach will contribute to understanding the dynamics of plasticity evolution.

Models of plasticity evolution continue to be elaborated and refined. They have the potential to drive empirical efforts and, in turn, be shaped by those efforts. We strongly encourage such synergies. Ultimately, these models will help shape to ongoing development of evolutionary theory. More than 50 years ago Schmalhausen (1949) attempted to integrate plasticity into the Modern Synthesis. Current modeling efforts are now making that linkage.

7

Integrated Solutions to Environmental Heterogeneity

Theory of Multimoment Reaction Norms

THOMAS J. DEWITT
R. BRIAN LANGERHANS

Environmental Heterogeneity

Environmental change drives evolution. If environments were constant, evolution rapidly would proceed to a rather humdrum, mostly static, equilibrium. The diversity of life we see today is the result of moderate but continual environmental challenges. Put simply, the seed of creation is the strife of organisms at odds with their changing environments. The premise of this book is that environmental variation is responsible for a spectacular suite of adaptations more intricate and labile than those for dealing with fixed environments. Phenotypic plasticity is one of those adaptations, but several others exist.

Among the many adaptations organisms have to cope with environmental variability are dormancy (i.e., seed banking or diapausing), to outlast problem environments; plasticity to produce relatively fit phenotypes for the demands of alternative environments; intermediate phenotypes (generalization) and bet-hedging, both of which reduce variance in performance across environments; and dispersal, to leave when environments are unfavorable. Much theoretical literature addresses the merits of each strategy, generally, compared with ecological specialization (e.g., Levins 1968; Lewontin and Cohen 1969; Cohen 1976; Lively 1986a; Seger and Brockmann 1987; Van Tienderen 1997; reviewed in Wilson and Yoshimura 1994; chapter 6).

Traditionally, evolutionary ecologists define and contrast four strategies for coping with environmental heterogeneity. (1) *Specialization*: one phenotype is produced that is optimal for a given environment, even though the specialist may find itself sometimes in alternative environments. (2) *Generalization*: an intermediate or otherwise general-purpose phenotype is produced, which is at least moderately successful in most environments. (3) *Bet-hedging*: an organism produces either (a) several phenotypes (e.g., among units in modular organisms or through producing diversified offspring) or (b) single phenotypes probabilistically. (4) *Phenotypic plasticity*: environmental factors trigger production of alternative phenotypes. Other strategies are conceptually similar and can be considered as forms of these four. Strategies such as dormancy or iteroparity, for example, can be viewed as a type of temporal bet-hedging (Philippi and Seger 1989). A generalized framework can accommodate all these possible strategies, such that each traditional strategy, as well as additional strategies, is a special case of the general construct.

The distinction between these strategies comes down to whether an organism adopts a single phenotype (specialists and generalists) or variable phenotypes (plasticity and bet-hedging). Plastic strategists produce variant phenotype based on the nature of the environment, whereas bet-hedging strategists produce phenotypic variation within single environments. What is strategic about bet-hedging is that it reduces variance in fitness across generations, and hence increases geometric mean fitness (Dempster 1955; Lewontin and Cohen 1969). Many theorists also require in their definition that bet-hedging reduce arithmetic mean fitness (Seger and Brockmann 1987; Philippi and Seger 1989), but we see no need for, and do not adopt, this restriction.

Understanding how reduced variance in fitness across generations increases geometric mean fitness is core to the models we present and central to all evolutionary thinking. The basic problem environmental variation presents is that it creates functional trade-offs between environments, a pattern of selection known as divergent natural selection (see chapter 1). Divergent natural selection is a problem for organisms because being equipped for one environment necessarily reduces performance in alternative environments. This type of trade-off is represented graphically in figure 7.1. In a given environment, a specialist (one who always produces the best phenotype for one of the environments) achieves maximum fitness in its specialized environment. Say, for example, that this maximum is 10, and it has fitness of 2 in the alternative environment. If variation occurs within generations, we can calculate fitness simply by summing environment-specific fitnesses, weighted by environmental frequencies. If environments are equally frequent, the average fitness is 6 [i.e., $(10 \times 0.5) + (2 \times 0.5)$]. Note that a generalist producing an intermediate phenotype (with fitness 5 in either environment) has an average fitness of 5. If variation were between generations, however, consider how these genotypes (and their subsequent lineages) do:

Specialist: $1 \times 10 \times 2 \times 10 \times 2 = 400$ progeny after four generations (average fitness = 6)
Generalist: $1 \times 5 \times 5 \times 5 \times 5 = 625$ progeny after four generations (average fitness = 5)
Bet-hedger: $1 \times 5 \times 6 \times 4 \times 5 = 600$ progeny after four generations (average fitness = 5)
Perfect plasticity: $1 \times 10 \times 10 \times 10 \times 10 = 10{,}000$ progeny after four generations

The clearest generality from this type of modeling is that plasticity, in the absence of constraints (i.e., perfect plasticity), is always the superior strategy in variable environments. The fact that plasticity is not ubiquitous (evident for all traits in all species) therefore suggests that constraints are ubiquitous in natural systems (reviewed in DeWitt et al. 1998). First, plasticity may be costly (Van Tienderen 1991; Via et al. 1995). Alternatively, plasticity could be constrained by developmental range limits, where plasticity cannot produce phenotypic extremes (because of the need to retain flexibility) (DeWitt 1998). Another possibility, one that frequently has been modeled (Moran 1992;

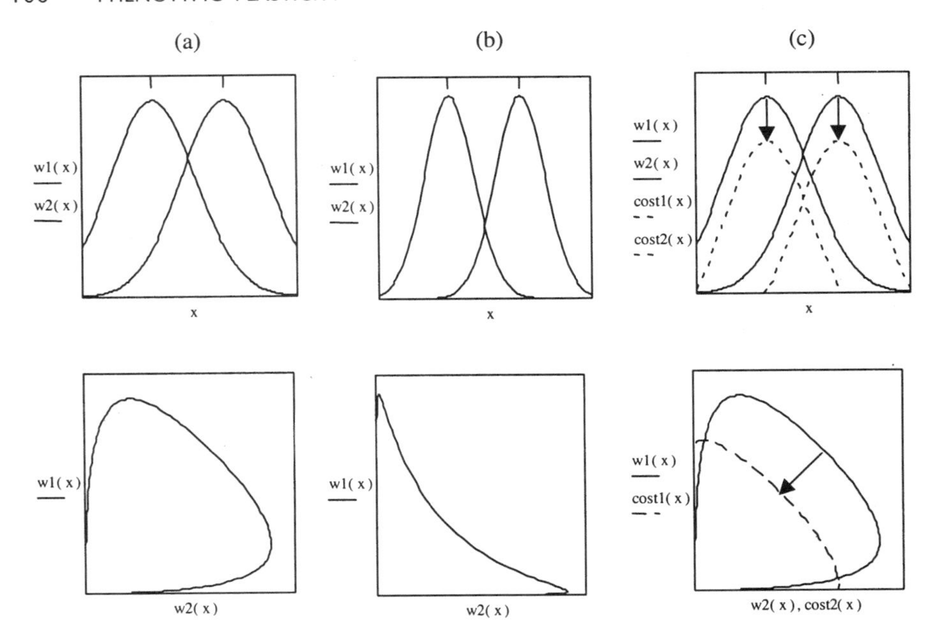

Figure 7.1. Fitness sets for the case of two environments. (a) Fitness functions (top) for two environments in which functions overlap broadly, producing a convex fitness set (bottom). (b) Slightly overlapping fitness functions (top), which produce a concave fitness set (bottom). (c) Fitness functions (solid lines) plus reexpressed functions with a value subtracted to represent the cost of phenotype production (dashed lines). When modeling costly plasticity, fitness for phenotypes produced through flexible development are taken from the dashed lines [Following Leon (1993).]

Getty 1996; Tufto 2000), is that plasticity is limited by accuracy of environmental cues used to guide development. Such constraints must be incorporated in any realistic model of the evolution of phenotypic plasticity.

Next we examine each of the traditional strategies (specialization, generalization, plasticity, and bet-hedging) and model integrated strategies that simultaneously combine elements of the foundational "pure" strategies. We evaluate whether integrated strategies provide better solutions to environmental heterogeneity than pure strategies. The goal is to find how closely each strategy approaches the ideal of perfect plasticity, given that certain constraints may be operating.

The Models

Following Levins (1962, 1968), the consequences of environmental shifts can be specified using a separate fitness function for each environment being modeled. Each fitness function describes the dependence of organismal performance (fitness) on phenotype for each specific environment. For simplicity, we use two environments in our models. Results from this simple scenario can easily be generalized, however, to environmental gradients or multiple categorical environments. When using only two environments, a fitness set is a useful way to summarize the fitness functions. For example, Figure 7.1, A and B, gives two examples of fitness sets, with the component fitness functions specified in the top row and the corresponding fitness sets mapped below. In the first scenario, fitness functions are highly overlapping and so produce a convex fitness set. The second scenario involves only slightly overlapping fitness functions (i.e., strongly divergent selection) and so produces a concave fitness set. An intermediate amount of overlap in fitness functions results in a flat fitness set (not shown). In general, convex fitness sets favor generalists and concave sets favor specialization (Levins 1962, 1968). However, the exact outcome depends upon factors such as environmental frequencies, whether environmental variation occurs within or between generations, and whether fitness is frequency dependent. Because these issues are trivial to the present goal, in the explicit models that follow we generally assume equally frequent environments, intergenerational variation, and no soft selection. However, the notation used allows for easy alteration of the first two assumptions.

Fitness functions define only the performance of a phenotype when in the specified environment. Calculating the actual fitness of genotypes must take into account (1) the probability of producing various phenotypes, (2) the frequency with which organisms experience each environment, and (3) whether alternative environments are encountered within single generations (fine-grained variation) or between generations (coarse-grained variation). Denoting the phenotype distribution produced by a genotype as $d(z;\, \mu_{i\,j}, \sigma_{i\,j})$, fitness within a generation is calculated by integrating the product of the phenotype distribution and fitness function (Yoshimura and Shields 1987, 1992, 1995):

$$w_j(\mu_{i\,j}, \sigma_{i\,j}) = + \int f(z)_j \cdot d(z;\, \mu_{i\,j}, \sigma_{i\,j})dz, \qquad (7.1)$$

where $w_j(\mu_{i\,j}, \sigma_{i\,j})$ is the fitness of genotype i in environment j. Fitness for the two environment case can be written for fine-grained variation as

$$W(\mu_{i\,j}, \sigma_{i\,j}) = w(\mu_{i,\,j=1}, \sigma_{i,\,j=1}) \cdot p + w(\mu_{i,\,j=2}, \sigma_{i,\,j=2}) \cdot (1-p), \qquad (7.2)$$

where p is the frequency of environment 1 and $(1-p)$ is the frequency of environment 2. For coarse-grained variation, the geometric mean is used to calculate fitness:

$$M(\mu_{i\,j}, \sigma_{i\,j}) = w(\mu_{i,\,j=1}, \sigma_{i,\,j=1})^p \cdot w(\mu_{i,\,j=2}, \sigma_{i,\,j=2})^{(1-p)} \qquad (7.3)$$

It is important to emphasize for each of the above equations that each genotype produces an alternative phenotype density function with mean μ_i and standard deviation σ_i unique for each environment j. For the models that follow, we apply the above equations assuming intergenerational variation—parents and offspring may encounter different environments but in any given generation only one fitness function holds.

Strategies

Now we can define the strategies: (1) A specialist genotype produces only z_1*, the optimum phenotype in environment 1. (Optimal values are denoted by asterisks.) Note that an environment 2 specialist is not defined to reduce redundancy. Because our fitness sets are symmetrical about their diagonal axis, specialists for environments 1 and 2 are equivalent in terms of fitness. (2) A generalist genotype produces phenotypes intermediate between the fitness peaks for the two environments. (3) A plastic genotype produces phenotypes near z_1* or z_2* depending on its interpretation of environmental cues and other logistic constraints. (4) A bet-hedger genotype produces phenotypes probabilistically, with mean μ_i and variance σ_i^2.

Traditionally the bet-hedging strategy involves adding variance around an intermediate phenotypic mean (i.e., a generalist who produces phenotypes probabilistically). However, such variance could be added to any of the first three strategies. Therefore, we equate the term bet-hedging with variance *per se* and not with a singularly unique strategy unto itself. So we now have three core strategies with bet-hedging being a possible attribute of each (box 7.1). Furthermore, the degree of bet-hedging (phenotypic variance) can differ across environments. Although phenotypic plasticity produces different mean phenotypes across environments, it is perfectly conceivable that the magnitude of variance could also differ across environments. Therefore, we could define several synthetic strategies, such as a trait-mean specialist with plasticity for trait-variances (i.e., $\mu_{i,j=1} = \mu_{i,j=2}$, but $\sigma_{i,j=1} \neq \sigma_{i,j=2}$; box 7.1, column 3, rows 1 and 2).

Constraints on Plasticity

To model constraints on phenotypic plasticity, we simulated costs, developmental range limits, and varied reliability of cues. For each constraint, equation (7.3) was modified to reflect the unique nature of the constraint at hand:

$$M(\mu_i, \sigma_{ij}) = [w(\mu_{i,j=1}, \sigma_{i,j=1}) - C]^p \cdot [w(\mu_{i,j=2}, \sigma_{i,j=2}) - C]^{(1-p)} \tag{7.4}$$

$$M(\mu_i, \sigma_{ij}) = w[(\mu + L)_{i,j=1}, \sigma_{i,j=1}]^p \cdot w[(\mu - L)_{i,j=2}, \sigma_{i,j=2}]^{(1-p)} \tag{7.5}$$

$$M(\mu_i, \sigma_{ij}) = [w(\mu_{i,j=1}, \sigma_{i,j=1})^a \cdot w(\mu_{i,j=2}, \sigma_{i,j=2})^{(1-a)}]^p \cdot [w(\mu_{i,j=2}, \sigma_{i,j=2})^a \cdot w(\mu_{i,j=1}, \sigma_{i,j=1})^{(1-a)}]^{(1-p)} \tag{7.6}$$

In equation (7.4), C is the cost of plasticity. When C is zero this equation reduces to equation (7.3). This cost is assumed to be equal in both environments (figure 7.1C), although this easily could be modified so that $C_1 \neq C_2$. L in equation (7.5) is the magnitude of developmental limits upon plasticity. Assuming $z_1^* < z_2^*$ (i.e., the fitness function for environment 1 is to the left of that for environment 2), L specifies the number of phenotype units by which the plastic genotype misses respective optima. If $z_1^* = 5$ and $z_2 = 10$, then a plastic strategist under $L = 1$ would produce phenotypes of 6 and 9, respectively, in environments 1 and 2. Finally, equation (7.6) specifies the geometric mean fitness when plastic strategists are constrained by cue accuracy. The probability of the cue being accurate is represented by a and the probability of the cue signaling the wrong environment is $1 - a$. Where the cue is completely reliable ($a = 1$), equation (7.6) reduces to equation (7.3). To more realistically reflect nature, we can specify environment-dependent cue accuracies ($a_1 \neq a_2$) in equation (7.6) to account for differences in reliability of cues between environments (see also Lively 1986a, 1999a; Moran 1992; Leon 1993; Getty 1996).

We proceed by modeling each strategy (i.e., those illustrated in box 7.1) under three fitness sets: concave (figure 7.1A; $\sigma = 1$), intermediate ($\sigma = 2$), and convex (figure 7.1B; $\sigma = 3$). First, fitnesses of the purely deterministic strategies (box 7.1, first column) are compared. We additionally examine how pure generalization compares with plasticity

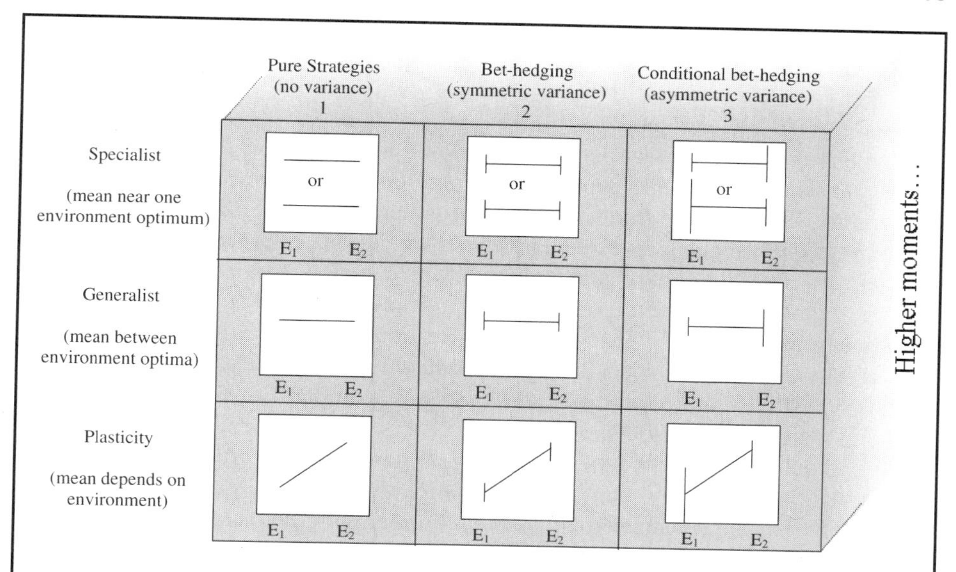

Box 7.1. Multimoment reaction norms, where z = trait values and E_i represent alternative environments. Column 1 illustrates the traditional "pure" strategies of specialization, generalization, and phenotypic plasticity. Bet-hedging is depicted in column 2, where variance is added to the former three strategies. (Note that the generalist strategy with added variance is the traditional "bet-hedger" in ecological modeling). Asymmetric bet-hedging, where variance differs by environment, is depicted in Column 3. That is, the amount of developmental noise might be conditional, perhaps based on the adaptive optimum level of variance prescribed by each environment.

In the two dimensions presented in this figure, we describe an expansion from the traditional suite of four strategies to nine. In theory the strategy space involves not only trait means and variances but all aspects of phenotype distributions. For example, a third dimension of this box could describe strategic flexibility in skewness, or kurtosis (see Discussion). The strategies shown here can be extended to include nonlinear reaction norms across environmental gradients. The most general framework for understanding phenotypic strategies is one of multimoment polynomial reaction norms.

with various constraints. Second, we evaluate bet-hedging strategies (box 7.1, second column). Finally, variance itself might be asymmetric across environments. Asymmetric (conditional) variance, as with simple (unconditional) bet-hedging, can be combined with any of the traditional pure strategies (box 7.1, third column). Strategies with asymmetrical variance are compared with those with symmetrical variance.

Computationally, the addition of bet-hedging to any pure strategy is achieved by adding a unique standard deviation σ to the phenotypic distribution for each genotype or genotype–environment combination. When testing effects of bet-hedging we varied σ from 0.5 to 6.

Model Outcomes

Pure Strategies

Under all fitness sets perfect plasticity was the optimal strategy. The relative fitness of specialists and generalists depends, as in previous models (reviewed in Wilson and Yoshimura 1994), on the frequency of environments and structure of the fitness set. Assuming equal frequencies of the two environments and intergenerational variance, specialization always fails as a strategy relative to generalization. However, specialists do quantitatively better under convex fitness sets under our assumption of intergenerational variance. If environmental variance were intragenerational and the fitness set were concave, specialists would outcompete generalists. Yet, because specialists typically fail in our models, the interesting comparison is between the generalist and constrained plastic strategists.

Our model confirmed that the advantages of plasticity were greatest under concave fitness sets. In general, plasticity was fallible relative to generalists only under severe levels of two of the constraints (figure 7.2). Furthermore, the nature of fitness differences between plasticity and generality depended on the constraint at hand (costs, developmental range limits, cue accuracy).

Costs

The structure of the fitness set had a large influence on the relationship of generalization and plasticity with costs (figure 7.2A). We expressed costs of plasticity as a percentage fitness decline relative to perfect plasticity. Under a convex fitness set, a relatively low level of plasticity cost (C = 29%) allowed generalists to beat plastic strategists. Under a concave fitness set, however, costs had to be particularly severe (i.e., C = 96%) for generalists to beat plasticity. Generally, more divergent selection requires commensurately strong costs to prevent the evolution of plasticity.

Developmental Range Limits

Even extreme developmental range limits failed to lower the value of plasticity below that of generalization. That is, plastic strategists can be severely constrained by developmental range limits and still outperform generalists. The degree of superiority of plasticity depended on shape of the fitness set and degree of developmental limits (figure 7.2B). Concave fitness sets resulted in an extreme advantage for plasticity even under high levels of constraint ($L < 80\%$). Plasticity's superiority was relatively modest under both convex and intermediate fitness sets.

Accuracy of Cues

The advantage of plasticity over generalization increased with cue accuracy. The fitness of plastic strategists converged on that of generalists at 75% cue accuracy (i.e., 25% inaccuracy; figure 7.2C). Interestingly, this effect was independent of fitness set shape and frequency of environments. As a consequence of using normal curves for fitness functions, this effect also remained invariant to changes in the individual fitness

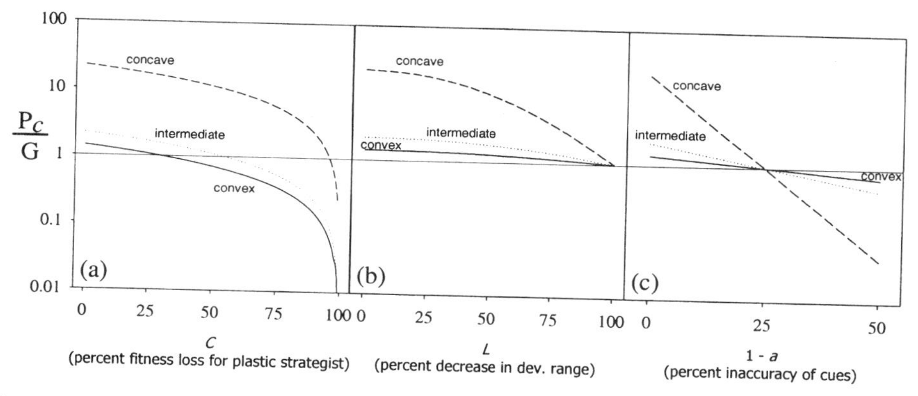

Figure 7.2. Fitness of plastic strategists relative to that of generalists under three types of constraint for three fitness sets. P_C/G is the ratio of fitness for plasticity under a given constraint to generalization. (a) Costly plasticity. (b) Plasticity with limited developmental range. (c) Ignorant plasticity.

function breadths or peak locations, as long as the generalist produced a phenotype exactly between the peaks.

Strategies with Bet-hedging

Generally, both specialists and generalists benefited from some level of bet-hedging (figure 7.3). The single exception was that generalists under convex fitness sets did not benefit from bet-hedging (figure 7.3C). And if fitness sets were concave, a specialist with a small amount of bet-hedging ($\sigma = 1.14$ in our model) overcame pure generalization (figure 7.3A). Bet-hedging decreased the fitness of unconstrained plasticity. Even when costs were imposed on plasticity bet-hedging awarded no benefits. So we tested to see whether plastic strategists constrained by developmental range limits and accu-

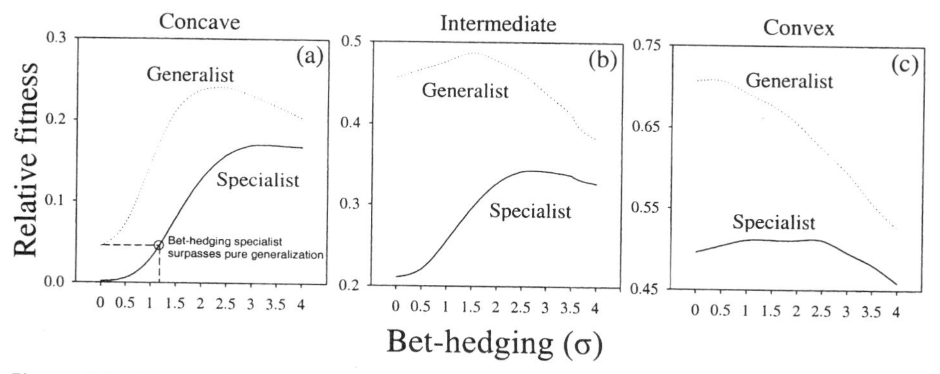

Figure 7.3. Fitness of bet-hedging specialists and generalists relative to perfect plasticity under three fitness sets (a–c). In (a), we have indicated the level of bet-hedging at which specialists surpass fitness of pure (non-bet-hedging) generalists.

racy of cues could benefit by bet-hedging. To accomplish this, we invoked a given level of constraint and compared fitness of plastic strategists to strategists using plasticity and bet-hedging simultaneously.

Developmental Range Limits

When constrained by limits on phenotypic range, plastic strategists benefited from bet-hedging only under concave fitness sets. Furthermore, the optimal degree of bet-hedging increased with the magnitude of developmental limits (figure 7.4). That is, the greater the developmental limits of plasticity the more bet-hedging is required to compensate for this limit.

Accuracy of Cues

When environments were inscrutable, to any degree, plastic strategists benefited from bet-hedging. The optimal degree of bet-hedging increased as cues became increasingly flawed (figure 7.5). Under intermediate and convex fitness sets, bet-hedging provided little benefit unless cues were very poor indicators of environmental conditions. Yet under a concave fitness set, even slight inaccuracy in cues (5% inaccuracy) made bet-hedging beneficial. Thus, just as with developmental range limits, cue inaccuracy required compensatory increases in the amount of bet-hedging.

Asymmetric Bet-hedging

Under assumptions that we have maintained so far (e.g., symmetrical fitness sets), only specialists gain from plasticity in trait variances (figure 7.6A). Under a concave fitness set, specialists that produce no variance in their specialized environment but bet-hedge in their nonspecialized environment (i.e., "conditional bet-hedgers") benefit greatly. Specialists with optimal asymmetric bet-hedging are 187 times more fit than pure spe-

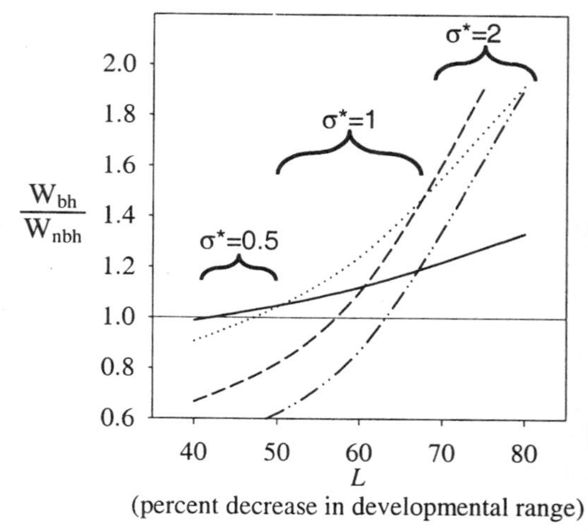

Figure 7.4. Fitness of bet-hedging plastic strategist relative to pure plastic strategists under varied developmental range limits. These results apply for the concave fitness set only; no benefits of bet-hedging were evident under intermediate or convex fitness sets. The optimal degree of bet-hedging (standard deviation of phenotypic mean) is depicted by σ^*. (solid line, $\sigma = 0.5$; dotted line, $\sigma = 1$; dashed line, $\sigma = 2$; dot-dashed line, $\sigma = 3$). W_{bh}/W_{nbh} is the ratio of fitness of bet-hedging plastic strategist to non-bet-hedging plastic strategist.

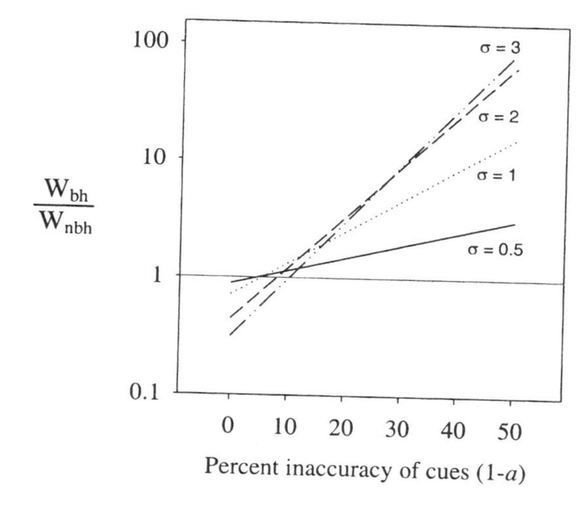

Figure 7.5. Fitness of bet-hedging plastic strategists under various cue accuracies, relative to pure plastic strategists. Depicted results apply for the concave fitness set. Minimal benefits of bet-hedging are awarded under intermediate and convex fitness sets. The optimal degree of bet-hedging (standard deviation of phenotypic mean) increased as inaccuracy of cues increased. W_{bh}/W_{nbh} is the ratio of fitness of bet-hedging plastic strategist to non-bet-hedging plastic strategist.

cialists. Furthermore, this conditional strategy (bet-hedge only in the nonspecialized environment) increases fitness nearly twofold over the best symmetrical bet-hedge. Although generalists as we defined them do not benefit from asymmetric bet-hedging, they would if their phenotypic mean was not exactly intermediate between the two optimal phenotypes.

Just as unconstrained plastic strategists never benefited from bet-hedging, they failed to benefit from asymmetric bet-hedging. Furthermore, plastic strategists constrained by developmental range limits or cue inaccuracy failed to benefit more from asymmetric bet-hedging than from symmetric bet-hedging. However, if cue accuracy differed between environments, then asymmetric bet-hedging was beneficial (figure 7.6B). Similarly, asymmetric bet-hedging provided fitness benefits when the developmental range limits differed by environment (not depicted).

Although the bulk of our modeling assumed symmetrical fitness sets, natural environments often differ in the shape of their fitness functions. In such cases asymmetric bet-hedging provided benefits to specialists, generalists, and constrained plastic strategists.

Integrated Strategies

An extensive literature compares specialization, generalization, bet-hedging, and phenotypic plasticity as evolutionary alternatives (reviewed in Wilson and Yoshimura 1994). Most of these publications demonstrate how certain factors (e.g., fitness set shape, habitat choice, density dependence, intergenerational environmental variance) favor one strategy over others. Rather than treat each foundational strategy as alternatives, our main result demonstrates integration of strategies as more often than not the optimal solution to environmental heterogeneity.

When comparing foundational strategies, our results concur with previous work. For example, pure generalists outcompete pure specialists given coarse-grained environmental variance, similar environmental frequencies, and convex fitness sets. Also, perfect plasticity (i.e., assuming no constraints) is always optimal. However, it is increasingly

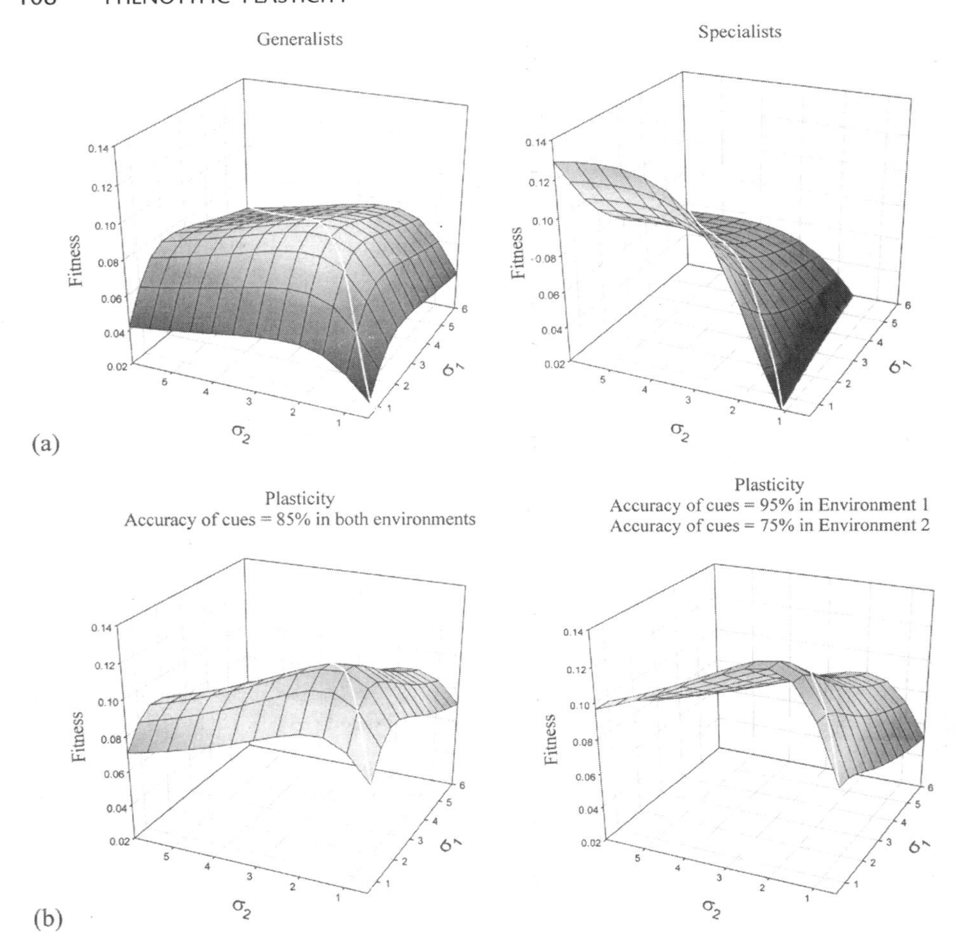

Figure 7.6. Fitness landscapes for variance in alternative environments under concave fitness sets (σ_i = standard deviation of phenotypes in alternative environments). The white line indicates symmetrical bet-hedging. (a) Symmetric bet-hedging is optimal for generalists (left), but asymmetric bet-hedging is optimal for specialists (right). (b) Symmetric bet-hedging is optimal when plasticity is equally constrained by cue accuracy in both environments (left). Asymmetric bet-hedging is optimal if cue accuracy differs for alternative environments (right).

clear that perfect plasticity is uncommon in nature. Constraints on plasticity must exist. Either plasticity has costs or logistic difficulties that limit its efficacy (e.g., developmental range limits or inaccurate cues; DeWitt et al. 1998). In theory, these constraints need to be severe to prevent the evolution of plasticity (e.g., Van Tienderen 1991). Empirical studies also support this idea, because plasticity seems to have evolved despite the demonstration of potentially severe constraints (Langerhans and DeWitt 2002). Only under a limited range of conditions were generalists ever superior to constrained (costly or ignorant) plasticity (figure 7.2A,C). Developmental range limits never lowered fitness of plastic strategists below that of generalists (figure 7.2B). Therefore, among the pure (invariant within environment) strategies, plasticity is very hard to beat.

In our models, bet-hedging (adding variance, or noise, around phenotypic means) was generally a good thing. Perhaps most remarkably, specialists increased fitness through bet-hedging and even surpassed pure generalists when fitness sets were concave (figure 7.3). Therefore, a specialist with a little noise can beat the traditional jack-of-all trades. However, a bet-hedging generalist (i.e., the traditional "bet-hedger" strategy) can always surpass a bet-hedging specialist. Bet-hedging generalists also surpass pure generalists under all but convex fitness sets.

When unconstrained or constrained only by costs, plastic strategists gained no advantages from bet-hedging. Yet when plasticity was constrained by developmental range limits or cue accuracy, integrating bet-hedging with plasticity proved beneficial, especially under concavity (figures 7.4 and 7.5). As these two constraints increased in strength, the optimal magnitude of bet-hedging increased commensurately. So, under most conditions plasticity should integrate bet-hedging to optimize performance. Interestingly, one way to view the adaptive benefits of integrating strategies is that bet-hedging can compensate for limits of plasticity. This type of compensatory relationship, which is one way to create integrated adaptations, has been termed trait compensation (DeWitt et al. 1999).

In theory, asymmetric bet-hedging can be integrated with any of the foundational strategies (box 7.1, column 3). If the optimal variance changes across environments, we can expect the evolution of conditional bet-hedging. Specialists, in particular, stand to gain enormously through asymmetric bet-hedging. The optimal bet-hedging specialist produces no variance in its specialized environment and some optimal level of variance in its nonspecialized environment. This strategy can be termed plasticity for variance. Thus, both trait means and trait variances can exhibit reaction norms. This implies that we should expect to find interesting forms of plasticity where both trait means and variances differ across environments in manners that transcend simple mean–variance correlations (see below).

Although our standard model was structured with many symmetries (equal probability of alternative environments, symmetrical fitness functions, same-shape fitness functions for alternative environments, and exactly intermediate generalists), many of these assumptions will not hold in natural systems. Therefore, some of these symmetries were relaxed to consider evolution under more realistic conditions. Of particular interest was that asymmetric bet-hedging provided benefits to plastic strategists if information reliability was greater in one environment than another ($a_1 \neq a_2$). Specifically, greater variance was favored in the more inscrutable environment. This may be a general phenomenon. Many plastic organisms experience either inducing or noninducing environments. Induction cues, when perceived, may more accurately reflect the true nature of the environment (i.e., reliably indicate the presence of the inducing agent) than failure to perceive a cue at all. For example, imagine a zooplankter that can chemically perceive planktivorous sunfish, but because of constraints of its sensory mechanism is unable to sense planktivorous minnows. In this case, the sunfish cue informs the zooplankter with certainty that a planktivore is present, but the absence of a perceived cue could mean that either no planktivores are present or minnows are present. Therefore, it may be a general rule that less phenotypic variance (bet-hedging) occurs in inducing environments.

Conditional bet-hedging may turn out to the rule in nature. Such a strategy should be selected not only by differentially reliable cues but also under many other situations.

For example, asymmetric developmental range limits or asymmetric fitness sets would favor conditional bet-hedging. Likewise, if the costs or logistic limits of canalization differed between environments, bet-hedging in the more constraining environment can partly compensate for plasticity's limits.

A General Conceptual Framework for Evolution

So what represents the most complete (integrated) adaptive solution to naturally variable environments? We have stressed the need to think about plasticity in terms of trait means and trait variances. We also stressed that bet-hedging can be used to compensate for the limits of plasticity, thereby providing an optimal integrated solution to environmental variation. Yet we have only touched upon the breadth of possibilities with our models, expanding the usual strategy space from a four-box paradigm to a nine-box paradigm (box 7.1). Rather than limit our thinking to only nine boxes (based on the first two moments of phenotype distributions), however, it is possible to expand further. Quite likely, skewness and other moments could be selected to differ in alternative environments. DeWitt and Yoshimura (1998) first suggested that higher moments of phenotype distributions may be under selection. And preliminary data have indicated that this may be so (A. E. Weis and T. J. DeWitt, unpublished data). What we need to do now is assess whether the expanded paradigm—especially plasticity in higher moments of phenotype distributions—is of use for explaining the distribution and abundance of phenotypes in nature.

We believe the new view is important. To illustrate, let us ask what evolution would look like in this new view. We can make several predictions about the early course of population divergence. Assume an initially flat reaction norm for a population whose members have a moderate level of normally distributed phenotypic variance within environments (figure 7.7A). Now suppose divergent selection arises: the phenotypic optima shift below the norm in one environment and above the norm in the other environment. Selection will now simultaneously affect trait means, variances, and higher moments, and the patterns for each moment will change as evolution proceeds. At first trait means and skewnesses will increase in the direction of the new optimum in each environment. Simultaneous with the mean and skewness changes, trait variance and kurtosis will increase in both environments (figure 7.7B). As the reaction norm approaches the optima, variance and skewness will begin to decrease, and selection for trait-mean plasticity will continue. If perfect trait-mean plasticity is achieved, then all other moments should theoretically evolve to zero. Realistically however, some developmental noise will always persist (figure 7.7C). Depending on the final constraints operating, some level of each moment may remain.

Concluding Remarks

In our theory of multimoment reaction norms, traditional strategies (specialist, generalist, bet-hedger, trait-mean plasticity) are special cases within a general strategy space. To maintain conditional strategies for multiple moments of phenotype distri-

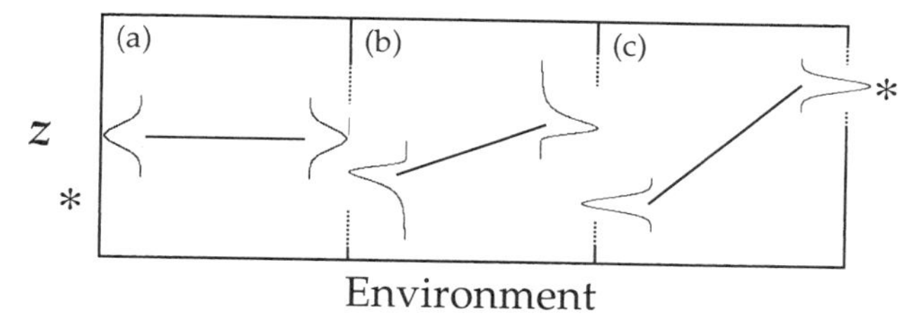

Figure 7.7. Evolution of multimoment reaction norms. Asterisks denote phenotypic optima in alternative environments, and the progression of from A to C indicates the evolutionary trajectory. (a) A hypothetical population initially produces a flat reaction norm with phenotypes distributed normally in all environments. (b) Selection produces divergence in trait means and increases variances, skewness toward optima, and kurtoses (leptokurtis). (c) When trait means reach phenotypic optima, selection produces symmetrical distributions with minimal variance. In the absence of constraints, the end point of evolution should be perfect plasticity (i.e., no variance within environments).

butions may at first seem unlikely, perhaps too sophisticated to expect. However, this should be an empirical matter—we must look for such patterns where the functional ecology warrants.

Also, during the course of divergent natural selection, it may be that shifts in trait means are accompanied by interesting and predictable evolutionary trajectories for higher moments of phenotype distributions (figure 7.7). Therefore, the concept of multimoment reaction norms may be the most general conceptual framework for describing the strategy space of organisms in naturally variable environments.

Acknowledgments Our work was supported by National Science Foundation grant DEB-9908528.

8

A Behavioral Ecological View of Phenotypic Plasticity

ANDREW SIH

In recent years, there has been a burst of new interest in understanding evolutionary forces that shape patterns of phenotypic plasticity, and in the effects of plasticity on evolutionary dynamics and ecological patterns (see reviews by Stearns 1989; West-Eberhard 1989; Newman 1992; Scheiner 1993a; Via et al. 1995; DeWitt et al. 1998). Most studies that apply the term phenotypic plasticity examine developmental plasticity; typically, environmentally induced changes in morphological, or life history traits (see above reviews). Behaviors, despite their obvious plasticity, are often explicitly excluded from the rubric of phenotypic plasticity. Based strictly on its definition (the change in the expressed phenotype of a genotype as a function of the environment), however, phenotypic plasticity could refer to any trait that is influenced by both genetic and environmental effects.

Here, I discuss (1) existing ideas on the distinction between behavioral and developmental plasticity, (2) the different approaches that dominate the study of phenotypic plasticity versus adaptive behavior, (3) some insights for the study of phenotypic plasticity that can be gained from the behavioral ecology view, (4) the joint evolution of behavioral and developmental plasticity, and (5) some insights for ecologists from genetically based studies of phenotypic plasticity. My primary focus will be on issue (3). I will suggest that the optimality approaches' emphasis on explicit biological mechanisms that relate plastic phenotypes to fitness costs and benefits can provide new insights on how to measure and how to think about relationships among realistic, complex environments, reaction norms, and fitness.

Is Behavior Fundamentally Different from Developmental Plasticity?

The usual idea is that behavioral plasticity differs from developmental plasticity in both the speed and reversibility of response. At one extreme, behavior might, in some cases, be infinitely plastic (cf. Houston and McNamara 1992); that is, capable of immediate and infinitely reversible changes in response to spatially or temporally varying environments. At the other extreme, developmental plasticity might be relatively slow to unfold and often irreversible. As noted in previous reviews, the differences between these extreme ends of the spectrum are important both for the evolutionary process and for the likely outcome of evolution (Scheiner 1993a; Via et al. 1995).

For the evolutionary process, a key question is whether an individual can express a reaction norm (the pattern of plasticity across environments expressed by a given genotype), or is a reaction norm a property of a group of individuals that share a genotype? If, for a given type of trait, individuals show rapid, reversible plasticity, then individuals can express a full reaction norm (provided that they experience a broad range of environments). In that case, reaction norms can be subject to selection, and given a genetic basis, reaction norms can evolve much like other traits. For highly labile traits, models predict that under a broad range of conditions, populations should evolve to the optimal reaction norm (e.g., Gabriel and Lynch 1992; Gomulkiewicz and Kirkpatrick 1992; Houston and McNamara 1992; Hammerstein 1996).

For example, generalist foragers typically have highly flexible diet breadths (number of food types in their diet) that depend on the relative value and availability of different food types (Stephens and Krebs 1986). That is, each individual can potentially express a full diet reaction norm (e.g., pattern of diet breadth as a function of total food availability). If variations in individual fitness depend on variations in diet choice, and if differences in patterns of diet choice are heritable, then adaptive diet choice should evolve.

In contrast, if plasticity involves a single, irreversible response to the organism's environment during development, then individuals cannot express a reaction norm. Instead, a reaction norm is a characteristic of a genotype. If groups can show a reaction norm, but individuals cannot, then a reaction norm as a distinct trait should evolve only by group selection (Via et al. 1995). An alternative view is that for developmental plasticity, reaction norms do not evolve as distinct traits, but instead represent the emergent byproducts of selection favoring different traits in different environments balanced against constraints imposed by genetic correlations among traits across environments (Via 1987, 1993a,b; Kirkpatrick and Lofsvold 1992).

With regard to evolutionary outcomes, if developmental plasticity is slow and irreversible, then plasticity is only favored if the future environment can be accurately predicted. If the environment varies rapidly and unpredictably, then a slow, irreversible response is likely to often result in traits that poorly match environmental optima. Models predict that under these circumstances, populations should evolve to a fixed phenotype (or a genetically constrained reaction norm) that makes the best of an unpredictable situation by balancing selection pressures in different environments (Moran 1992; Gavrilets and Scheiner 1993a).

The upshot is that if behavioral plasticity is rapid and reversible, whereas developmental plasticity is slower and irreversible, then these different sorts of plasticity should often show different evolutionary dynamics and outcomes. If behavioral plasticity often evolves to an optimum, then the optimality approach should be the most valuable way to study behavior. In contrast, if developmental plasticity is constrained, it might be most relevant to study genetic and developmental constraints that limit plasticity. Insights on the evolution of one sort of plasticity might not apply to the other.

The problem with this view, of course, is that it oversimplifies the nature of the two sorts of plasticity. Ecologically interesting behaviors are rarely infinitely rapid and reversible. Species that form long-term mating bonds, by definition, make mate choice decisions in one environment that carry over into future, often changing, environments. Foragers can develop dietary specializations or modes of foraging depending on their rearing environment, that subsequently change only slowly, if at all (Curio 1973). Prey often stay in refuge long after predators have left an area (Sih 1992).

Conversely, developmental plasticity need not be fixed and irreversible. Many life history decisions (that have often been studied using the framework of phenotypic plasticity) are repeated many times over a lifetime and can be reversible (Roff 1992; Stearns 1992). Even induced morphological traits, which are often thought to be developmentally fixed, can, in some systems, be reshaped throughout development. For example, although the size and shape of a given leaf might be fixed after development, for an entire plant, the size and shape of new leaves can continue to change throughout the plant's lifetime. That is, rather than being fundamentally different, both behavioral and developmental plasticity encompass broad and overlapping ranges of a continuum of plasticity. Both should be considered in the study of phenotypic plasticity.

Contrasting Approaches to Studying Plasticity

Although one can argue that behavioral and developmental plasticity overlap enough to deserve to be studied under a unified framework, the perceived differences between the lability of behavior versus developmental plasticity have contributed to fundamental differences in the conceptual approaches used to study the two.

In recent years, many of the advances in the study of developmental plasticity have been generated by the quantitative genetic (QG) approach (reviewed by Scheiner 1993a; Via et al. 1995). This approach focuses on quantifying three complementary factors that influence the evolution of plasticity: (1) reaction norms *per se*, (2) natural selection on plasticity, and (3) the genetics of plasticity. Although the statistical details on how to address each of these issues has been the subject of some debate (Scheiner 1993a; van Tienderen and Koelewijn 1994; de Jong, Via et al. 1995; see chapter 3), the basic task of quantifying these parameters remains a central goal of this field.

In contrast, the main conceptual approach for studying patterns of behavioral plasticity has been the optimality approach (Mangel and Clark 1986; Krebs and Davies 1996). This approach uses optimality models that make explicit assumptions about costs and benefits of different behaviors to predict the optimal behaviors for a range of different environments. These predictions are then tested by comparing predicted and observed patterns of behavioral plasticity.

Although the optimality approach, in theory, generates predictions on optimal reaction norms, behavioral ecologists rarely couch their studies in those terms. More important, perhaps, behavioral ecologists rarely quantify the two major mechanisms of evolution (natural selection and inheritance) that occupy the efforts of quantitative geneticists. Although the optimality approach implicitly assumes that natural selection is the dominant force in evolution, optimality-based studies of behavior rarely attempt to measure natural selection. Similarly, behavioral ecologists typically do not attempt to quantify genetic variances and covariances for ecologically interesting behaviors (but see Boake 1994). Instead, behavioral ecologists usually implicitly or explicitly assume that heritabilities are high enough and genetic covariances low enough that genetic constraints do not prevent the evolution of optimal behaviors. Although these assumptions might apply in some systems, they appear to be violated in others. Accordingly, recent reports have noted valuable insights that optimality-based behavioral ecologists might gain from quantitative genetics (e.g., Moore and Boake 1994).

The above discussion implies a largely, one-way flow of insights from the quantitative genetics of reaction norms to the study of behavioral plasticity. My theme in the next few pages, however, will take the opposite tack. I will not dispute the value of studying evolutionary mechanisms in behavioral evolution. I will, however, argue that the optimality approach has potentially important insights for quantitative geneticists studying reaction norms. In particular, I suggest that the optimality approach can provide a mechanistic, functional basis for characterizing reaction norms and for understanding natural selection on reaction norms. That is, the optimality approach can generate predictions on shapes of reaction norms and on fitness functions for reaction norms. This complements the QG approach to studying reaction norms. In broad strokes, for reaction norms and natural selection on reaction norms, the QG approach asks what the observed patterns are, whereas the optimality approach asks why the patterns are the way they are. The optimality approach can thus bring a new level of hypothesis testing into the study of phenotypic plasticity.

An Optimality-Based View of Characterizing Reaction Norms

The main point of this section is to suggest that the shape of reaction norms should be characterized by fitting them to explicit functions generated by optimality theory, rather than by statistical models. The advantage of using optimality functions is that they should have biologically meaningful parameters, rather than statistical parameters *per se*.

The QG approach offers three main ways of quantifying reaction norms (Scheiner 1993a; Via et al. 1995). The approaches yield the same results in the simple two-environment situation; however, they differ in their ability to describe patterns of plasticity in multiple environments, particularly across an environmental gradient (Scheiner 1993a). In the character state approach, the reaction norm is characterized by correlations between trait values across pairs of environments (Via and Lande 1985). This approach, however, does not provide a simple index of the pattern of plasticity across an environmental gradient (Scheiner 1993a).

Two other methods use continuous functions to better characterize the pattern of plasticity. One approach fits a polynomial function to trait values across the environmental gradient (Gavrilets and Scheiner 1993a,b; van Tienderen and Koelewijn 1994). The pattern of plasticity is measured by the coefficients of the polynomial function [e.g., $z(E) = b_0 + b_1E + b_2E^2 + b_3E^3 + \ldots + b_nE^n$], where z is the phenotype as a function of the environment E. Polynomial regressions are used to estimate the coefficients (linear, quadratic, cubic, etc.). Genetic variances and covariances are estimated for the coefficients, and evolution of reaction norms involves evolution of the coefficients. Although polynomials can, in theory, account for virtually any shape of reaction norm, a conceptual problem with this approach is the lack of an obvious biological basis for the nonlinear terms. What mechanistic basis (e.g., genetic, developmental or functional) underlies the quadratic, cubic or quartic terms? What are the biological implications of evolution of each of these terms?

The other continuous approach is the infinite dimensional approach (Kirkpatrick et al. 1990; Kirkpatrick and Lofsvold 1992), which characterizes the pattern of genetic variation and covariation across environments by a continuous eigenfunction. Although this method can also accommodate any shape of reaction norm, it also suffers from a lack of a clear biological basis. In my experience, few evolutionary biologists can interpret eigenfunctions, much less understand the underlying mechanistic basis or meaning of evolutionary change in eigenfunctions.

What can the optimality approach add to this issue? The optimality approach, of course, generates optimal reaction norms. This is true both for the literature (primarily on life history traits) that explicitly invokes optimal reaction norms (Stearns and Koella 1986; Kawecki and Stearns 1992; Houston and McNamara 1992; Sibly 1995) and for optimality analyses in behavioral ecology that usually do not refer to the genetics of reaction norms (Sih 1984; Mangel and Clark 1986; Krebs and Davies 1996). The basic approach requires the modeler to make explicit assumptions (in some cases, derived empirically) about cost–benefit functions that link environmental factors and phenotypes to fitness. Ideally, parameters in these cost–benefit functions have a biological basis. As a result, the predicted relationship between an environmental gradient and the optimal reaction norm involves a function with parameters that have an interpretable biological basis.

Consider the following example involving behavioral plasticity. Foraging theory generates a specific reaction norm function for the optimal time that foragers should spend in patches (the phenotype) as a function of distance between patches, as reflected in the mean transit time between patches (the environmental variable). For example, a forager's cumulative net energy intake in the average patch, $g(T)$, as a function of time spent in the patch, T, can be expressed as $gT/(c + T)$, where g is the maximum net energy extracted from a patch, and c is a measure of foraging rate (lower values of c correspond to higher foraging rates). Following the marginal value theorem (Charnov 1976), the optimal time to spend in patches, $T^* = (ct)^{1/2}$, where t is the mean transit time between patches (Sih 1980; figure 8.1). The expected form of observed reaction norms is thus not a polynomial function, but a square-root function. Differences among individuals or genotypes might reflect differences in either their foraging rate, c, or in their ability to match the predicted optimal reaction norm.

The suggestion, thus, is to use the optimality approach to generate a specific, predicted function for an optimal reaction norm that has parameters that have a plausible biological basis. Observed reaction norms should be fit statistically to a function that follows the form of the optimal reaction norm function, rather than to one chosen for statistical purposes (e.g., a linear or polynomial function). Observed reaction norms should, of course, be tested for their fit to predicted optima. The approach, however, does not require individuals or genotypes to actually show optimal reaction norms. Instead, the suggestion is that the statistical characterization of reaction norms, including deviations from the optimum, can be assessed by variations in parameters with a biological basis. The evolution of reaction norms might be best described in terms of the evolution of biologically based parameters.

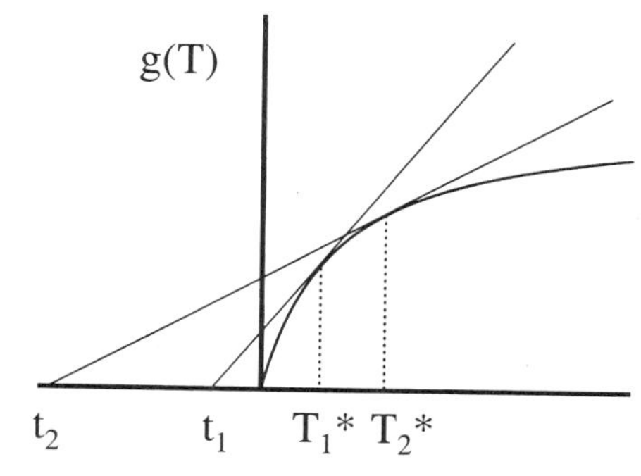

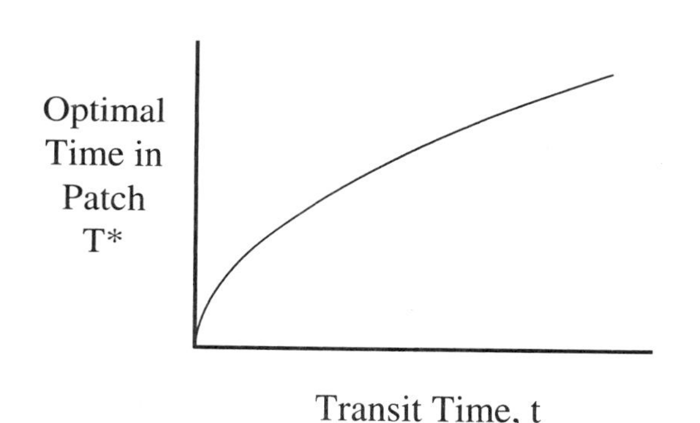

Figure 8.1. (Top) Cumulative net energy intake rate $g(T)$ as a function of time spent in a patch. For two values of mean transit time, t_1 and t_2, the tangent method derived from the marginal value theorem identifies the optimal time to spent in patches, T^*. (Bottom) A foraging optimal reaction norm: the optimal time spent in the average patch (the phenotype) as a function of mean transit time between patches (the environmental variable).

I see two important challenges that must be met to implement the above suggestion. First, for some plastic traits, we might not understand the mechanisms underlying fitness costs and benefits of plasticity well enough to build a useful explicit, optimality model. Second, we will likely need new statistical methods (probably, new randomization tests, Crowley 1992) to test hypotheses about parameters in reaction norm functions generated by optimality models. For adaptive patterns of plasticity, the extra effort should prove worthwhile.

Optimality Theory and Quantitative Measures of Natural Selection

A second goal of QG studies of phenotypic plasticity is to quantify natural selection on reaction norms. In theory, the basic method is an extension of the standard, multivariate statistical techniques for analyzing selection that can be applied to any trait (Lande and Arnold 1983; Mitchell-Olds and Shaw 1987; Brodie et al. 1995). Optimality theory, of course, also invokes natural selection in the sense that it identifies fitness-maximizing traits under the assumption that natural selection has shaped traits (including patterns of plasticity). Given their shared focus on natural selection, it would seem appropriate that there should be numerous bridges between optimality theory and the quantitative measurement of natural selection. In fact, relatively few studies bridge these two fields. Next, I describe a particular form of the optimality approach in more detail and discuss some benefits of integrating optimality and quantitative analyses of natural selection (QNS). The insights on this integration could apply to any trait—plastic or not.

Figure 8.2 shows a series of steps that link trait Z to fitness W. The trait has benefits and costs that influence fitness components that contribute to fitness. Fitness components include survival, fecundity, and growth or development (if survival or fecundity is size or stage dependent). Fitness components can be defined in terms of transition probabilities among such categories as age, size, or stage (Caswell 2001). Transition elements, a_{ij}, represent the probability of moving from category j to category i in a given time period. If category 1 is new offspring, then a_{1j} is the fecundity of category j. The probability of survival in category j without growth or development is a_{jj}, and the probability of survival with growth or development to a new stage is a_{ij}.

In this framework, to implement the optimality approach, one must make explicit assumptions about how a given trait, Z, influences each of the fitness components (transition elements) and how the fitness components influence fitness. Given these assumptions, optimality theory predicts the fitness-maximizing trait, Z^* in a given environment, and the optimal pattern of plasticity, $Z^*(E)$, as a function, for example, of the food, predation risk, or social environment.

To find the maximum (or minimum) for a function, $W(Z)$, one finds the value of Z that results in $dW/dZ = 0$ [where dW/dZ is the derivative of $W(Z)$ with respect to Z; or partial derivatives if there are multiple traits influencing fitness]. Of course, to implement this process, the modeler must first derive $W(Z)$. That is, as part of the process of identifying an optimum, optimality models typically must generate a predicted fitness function (the quantitative function for the relationship between variation in the trait and fitness or fitness component) for every environment that fits the model's scenario. Although the optimality approach has focused primarily on optima, en route the method also predicts the strength of selection (the selection gradient, dW/dZ) for each trait value in every environment.

A few analyses have noted that information on the relative strength of selection can be used to predict when we would most expect to see adaptive behavior (i.e., when selection is strong), and when we might not be surprised to find suboptimal behavior (i.e., when selection is weak; Sih 1982; Crowley et al. 1990). In addition, according to risk sensitivity theory, nonlinearities in the fitness function can be important in an uncertain environment (Real and Caraco 1986).

Although the optimality approach implicitly generates predictions on fitness functions and on the strength of selection, relatively few optimality studies have quantified these functions. Multivariate statistical analyses of selection can yield several benefits that complement the optimality approach. First, explicit measurements of selection test critical assumptions about expected relationships between traits and fitness. Behavioral ecologists often assume that if observed behaviors resemble optimal behaviors, this confirms that behavior maximizes fitness, when in fact they have not measured relationships between behavior and fitness. Direct measurements of complex, nonlinear selection landscapes can clarify relationships that should be incorporated into optimality models. Second, statistical analyses of selection emphasize the need to collect data on

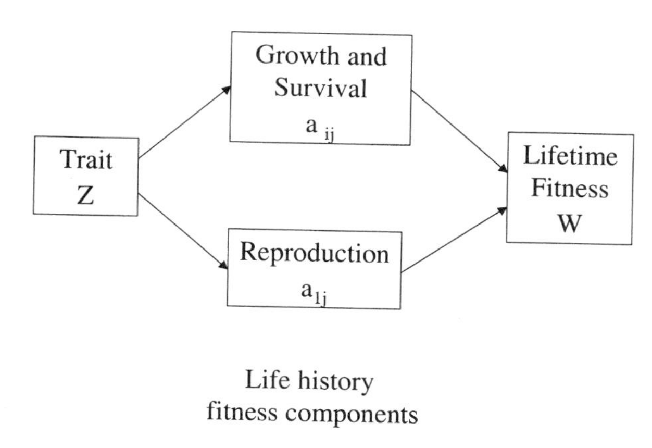

Figure 8.2. The relationship between a trait, *Z*, and fitness, *W*, via effects of the trait on various life history components that, in turn, influence lifetime fitness. Selection studies typically measure relationships between the trait and one component of fitness in one life history stage. Elasticity analyses from projection matrices can be used to relate variation in the life history component to lifetime fitness.

individual variation. The process typically involves following individuals and measuring their traits and associated fitness effects in different environments. In contrast, many evolutionary ecologists treat individual variation as noise. An insight for the optimality approach therefore is the need to use existing QNS methods to quantify natural selection.

This knife, however, cuts both ways. Although numerous studies have quantified selection (e.g., Simms and Rausher 1989; Weis and Gorman 1990; Brodie 1992; Dudley and Schmitt 1996), to my knowledge, few have compared observed fitness functions to a predicted fitness function. That is, although evolutionary biologists often quantify natural selection, presumably with an eye toward understanding adaptation, they generally do not attempt to relate observed fitness functions to an explicit mechanistic (e.g., optimality) model. Without an optimality model, studies of selection test statistical hypotheses about whether the strength of selection is greater than zero, or whether selection is linear or nonlinear, but they do not test hypotheses about biological mechanisms that underlie selection. Without some sort of mechanistic model, each labor-intensive measurement of a fitness function is done in a vacuum, with no explicit way of relating the results to expected fitness functions in other environments or with other related systems. As noted above, optimality models offer a unifying format for relating fitness functions observed in different situations, and they predict fitness functions in situations that have not yet been observed.

Projection Matrices and Measuring Natural Selection

An offshoot of the optimality approach also offers a quantitative method for dealing with an oft-ignored problem that affects most existing studies that measure fitness functions. Because of logistical constraints, in most cases, studies that purport to measure natural selection do not actually measure lifetime fitness. Instead, the dependent variable measured is a fitness component, such as short-term survivorship or fecundity (i.e., a transition probability; figure 8.2). The problem is that transition probabilities differ in their effects on fitness (e.g., Crouse et al. 1987; Caswell 2001; Byers and Meagher 1997). Typically, some transition elements (often, survival through very early stages or in very late stages) have very little effect on lifetime fitness, whereas other components of fitness (often, survival around the age of first reproduction) have important effects on overall fitness.

Relationships between traits, components of fitness, and lifetime fitness can be examined quantitatively by using a hybrid of projection matrix (PM) analyses and quantitative measures of natural selection (QNS). A PM consists of a matrix of transition probabilities (a_{ij}, as defined for figure 8.2). Given a particular matrix, standard methods can be used to calculate the sensitivity ($\delta W/\delta a_{ij}$) or elasticity ($(1/a_{ij})(\delta W/\delta a_{ij})$) of each element as a measure of selection on that fitness component. Studies that purport to quantify natural selection (e.g., selection gradients, or the entire fitness surface) typically measure the effects of a trait on a component of fitness (i.e., $\delta a_{ij}/\delta Z$). A true measure of natural selection ought to assess the effect of the trait on lifetime fitness (dW/dZ), which depends on the products of the effect of the trait on a transition element multiplied by the sensitivity of that element, summed up for all components of fitness that are influenced by that trait [$\Sigma(\delta a_{ij}/\delta Z)(\delta W/da_{ij})$; Caswell 2001].

Projection matrices are therefore necessary for comparing the relative strength of selection on two or more traits that influence different fitness components, or for contrasting the effects of a given trait on fitness through different fitness components. Conversely, PM analyses of the effects of a trait on each fitness component require multivariate statistical methods to actually measure these relationships. A hybrid of the two methods, where logistically feasible, should prove highly useful.

Lessons from Behavioral Ecology for the Study of Reaction Norms

Another important insight from optimality-based behavioral ecology for the study of reaction norms is a recognition of the complexity of the environment, and the resulting complexity of the relationship between the phenotype, the environment, and fitness. Although this insight will prove difficult to deal with using the QG approach, it is, I believe, a real fact about nature, and therefore should not be ignored.

Most QG-based studies of phenotypic plasticity take a simplistic view of the environment; that is, most studies quantify reaction norms relative to one or two environmental gradients, often comparing only two or three environments along a particular gradient. The scope of these studies is typically limited by logistical constraints associated with the QG approach (e.g., the requirements of having large numbers of sibships, sufficient individuals per sibship per environment, multiple measurements on each individual). To further study natural selection on reaction norms, an investigator must cope with the additional logistical problems associated with measuring selection. As a result, although measurements of selection on reaction norms should, in theory, be critical for understanding the evolution of reaction norms, in reality, relatively few studies have attempted this task (Weis and Gorman 1990; Dudley and Schmitt 1996). Thus, the QG approach has yielded relatively little empirical insight on natural selection on reaction norms.

In contrast, because behavioral ecologists address selection in a more superficial way, they can examine the effects of a greater range of fitness factors. One clear insight from behavioral ecological studies is that fitness often depends on a complex interplay of several conflicting selection pressures (Sih 1987, 1994; DeWitt and Langerhans 2003). Fitness functions and optimal behaviors along one environmental gradient depend on one or more other environmental factors. For example, optimal patch use as a function of food availability depends on predation risk (Lima and Dill 1990), competition (Milinski and Parker 1991), and abiotic factors, as well as interactions among these factors (McNamara and Houston 1990; Sih 1998). Any one of these broad factors actually includes several, if not many, distinct subfactors. For example, predation risk typically involves a suite of multiple predators that can exert different, conflicting selection pressures (Sih et al. 1998, DeWitt and Langerhans 2003). A similar statement about the complex interplay of environmental effects can be made about factors affecting plasticity in life history or morphological traits (Crowl and Covich 1990; Clark and Harvel 1992; Skelly and Werner 1990; Sih and Moore 1993).

Effects of the social environment on optimal reaction norms can be particularly complex because they depend not only on the phenotypes of other individuals, but also on the frequency of these phenotypes in the population. This is well known to apply for various social situations including cooperation, competition for food or mates, and male–female sexual conflicts. In behavioral ecology, these effects are analyzed via game theory (Maynard Smith 1982; Dugatkin and Reeve 1998). Recent models of indirect genetic effects have used the QG approach to analyze effects of the social environment on the evolution of phenotypes and, potentially, on reaction norms; however, to date, these models have not included the frequency-dependent aspects that are at the heart of game theory (Wolf et al. 2000). Frequency-dependent considerations could also play a major role in the evolution of patterns of developmental plasticity in competitive (and other social) contexts.

A further, general complexity involves the effects of individual state (e.g., size, condition, energy reserves) and time horizons (e.g., time until the end of the season) on fitness functions and optimal behaviors (Mangel and Clark 1986; McNamara and Houston 1986). As a generality, for example, individuals that are in poor condition often take greater risks to feed than do individuals in better condition, particularly as the time horizon approaches. Although the variations in individual state might not be inherited, it is plausible that there is a genetic basis to behavioral reaction norms as a function of the external environment, the individual's current state, and the time remaining before a time horizon. That is, the individual's state and the time horizon can be considered additional environmental gradients for a reaction norm, with the complication that the phenotype (behavior) influences the future environment (individual state). The branch of optimality theory that addresses these effects is dynamic optimization. It has been applied to the evolution of life histories (Werner 1986; Rowe and Ludwig 1991) and of induced morphological plasticity (Clark and Harvell 1992) but, by and large, is ignored by QG-based analyses of reaction norms.

Overall, the message is that we already know that both reaction norms and selection on reaction norms should depend heavily on a complex mix of biotic and abiotic environmental factors, including the social environment, as well as the individual's state, and the time until a time horizon. Given the considerable effort required to apply the QG approach to address even simple environmental scenarios, it seems unlikely that we will be able to account for these important selective effects within the QG framework, except perhaps in particularly tractable model systems. This poses a dilemma that, at minimum, should be taken as a cautionary note in studies of phenotypic plasticity.

Adaptive versus Nonadaptive Plasticity

I have argued that the optimality approach has important insights for future studies of phenotypic plasticity. However, because the optimality view hinges on the assumption that traits are adaptive, this view is not likely to be useful for understanding patterns of plasticity that are nonadaptive. As a generality, nonadaptive plasticity tends to involve traits that are either too close to fitness or too far removed from fitness. Obviously, the study of traits that are plastic but have little or no effect on fitness should not be guided by optimality theory. More interesting, plasticity in fitness itself might also be generally nonadaptive (Scheiner 1993a). With fitness, the simple generality is that "more is better." Therefore, plasticity in fitness across environments likely represents constraints generated by variations in environmental quality, and not adaptive plasticity. That is, low fitness in poor environments is a constraint, rather than an adaptive decision.

Moving one step away from fitness *per se* to components of fitness (e.g., life history traits such as age-specific survival or fecundity) allows for the possibility that plasticity reflects both environmental constraints and adaptive trade-offs (e.g., Roff 1992; Stearns 1992). With trade-offs, "more is always better" no longer applies. Thus, low fecundity in a poor environment might be due both to environmental stress *per se* (low total energy availability) and to an adaptive decision to reduce current reproductive effort so that energy can be diverted to growth, survival, or future reproduction. The optimality approach focuses, of course, on adaptive plasticity. Indeed, in my experience, optimality-based evolutionary ecologists often think that much of nonadaptive life history variation is not interesting, that is, that it is simply not interesting that stressful environments or poor diets result in slow growth, low survival, or low fecundity. Unfortunately, studies of phenotypic plasticity in life history traits often do not attempt to distinguish between adaptive and nonadaptive aspects of plasticity.

Plasticity in phenotypic traits that are not direct fitness components (e.g., behavior, physiology, morphology) can also be either adaptive or nonadaptive. With these traits, the "more is better" generality is even less applicable. In that case, both adaptive and nonadaptive plasticity can be interesting in the sense that they both result from interesting biological mechanisms. In these traits, adaptive plasticity often appears to fit functional considerations, whereas nonadaptive plasticity might reflect underlying genetic and developmental constraints (Smith-Gill 1983; Cowley and Atchley 1992). To fully understand reaction norms, both adaptive and nonadaptive, we need QG methods to quantify patterns of plasticity, and explicit mechanistic models (functional, genetic and developmental) along with tests of these models to gain insights on mechanisms. The optimality approach should play a valuable role in this integrative approach.

Joint Evolution of Behavioral and Developmental Plasticity

Up to this point, I have contrasted aspects of behavioral versus developmental plasticity. In this section, I emphasize that organisms often show both types of plasticity in response to the same selective pressures. For example, recent studies have shown that individual freshwater snails (*Physa*) show behavioral (refuge use), life history (age and size of first reproduction), and induced morphological (shell shape) responses to predatory fish and crayfish (Crowl and Covich 1990; Turner 1996; DeWitt 1998). Similar joint responses to predators have been observed for several other species [e.g., for *Daphnia* (Dodson 1988, 1999; Spitze 1992), freshwater isopods (Sparkes 1996), and guppies (Endler 1995)].

Multiple responses to a given selective agent are particularly interesting and important when they have interacting effects on fitness. In that case, understanding selection and optimal reaction norms for one type of response requires knowledge of the other responses. For example, if prey are well defended morphologically, then they experience less selection favoring a strong behavioral response. Indeed, if morphological adaptations make prey invulnerable to attack, then if predator avoidance has costs (e.g., reduced foraging rates or access to potential mates), prey should not avoid predators. Conversely, if prey show strong behavioral responses to predators, then they have less need for costly morphological defenses. That is, prey can behaviorally compensate for having a poor antipredator morphology. Recent work on physid snails revealed that they indeed show behavioral compensation for morphological vulnerability (DeWitt et al. 1999). This effect (behavioral compensation) seems also to apply across species (Rundle and Brönmark 2001).

Figure 8.3 shows a way of visualizing joint (behavior-morphology) reaction norms for two genotypes in four environments. The specific scenario illustrated represents behavioral versus morphological responses of physid snails to fish versus crayfish (see discussion above). The four environments involve the combination of presence versus absence of the two types of predators. The stars represent hypothesized optima in each environment. The antipredator behavioral response (use of refuges) is effective against both predators but is costly. Because of the cost of predator avoidance, snails should respond less when neither predator is present. In the presence of shell-crushing predators (e.g., fish), snails develop a rounder shell that is more difficult to crush. In contrast, in the presence of crayfish that extract snails through their aperture, snails develop a narrower aperture (that is correlated to narrower shell). A default, intermediate shell shape might be expected in habitats with neither predator, and perhaps in environments with both predators.

By following standard QG methods (i.e., rearing split sibships with individuals in each environment; see chapter 3), it is possible to generate the joint reaction norms shown in figure 8.3. By following standard QNS methods, it is, in theory, possible to generate fitness functions for a range of behavior–morphology combinations in each of the four predation regimes. Thom DeWitt and I are now in the process of using these basic methods to gather data on physid joint reaction norms in the scenario outlined above. Below, I briefly note, using hypothetical data shown in figure 8.3, some of the novel issues that we should be able to address.

Given two types of responses to predators, genotypes can be behavioral specialists (use refuges but show little induced morphological response), morphological specialists (induce an effective shell size and shape, but show little behavioral response), or behavior–morphology generalists (show both types of response). Behavioral compensation within each environment can be assessed by examining genetic correlations between traits. In the fish environment, the behavior–morphology correlation should be positive, whereas in the crayfish environment, it should be negative (see figure 8.3). The relative plasticity exhibited by each of the three genotypes is measured by the areas inside the polygons. If the sizes of these polygons are generally smaller than the area of

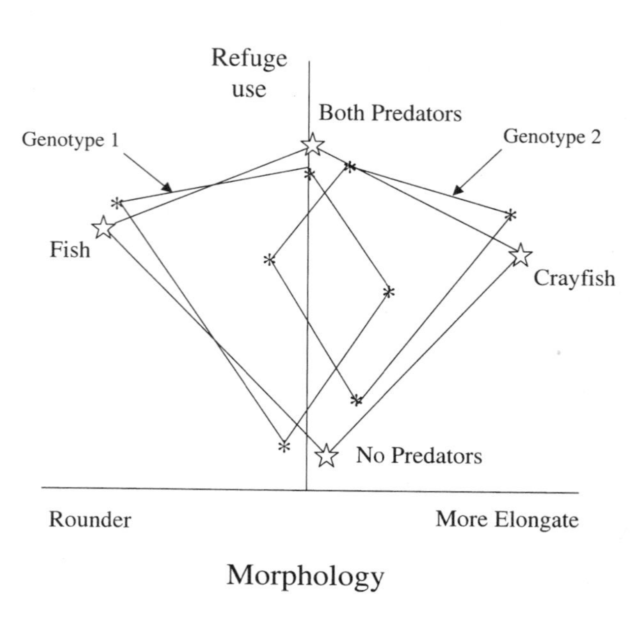

Figure 8.3. A two-dimensional hypothetical reaction norm. Each point represents a genotype's behavior–morphology combination in a given environment. Stars represent the optimal behavior–morphology combinations in each of four predation regimes. The area of the polygon linking the four stars measures the degree of plasticity required to show an optimal reaction norm. Both genotypes have smaller polygons than the optimal reaction norm; thus they show suboptimal phenotypes because of limits to plasticity. Genotype 1 is well adapted to fish but poorly adapted to crayfish, whereas genotype 2 is well adapted to crayfish but poorly adapted to fish.

the polygon joining the four optimal points, then fitness should be reduced by limited plasticity (perhaps because of genetic and developmental constraints). Further statistical analyses can reveal the relative effects of limited behavioral versus morphological plasticity on fitness.

The location of each polygon reflects relative adaptation to each predation regime. For example, genotype 1 is well adapted to the fish environment, but because of limited morphological plasticity, it is poorly adapted to the crayfish environment. It partially compensates for its limited morphological plasticity by exhibiting increased crawl-out in the presence of fish. In contrast, genotype 2 is well adapted to crayfish but shows limited plasticity and thus poor performance with fish.

Overall, the suggestion is that, given that organisms often show multiple types of responses to any given selective agent, studies of the joint evolution of multiple forms of plasticity should yield new, important insights. In particular, it should be interesting to contrast patterns of joint plasticity for more rapid, behavioral and slower, developmental plasticity.

Summary and Suggestions for Future Work

1. Reaction norms are usually characterized by statistical models whose parameters might not have a clear biological basis. To add more biological meaning to quantitative descriptions of reaction norms, workers can use optimality models to iden-

tify classes of reaction norm functions that are built on biologically meaningful parameters.

2. Optimality models generate fitness functions as part of the process of predicting optima. Furthermore, optimality models predict how fitness functions should change depending on the environment. To date, behavioral ecologists have rarely used quantitative natural selection (QNS) methods to quantify predicted fitness functions. In turn, QNS studies rarely contrast and attempt to understand how and why fitness functions differ in different environments. More studies should aim to integrate studies of optimality and QNS.
3. QNS studies rarely address lifetime fitness; instead, they typically correlate traits to a component of fitness in one life history stage. PM analyses offer a quantitative method for relating a component of selection in one life history stage to overall selection on lifetime fitness. In particular, PM elasticity analyses suggest that the strength of selection can vary across life history stages. It should be useful to incorporate this life cycle perspective more actively into QNS studies.
4. The theme of much of modern behavioral ecology is that selection is a complicated outcome of conflicting selection pressures that can depend on individual state, time horizons and on variation in abiotic, biotic, and social conditions. These complexities need to be included into QG studies of plasticity, or at least recognized as cautionary comments.
5. Studies of plasticity, particularly life history plasticity, should be more explicit about adaptive versus nonadaptive aspects of plasticity.
6. Organisms often show both behavioral and developmental plasticity. Studies on joint plasticity and limits to joint plasticity across environments should prove insightful.

9

Patterns and Analysis of Adaptive Phenotypic Plasticity in Animals

PAUL DOUGHTY
DAVID N. REZNICK

The most obvious ecological difference between animals and plants is that most animals move through space in response to changing conditions whereas plants do not. The evolution of a complex neuromuscular system that enables animals to receive, process, and respond to information with coordinated movement through space is perhaps the most impressive feat of adaptive phenotypic plasticity, broadly defined. However, this ability to move through space cannot resolve all environmental challenges. There will therefore be selection for adaptive phenotypic plasticity in morphology, physiology, behavior, and life history characters in response to the many diverse and risky situations animals face. Two main questions of interest to evolutionary biologists investigating this topic are (1) what kinds of environmental variation select for adaptive plasticity, versus other evolutionary responses such as local adaptation or polymorphism, and (2) what constrains the evolution of plastic responses?

Empirical evaluation of adaptive plasticity in animals over the past two decades has revealed a wide diversity of solutions to an equally wide array of environmental problems (Scheiner 1993; Travis 1994; Gotthard and Nylin 1995; Skulason and Smith 1995; Via et al. 1995; Roff 1996; Nylin and Gotthard 1998; Schlichting and Pigliucci 1998; Tollrian and Harvell 1999; chapter 10). In many areas, theoretical explorations have far outpaced empirical testing (Via et al. 1995; Agrawal 2001; chapter 6). For example, testing some of the more recent models will require information on the spatial and temporal distribution of alternative environments, whether they are population sources or sinks, and movement among patches (e.g., Houston and McNamara 1992; Sibly1995; Whitlock 1996; Sasaki and de Jong 1999). This imbalance between conjecture and data will probably persist because rigorously testing plastic adaptations can be time-consuming and logistically difficult. As we hope to demonstrate in our review of empirical studies, progress in this field to date has resulted from testing hypotheses in relatively simple systems. Most studies, however, have tended to rely on the argument from design (Williams 1992; Lauder 1996) rather than on detailed functional and ecological studies. Testing adaptations is an onerous task even for nonplastic phenotypic characters. Here we discuss additional issues involved in identifying phenotypic plasticity as an adaptation to heterogeneous environments. We focus on such traits as morphological structures used for defense or feeding and life history traits such as the age and size at metamorphosis or maturation. Aside from a few examples, we do not address the evolution of behavioral complexity in animals, despite its obvious importance for the success and diversification of this group (see chapter 8).

In the next section, we discuss the nature of phenotypic variation among and within taxa and how adaptive plasticity differs from nonadaptive plasticity. In the third section, we consider the criteria empiricists might use to identify phenotypic plasticity as an adaptive solution to environmental heterogeneity. In the fourth section, we review case studies of adaptive plasticity in animals to highlight the diversity of situations where adaptive phenotypic plasticity seems to have evolved. These examples are discussed with reference to the criteria set out in the preceding section. We conclude with a brief discussion of current gaps in knowledge.

When Is Phenotypic Plasticity an Adaptive Solution?

Phenotypic variation exists at all levels of the biological hierarchy. Phenotypic differences among higher taxa (e.g., genera, families) are usually interpreted as resulting from accumulated genetic differences driven by natural selection and phylogenetic history. Although differences in the environments occupied by taxa may contribute to species differences in phenotype, these environmental sources of variance are generally considered to be small compared with accumulated genetic differentiation. At a practical level, the scale of phenotypic differences among taxa is usually substantial enough for systematists and paleontologists to comfortably ignore environmental sources of variation in their analyses of phylogenies and macroevolutionary patterns of trait evolution. Investigators who carry out comparative studies of interspecific differences of potentially adaptive phenotypic characters also tend to ignore environmental sources of intraspecific variation.

In contrast to fixed macroevolutionary traits, environmental sources of phenotypic variation can profoundly affect phenotypic expression within and among populations. Accordingly, analysis of intraspecific phenotypic variation requires explicit inclusion of both genetic and environmental effects. Nearly all traits will show some plasticity because phenotypes result from the interaction of information in the genome with the developmental environment—the simple equation "genotype × environment = phenotype." Demonstrating that the environment affects phenotypic expression leaves unanswered the question of whether the response is adaptive and arose through natural selection, or if the response is merely the reflection of environmental variation on the phenotype (e.g., Smith-Gill 1983; Stearns 1989; Newman 1992; Doughty 1995). Genuine adaptive phenotypic plasticity involves the same lock-and-key matching seen when adaptation is achieved by a change in the mean phenotype of the population over many generations. With adaptive plasticity, however, phenotypic variation occurs in response to the developmental environment experienced during an individual's lifetime (or sometimes across one generation; e.g., Agrawal et al. 1999). Fortunately, empiricists can use many of the same principles and methods to study these two different types of adaptive phenotypic changes.

What kinds of conditions select for adaptive phenotypic plasticity? Both external (environmental) and internal (genetic variation and functional integration) factors play a role in determining whether adaptive plasticity will evolve. Although animals are capable of impressive movements through space, they cannot always move toward resources or mates or away from risky situations, especially when they occupy specific ecological niches and have limited dispersal abilities (e.g., tadpoles in ponds, caterpillars on trees). Temporal heterogeneity in the kind of environments that animals experience can also generate strong selection for adaptive plasticity. For example, short-lived animals can benefit from plasticity by deploying the strategy with the highest fitness for the time of year in which they emerge (e.g., Dixon and Kundu 1998).

Heterogeneity in the environment can select for either reversible or nonreversible plasticity depending on the scale over which heterogeneity occurs. Reversible plasticity can be favored by spatial heterogeneity if the animal moves through more than one kind of selective environment during its lifetime. For example, flatfish can rapidly change their dorsal coloration for camouflage when they settle on different backgrounds (Ramachandran et al. 1996). Spatial heterogeneity can select for nonreversible plasticity if patch size is greater than dispersal capabilities of individuals, but fine-grained enough so that the evolution of genotypic specialization will not ameliorate the need for plasticity (Levins 1968; Futuyma and Moreno 1988; Whitlock 1996). For example, tadpoles of several species show morphological responses (increases in fin size and bright colors in the tail) to the presence of dragonfly larvae and fish predators (McCollum and Van Buskirk 1996; McCollum and Leimberger 1997; Van Buskirk and Relyea 1998; Relyea 2001). Adaptive plasticity in the tail is a solution to this problem because there is variation among pools in the presence of predators, and tadpoles in small pools cannot move to another pool. Possession of larger fins increases survival by increasing escape speed or by directing predator strikes away from the more vulnerable body. Flexible patterns of development can be seen as nonreversible plasticity (Raff 1996; Schlichting and Pigliucci 1998).

Temporal heterogeneity can occur within or across generations, and theoretical models predict that adaptive plasticity is more likely to evolve when animals encounter temporal, versus spatial, heterogeneity (Moran 1992). Temporal heterogeneity that occurs within an individual's lifetime may favor the evolution of reversible plasticity. Tadpoles exposed to predators reduce activity levels and resume activity when predators are removed (reviewed in Anholt and Werner 1999), demonstrating the advantages of rapid reversible behavioral changes in response to temporal changes in the selective environment. One prominent aspect of temporal variation is seasonality. Most animals time reproduction and other important life history events to the seasons. When animals experience more than one season in their lives, plastic phenotypic changes are generally reversible (e.g., winter coats of fur or feathers, hibernation, migration; Piersma and Lindstrom 1997; Wilson and Franklin 2002). Short-lived animals, in contrast, are likely to encounter only one or a few seasons in their lifetimes, and these taxa frequently display across-generation irreversible morphological plasticity (e.g., Greene 1989a,b; Alekseev and Lampert 2001).

A distinction that is often not made explicit in discussions of adaptive plasticity is that between developmental versus selective environments (but see Levins 1968; Moran 1992; Scheiner 1993a; Sasaki and de Jong 1999). Developmental environment refers to the environment that results in expression of different phenotypes. For example, female aphids exposed to chemicals from ladybugs produce more winged offspring than do conspecifics not exposed (Dixon and Agarwala 1999; Weisser et al. 1999). In this case, detection of ladybug chemicals by aphids results in an adaptive phenotypic response because offspring can fly away from the plant, but the chemicals themselves do not kill the aphids and therefore are not the selective agent *per se*. In contrast, selective environment refers to the phenotypic selection favoring the evolution of a plastic response—in this case, predation by ladybugs. Thus, the presence of chemicals from the predator accurately predicts the risk of predation. Other examples of selective environments in adaptive analyses of plasticity include prey survivorship with or without defensive structures in the presence of predators (e.g., studies reviewed in Tollrian and Harvell 1999), and reproductive success as a function of the age and size at maturation (Stearns and Koella 1986; Smith 1987; Bernardo 1993).

Developmental and selective environments may be highly correlated. For example, lower temperatures induce darker pigmentation in *Drosophila*, and darker flies warm faster at lower temperatures than do lighter flies (chapter 4). The distinction between the two kinds of environments, however, should be made clear when describing a case of putative adaptive plasticity.

We emphasize a further distinction within developmental environments: the cue may be information based or resource based. This distinction is important in studies of life histories because adaptive reaction norms often involve shifts in energy allocation in response to variation in resource availability, rather than to a change in the selective environment. In this case, the resource-based cue might be propagule size from the offspring's perspective (Kaplan 1989; Parichy and Kaplan 1995; Bernardo 1996), growth rate (Wilbur and Collins 1973; Bernardo 1994), or fat levels in female reproductive allocation decisions (Reznick and Yang 1993; Doughty and Shine 1998; Bonnet et al. 2001). In this kind of adaptive plasticity, a resource-based cue triggers a change in energy allocation as a function of resource availability. For example, optimized reaction norms result in the initiation of metamorphosis or maturation at an age and size that yields the highest fitness possible given an individual's access to resources (Stearns and Koella 1986; Bernardo 1993). An adaptive life history reaction norm is an evolved response to help individuals cope with variation in resources in the developmental environment. This can apply to intraspecific variation in propagule size or to food availability among patches. All individuals, however, compete for reproductive success in the same selective environment. Emlen and Nijhout (2000) view scaling relationships of traits in a similar fashion.

In contrast to resource-based cues, information-based cues induce phenotypic effects because they predict the future selective environment. For example, changes in photoperiod can induce major phenotypic changes in diapause and reproduction in butterflies (Nylin 1992; Gotthard et al. 1994; Nylin et al. 1996). However, the plastic responses are not caused by photoperiod acting as an agent of selection *per se*, but by the evolved mechanisms that exploit photoperiod as an indicator of change in season and hence selective pressures. Detection of salient cues can therefore trigger appropriate developmental responses. For example, with inducible defenses, there is an adaptive change in energy allocation according to information in the developmental environment (e.g., predator chemicals or attack) that predicts the future selective environment (e.g., risk of predation).

Even when selection favors the evolution of adaptive plasticity, it may still not evolve for several reasons. As for nonplastic traits, lack of sufficient genetic variation can prevent the evolution of adaptive plasticity (Via and Lande 1985; Scheiner 1993a; Kirkpatrick 1996). Trade-offs with other traits or across life history stages (e.g., Stevens et al. 1999) may also constrain the evolution of plastic responses if induced structures are too costly to produce relative to their benefits (Scheiner and Berrigan 1998; DeWitt et al. 1998). Although it is theoretically possible for macromutations to result in polyphenisms (Orr and Coyne 1992; Skulason and Smith 1995), we believe polyphenisms are more likely to evolve via disruptive selection working over a range of continuous phenotypic variation in different selective environments. The evolution of information-based adaptive plasticity is not possible when there are no reliable cues available that are correlated with the future selective environment (Moran 1992; Getty 1996). Plasticity may not evolve in rapidly changing environments if there are long delays between detecting a cue and deploying a response (Padilla and Adolph 1996). Additionally, environments with different selection regimes must be encountered frequently enough for a plastic solution to evolve (Houston and McNamara 1992; Moran 1992; Kawecki and Stearns 1993; Whitlock 1996; Sasaki and de Jong 1999).

Criteria for Recognizing Phenotypic Plasticity as an Adaptation

What evidence is necessary to argue that a plastic trait is an adaptation? In general, labeling a trait as an adaptation implies that it evolved in response to a specific form of selection, or that there is a cause-and-effect relationship between the trait and the environment in which it is found. Because the history behind the evolution of a trait is generally not available (Reeve and Sherman 1993; Pagel 1994; Doughty 1996), we suggest using a diverse number of indirect approaches for testing whether plasticity is an adaptation. The argument for adaptation is most convincing when several lines of indirect evidence are brought to bear on a specific case (Curio 1973; Endler 1986; Reznick and Travis 1996). For phenotypic plasticity, these resolve to six questions presented in table 9.1.

Production of Different Phenotypes in Different Developmental Environments

The first step in arguing for adaptive plasticity is describing the nature of the phenotypic variation and the factors that cause it. Plastic traits can have distributions that are either qualitative (polyphenisms) or quantitative (reaction norms). The identification of two or more elaborate morphs that develop in response to the environment can strongly suggest adaptive phenotypic plasticity (e.g., Smith-Gill 1983; Gotthard and Nylin 1995). Such complex plastic traits are more counterintuitive than are traits that vary continuously, and therefore are putative adaptations based on the "argument from design" commonly used in evolutionary biology (Williams 1992; Gotthard and Nylin 1995; Lauder 1996). For example, Greene (1989a,b) described the caterpillars of a bivoltine moth that

Table 9.1. Six questions that can be addressed to evaluate whether phenotypic plasticity is an adaptation.

1. What are the plastic responses to the developmental environment?
2. What is the nature and extent of environmental heterogeneity?
3. Is there a reversal in relative fitness of the alternative phenotypes in different environments? This question includes understanding how selection acts on the plastic trait(s), trade-offs with other traits, and costs of plasticity.
4. For information-based plasticity, does the animal exploit a reliable cue that predicts the future selective environment?
5. Is there genetic variation for plasticity and does it evolve in response to selection?
6. Is there comparative evidence that plasticity is correlated with environmental heterogeneity?

appears as either a catkin (oak flower) morph in late winter and early spring or a twig morph that develops in summer. In this example, caterpillars from the two cohorts differ strikingly in morphology, coloration, and behavior; moreover, there is a close match between phenotypes of the two cohorts and the seasonal change in oaks. Therefore, it seems reasonable based on the argument from design to regard the plasticity as a putative adaptation, subject to further evaluation (Lauder 1996; Reznick and Travis 1996).

It is more difficult to interpret plastic traits that vary quantitatively (reaction norms) as adaptations because reaction norms can be generated from passive or active processes of development. Passively generated reaction norms can reflect phenotypic responses to basic physical and chemical features of the environment (most notably, food and temperature), and may be adaptations if selection has changed the form of an ancestral reaction norm (Doughty 1995; Gotthard and Nylin 1995; see below). Although traits may show plastic responses to novel environments, such responses would not be considered adaptations because they have not evolved in that environment (although the plasticity could be fortuitously beneficial to the animal) (Smith-Gill 1983; Stearns 1989; Gotthard and Nylin 1995). In contrast, more convincing evidence for adaptive plasticity is provided by studies demonstrating an active developmental response to environmental variation.

After the description of a polyphenism or reaction norm, further hypothesis testing is necessary. One way of looking at these kinds of data is just describing the phenotypic pattern that needs to be explained. The analogous case for an adaptationist analysis of a nonplastic trait would be to describe the trait (e.g., mean, variance) and then present a plausible scenario for its evolution. The scenario could be based on similar examples or devised by imagining the relevant selection pressures and the genotypic and functional constraints operating on the trait. Clearly, different kinds of evidence are necessary to go beyond the initial observation and scenario in support of an adaptive hypothesis of phenotypic plasticity (Newman 1992; Doughty 1995, 1996; Gotthard and Nylin 1995; Reznick and Travis 1996).

Environmental Heterogeneity

Demonstrating environmental heterogeneity is a necessary and early step toward establishing phenotypic plasticity as an adaptation because such variation is believed to be what selects for adaptive plasticity. It is important to document environmental heterogeneity in natural environments because experiments can induce phenotypic changes in response to environments that are rarely encountered in nature, and hence are unlikely to drive the evolution of adaptive plasticity. Such potentially misleading results are especially a concern for quantitative life history traits that vary in response to food or temperature (see above).

Most studies present only a qualitative description of environmental heterogeneity, and this is sufficient to initiate an investigation of adaptive plasticity. However, mathematical models have shown that the degree to which environments vary can affect the evolution of plasticity as an adaptation, versus the evolution of a change in the population mean (e.g., Moran 1992; Houston and McNamara 1992; Kawecki and Stearns 1993; Whitlock 1996; Zhivotovsky et al. 1996; Scheiner 1998; Sasaki and de Jong 1999). Accordingly, field studies that document heterogeneity in space or time are especially useful in gaining insight into how environments select for adaptive plasticity. Well-documented examples in animals are the effects of pond drying or food availability on larval phenotypes of spadefoot toads (Newman 1987, 1989; Pfennig 1990; Morey and Reznick 2000; Pfennig and Murphy 2000) and the association between gastropod predators and shell morphology in barnacles (Lively 1986b, 1999a; Lively et al. 2000).

Reversal of Fitness of Alternative Phenotypes in Different Environments

For adaptive plasticity to evolve, there must be trade-offs among traits that cause one phenotype to have higher fitness in one environment and an alternative phenotype to have higher fitness in an alternative environment. If this is not the case, then selection will favor the evolution of a single all-purpose phenotype that has high fitness in all possible environments. Ideally, one wishes to know how selection favors differential trait expression across environments. Accordingly, many of the same methods evolutionists use to study nonplastic traits can also be used to evaluate the fitness of plastic traits in much the same way (Endler 1986; Gotthard and Nylin 1995; Reznick and Travis 1996). These include quantitative estimates of selection surfaces (reviewed in Brodie et al. 1995) on natural or manipulated phenotypic distributions (Sinervo and Basalo 1996; Sinervo and Doughty 1996).

Manipulative experiments are often used to evaluate what effects alternative phenotypes have on fitness. For example, to evaluate how tadpole mobility affected predation by dragonfly larvae, Skelly (1994) placed anesthetized (immobile) and control (mobile) tadpoles into aquaria with dragonfly larvae predators. Mobile tadpoles suffered four times the predation rate experienced by immobile tadpoles; he therefore concluded that reducing activity in the presence of dragonfly larvae enhances tadpole survivorship in ponds with predators. A limitation with such studies, however, is that the benefits of plastic responses are often evaluated without proper estimation of their costs. Such functional studies are often one-sided, demonstrating an advantage of induced phenotypes in the presence of the inducing agent without showing a disadvantage of the induced phenotype in the noninducing environment. Again, this may be another way the argument from design is used presumptively in studies of plasticity. In this example, other studies have shown that reduced activity in the presence of predators results in lower growth rates (reviewed in Anholt and Werner 1999). Nevertheless, experiments designed to isolate specific hypothesized costs and benefits of plasticity can provide a more complete picture of the net fitness advantages to alternative morphs in contrasting environments (Scheiner and Berrigan 1998; DeWitt et al. 1998; Anholt and Werner 1999).

Reciprocal transplant experiments (the translocation of individuals with alternative phenotypes among alternative environments) test for reversals in phenotypic optima because they incorporate whatever costs and benefits accrue to alternative phenotypes in the alternative environments (discussed in Gotthard and Nylin 1995). This method has been used most often for studies on plants, but there are good examples for a diversity of animals (e.g., Berven 1982a,b; Niewiarowski and Roosenburg 1993; Bernardo 1994; Trussell 2000; Yeap et al. 2001).

Cue Reliability (for Information-Based Plasticity)

When genotypes respond to neutral cues in the environment, the cue must be highly correlated with the future environmental conditions (e.g., Moran 1992; Getty 1996). Such cues can be considered neutral because they had no major physical or chemical effects on phenotypic expression before the evolution of adaptive plasticity. Information-based cues, however, must have had some effect; otherwise, it would not be possible for natural selection to evolve mechanisms to sense and respond to them. If the correlation between environmental signal and future conditions is low, selection will favor the evolution of more sophisticated sensory mechanisms to exploit more reliable cues. For example, photoperiod is a cue that often induces phenotypic change in animals because it more reliably reflects the time of year than temperature or food availability (Shapiro 1976; Nylin et al. 1996). In Wilbur and Collins's (1973) model of amphibian metamorphosis, a decrease in growth rate is the "cue" that predicts a decrease in future growth opportunity, and thus in turn triggers metamorphosis into the terrestrial environment (Morey and Reznick 2000; Doughty 2002). An animal's response to a cue can be very coarse. For example, Langerhans and DeWitt (2002) found that snails respond to harmless species of fish in addition to those that actually eat snails. Presumably the sensory mechanisms have not been refined owing to a lack of strong selection or constraints on chemical detecting mechanisms to discriminate between the different kinds of fish.

Genetic Variation and Response to Selection

Establishing that there is an underlying genetic basis to plastic traits is necessary for understanding the plastic response as an adaptation. Genetic variation in plastic traits is of interest for two reasons. First, genetic variation for adaptive plasticity must be present for it to evolve. Second, when populations are found to differ phenotypically, this could be caused by genetic differences between them, or by the same genotype reflecting variation in the developmental environment. We briefly review the first point below, and then discuss the latter in the next section.

Genetic variation for phenotypic plasticity has been evaluated from two perspectives: allelic sensitivity (the differential expression of genes in different environments; see Via and Lande 1985) and plasticity genes (genes that affect the reaction norm directly; see Scheiner 1993a). This continues to be an active area for discussion in evolutionary genetics. See Via et al. (1995), Schlichting and Pigliucci (1998), and chapter 3 for reviews of these issues.

Although demonstrating genetic variation for overall levels of plasticity is important, what is probably more relevant is genetic variation for specific aspects of the animal's response to its environment. For instance, there may be genetic variation in reaction norms for intercept, slope, or shape. For example, reaction norms for horn size allometry in dung beetles (see example 4, below) have been artificially selected to change their intercept but not their shape. Hence, some aspects of an animal's response to the environment may be more genetically malleable than are other aspects. Local phylogenetic analyses (see below) would aid in identifying aspects of plastic responses that were relatively labile versus constrained evolutionarily.

Experiments with *Drosophila melanogaster* demonstrate that plasticity in body size and weight can respond to selection and evolve (Hillesheim and Stearns 1991; Scheiner and Lyman 1991; reviewed in Scheiner 1993a, 2002). In cases of adaptive phenotypic plasticity in vertebrates, genetic variation and response to selection are rarely measured. In many interesting systems, almost nothing is known of the genetics of plasticity (e.g., crucian carp: Bronmark et al. 1999; guppies: Reznick 1990). However, there are some notable exceptions based on analyses of parents and their offspring (tadpoles: Newman 1988a,b, 1994b; sailfin mollies: Trexler et al. 1990, but see Trexler and Travis 1990). There have been few selection experiments of plasticity conducted on vertebrates in the lab or in nature (but see Semlitsch and Wilbur 1989).

Comparative Evidence

Another line of evidence that can strengthen a case for adaptive plasticity is the finding that plastic phenotypic expression across species or populations is correlated with environmental heterogeneity (Newman 1992; Doughty 1995; Gotthard and Nylin 1995). Such data indicate that (1) genetic variation for plasticity existed in the past, and (2) there is an adaptive match between plasticity and environmental heterogeneity. If there are evolved differences in the genetic control of plasticity among populations or species, then there must have been genetic variation for plasticity in the past. Thus, differences in plasticity among populations or taxa are evidence for genetic variation in plasticity.

Comparative evidence is especially valuable for evaluating quantitative variation in reaction norms because such norms may result from inevitable nonadaptive responses to environmental factors, such as food availability or temperature (Smith-Gill 1983; Stearns 1989). If ecological differences among taxa are correlated with phenotypic plasticity, this variation is likely to have been shaped by selection in each species' or population's environment (Doughty 1995; Gotthard and Nylin 1995). Identifying correlations in quantitative characters provides a strong empirical basis for rigorous testing of an adaptive hypothesis (Pagel 1994; Doughty 1996; Larson and Losos 1996; Reznick and Travis 1996).

Most comparative tests to date have tested two populations or species that occur in different environments (e.g., Bernardo 1994; Leips and Travis 1994; Nylin et al. 1996). Wider taxonomic sampling may reveal complex spatial patterns of plasticity and avoid the pitfalls of two-taxon comparisons (Garland and Adolph 1994; Doughty 1996; Jockusch 1997).

Some Case Studies of Adaptive Phenotypic Plasticity in Animals

Here we review six case studies of adaptive phenotypic plasticity in animals. These examples highlight the diversity of plastic solutions to environmental heterogeneity and illustrate the criteria outlined above for identifying adaptive plasticity. We have tried to avoid overlap with other recent reviews. Chapter 4 illustrates several points on reaction norm evolution with the *Drosophila* model, and although we focus on studies by Palmer and others on snails here, Lively and colleagues (see Lively 1999a) have also made a strong case for adaptive plasticity in barnacle shape to predators.

We emphasize that meeting all the criteria for a specific case requires a tremendous amount of work and, in most cases, will take many years and perhaps several investigative teams to carry out. Hence, our indication of the lack of certain kinds of data does not reflect negatively on the positive results of the studies presented below, but rather represent opportunities for future tests of adaptationist hypotheses. This state of affairs underscores the relative youth of adaptive analyses of phenotypic plasticity in animals and the lack of studies that go beyond a description of phenotypic plasticity and an argument from design.

Example 1: Plasticity in Claw Size of Crabs in Response to Prey Hardness

Crabs use their claws to grasp and process prey. This simple interaction has resulted in a predator–prey arms race, primarily between crabs and mollusks, that has been documented in both the fossil record and contemporary communities (Vermeij 1987). There may also be sufficient variation in prey availability and composition for crabs to have evolved developmental plasticity in claw formation. Smith and Palmer (1994) conducted laboratory experiments to determine if the claw sizes of crabs (*Cancer productus*) changed as a function of the hardness of their prey (mollusks of the genus *Mytilus*). They fed shelled and unshelled mollusks to crabs with both claws free or one claw glued closed. In both shelled and unshelled treatments, closed claws decreased substantially in size (figure 9.1). In crabs that had to crush shelled mollusks, claw volume increased 5% when claws were free to manipulate prey. In contrast, claws of crabs fed mollusks without shells all decreased in volume by up to 10%. The largest differences in claw size were between the free and glued claws of crabs fed shelled prey, and these changes in claw size were most pronounced on the second molt after the initiation of the experiment. To assess the performance capabilities of different-sized claws, crabs were coaxed to grasp a device that measured crushing force. The results indicated that claw volume was a good index of crushing force, and hence the ability of crabs to process prey.

According to our first criteria, this study suggests that use-induced changes in feeding morphology are adaptive. The experimental treatments controlled for differences in nutrition between shelled and unshelled prey, although crabs that crushed shelled mollusks might have expended slightly more energy. A reversal in fitness of alternate phenotypes in alternative environments was not measured directly. However, we think it reasonable to presume that increased crushing force of large claws would increase feeding efficiency and hence fitness in an environment with hard prey. Conversely, in environ-

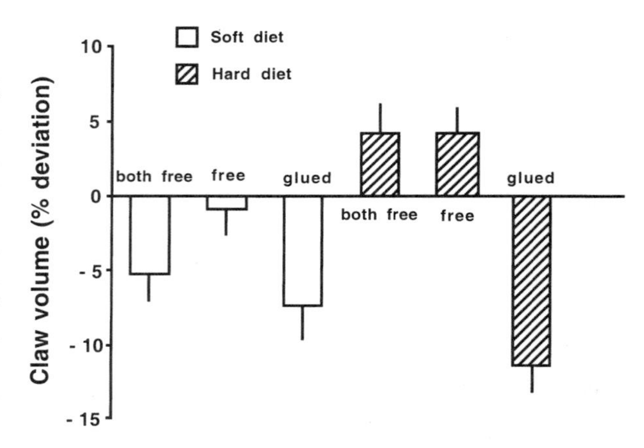

Figure 9.1. Size-corrected changes in claw volume of *Cancer productus* in response to soft (opened) or hard (unopened) mollusks (*Mytilus* sp.). Crabs had both claws free, or one claw free and one claw glued shut. Data shown are from the second molt after the treatments began, and error bars represent ±1 standard error. [From Smith and Palmer (1994)]

ments with soft-bodied prey, crabs would not need to produce larger claws and therefore could conserve energy for use in other energetically expensive functions.

Data to satisfy the rest of our criteria were not presented in their report, although some predictions can be derived from their adaptive scenario. Because crabs are generalist feeders, they are likely to encounter a wide variety of prey that differ in hardness. The adaptive significance of the plasticity would also depend on the time course of the response relative to the patterns of variation in prey availability. The authors discuss the possibility that the environmental heterogeneity may actually be the loss of a claw, rather than a change in available prey. In crabs, feeding efficiency is sometimes optimized when one claw crushes prey and the other manipulates it. However, if crabs commonly lose their crushing claw to predators or in agonistic encounters, then selection may favor individuals that can increase the size of their manipulating claw in order to crush prey more effectively. The environmental cue in this case is hardness of prey. This would be a reliable cue if crabs encounter prey of similar hardness over time scales that lasted for at least two molts.

Example 2: Plasticity in Morphological Defenses of Marine Snails in Response to Predation Risk from Crabs

Although selection has improved the ability of crabs to crush prey, there also has been selection to increase the resistance of mollusk shells to crushing (Vermeij 1987) either as a fixed feature of the genotype or as developmental plasticity. Plasticity in the form of induced defenses in response to the attack or presence of predators has been documented in numerous systems (Tollrian and Harvell 1999). Here we review morphological and behavioral changes of marine snails in response to water-borne chemicals that indicate the risk of predation by crabs. We focus on studies examining the responses of *Nucella lamellosa* (Appleton and Palmer 1988), *N. lapillus* (Palmer 1990), and *Littorina obtusata* (Trussell 1996) to chemicals released from crabs and the crab's prey. All three studies employed similar experimental procedures and produced qualitatively similar results (with some exceptions noted below). Therefore, we focus on the study of Appleton and Palmer (1988).

Snails were taken from the field and placed in aquaria in a 2 × 3 × 3 factorial design: source population, feeding regime, and predation risk. The source populations were from two contrasting environments: a protected shore where risk of predation by crabs is high and an exposed shore where few crabs are present because of high densities of fish that prey on crabs (R. Palmer, personal communication). The predation factor had three treatments: no crabs, crabs fed fish, and crabs fed conspecific snails. Test snails were not exposed directly to predators but rather to water that was occupied by crabs and their prey. The three snail feeding treatments were no food, fed 2 of 6 days, and fed 4 of 6 days. The dependent variables were the height of apertural teeth (ridges on the inside opening of the shell) and, in the treatments with food, the rate of the snail's consumption of prey (barnacles).

After 76 days, snails in the crabs + snails treatment had the highest apertural teeth, and as a result of eating the fewest barnacles, they grew the least compared with snails in the other treatments (figure 9.2). Snails in the crabs + fish treatment showed a pattern of response that was similar but not as pronounced as that toward the crabs + snails treatment. Starved snails produced relatively high apertural teeth regardless of the predator treatment. This response may explain some of the increase in height of snails from the crab treatments, because these snails ate less than those in the crabless treatments.

In this case study, snails responded to the perceived risk of crab predation by reducing feeding activity and producing high apertural teeth. Palmer (1990) found similar changes in *N. lapillus* using the same experimental procedures, but also found that snails from crab treatments had heavier shells, a smaller aperture, thinner apertural lips, and greater retractability. Trussell (1996) found that *L. obtusata* exposed to crabs had thicker shells than did snails not exposed to crabs. All of these traits reduce the effectiveness of crab predation (Bertness and Cunningham 1981; Vermeij 1987). The nature of environmental heterogeneity is the presence of predatory crabs. Palmer (1990) notes that marine snails attempt to flee from slowly moving starfish because morphological responses are ineffective against this predator (see also Marko and Palmer 1991). A morphological response to the faster-moving but less powerful crab predator may be adaptive if the abundance of crabs varies among seasons or localities, but no such data were presented. As with the construction of larger crab claws, producing thicker shells and apertural teeth in response to predation risk may be energetically expensive. Snails in crab treatments had decreased growth rates that probably reflect a decrease in activity (hence feeding) rather than a direct trade-off with producing a more predator-resistant

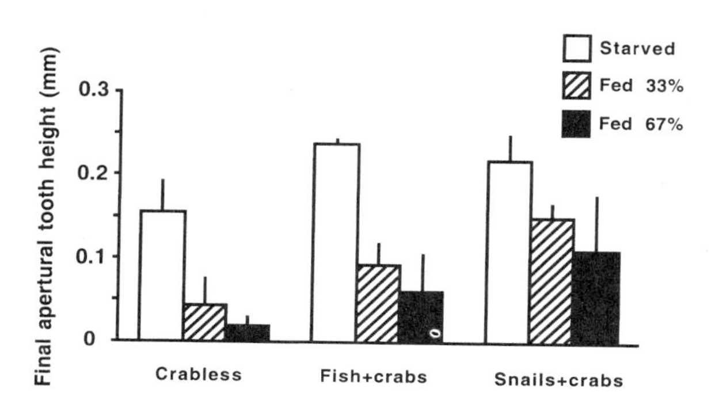

Figure 9.2. Apertural tooth height of a marine snail (*Nucella lamellosa*) as a function of three predator treatments, crossed with three food treatments after 76 days. Data shown are for the initially small and thick-shelled group only. Error bars represent ±1 standard error of the mean. [From Appleton and Palmer (1988).]

shell (see also Kemp and Bertness 1984; Trussell 2000; Yeap et al. 2001). Therefore, in this example, net fitness costs cannot be estimated without further experiments. Waterborne chemical cues from crabs probably would be carried more effectively in the less disturbed waters of sheltered sites compared with the rougher water of exposed sites. No evidence was presented for a genetic basis of the response.

All three studies compared the plastic response of snails from exposed and protected populations. Exposed populations are presumably less susceptible to crab predation because of predation of crabs by fish. Appleton and Palmer (1988) found that snails from the protected population showed greater plasticity in tooth height in absolute terms, which is consistent with an adaptive hypothesis (although in proportional terms the plasticity was similar; R. Palmer, personal communication). However, Palmer (1990) demonstrates the reverse pattern for defensive structures, and Trussell (1996) documents parallel slopes (reaction norms) of shell thickness as a function of length of snails from exposed and protected shores. Therefore, only one of three studies indicated a comparative pattern of local adaptation that is consistent with adaptive plasticity.

Example 3: Plasticity in Maternal Provisioning to Offspring in a Seed Beetle on Different Host Plants

Fox and colleagues (Fox et al. 1994, 1997, 1999; Fox and Mousseau 1996) demonstrate adaptive plasticity in egg provisioning by a seed beetle (*Stator limbatus*). These beetles deposit eggs directly on seeds from a variety of different host plants. When larvae hatch, they burrow through the coat of the seed and feed directly on its flesh. After pupating within the seed, the beetles emerge as adults, disperse, and commence mating within 2 days of settling on the plant on which they will lay their eggs. Fox and colleagues observed that females from the same population laid large numbers of small eggs on one host plant (cat-claw acacia, *Acacia greggii*) and small numbers of large eggs on an alternative host plant (blue paloverde, *Cercidium floridum*). Moreover, on these two host plants, beetles show different levels of egg survivorship and relationships between egg size and survivorship (Fox et al. 1994; Fox and Mousseau 1996). Egg survivorship on acacia seeds is relatively high regardless of egg size. In contrast, egg survivorship on paloverde seeds is relatively low (<50%) and positively related to egg size. This difference is probably due to the higher concentration of allelochemicals in seeds from paloverde and possibly the thicker coat that larvae must burrow through to inhabit seeds.

As in example 2 above, the difference in allocation pattern is consistent with two scenarios: genetic substructuring within the population, and adaptive plasticity in maternal reproductive allocation. To discriminate between these hypotheses, Fox et al. (1997) raised beetles from two populations on acacia plants for three generations in the laboratory. This procedure should have eliminated most potential environmental maternal effects. Next, they assigned recently mated beetles to petri dishes with paloverde or acacia seeds, and measured the number and size (length and width) of eggs laid directly on to the seeds. They found that females from both populations laid small numbers of large eggs on paloverde and large numbers of small eggs on acacia (figure 9.3A,B). In a second series of experiments, reproductive females that laid eggs on one host plant were switched to the other host plant to test for reversible plasticity in maternal allocation strategy. Females switched from acacia seeds to paloverde seeds depos-

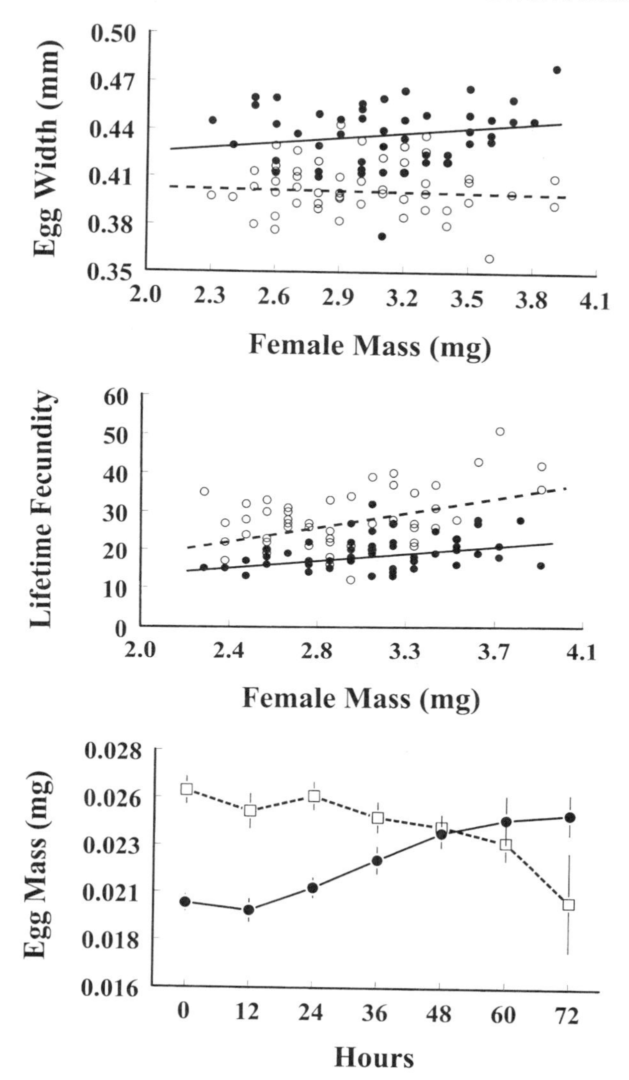

Figure 9.3. Plasticity in maternal reproductive allocation in the seed beetle *Stator limbatus*. (Top) Females lay larger eggs on blue paloverde (*Cercidium floridum*) than on *Acacia greggii*. (Middle) Females have lower fecundity on blue paloverde. (Bottom) When host plants are switched, females adjust egg size (shown) and clutch size (not shown) to the new host. [Courtesy of C. Fox.]

ited relatively larger eggs; however, females switched from paloverde to acacia did not substantially decrease egg size (figure 9.3C; although later studies have found the plasticity to be symmetrical; C. Fox, personal communication). Fox and Savalli (2000) also showed that females on different hosts change egg composition as well as egg size. Overall, these results indicate an adaptive change in the maternal allocation of energy to offspring as a function of host plant.

In this group of studies, seed beetles were observed to lay eggs on multiple host plants. High mortality of small beetle larvae on paloverde as a result of difficulty in penetrating the seed coat generates strong selection for large egg size. In contrast, acacia seed coats are easily penetrated by larvae regardless of their body size. Field studies of the distribution of beetles on the different host plants were not conducted but were qualitatively discussed. However, the presence of larvae on both hosts in at least one population indicates that plasticity in allocation strategies can be adaptive because there is temporal variation for when the hosts set seed. Although larvae on paloverde fared better if they were from larger rather than smaller eggs, the benefit of producing large eggs is offset by a genetically based trade-off with fecundity. The cue that induces females to change energy allocation is not known. However, because females remain on seeds for at least 24 hours before oviposition, it is likely that they discriminate among hosts using chemical cues. Quantitative genetic studies (Fox et al. 1999) indicate heritabilities for egg size ranging from 0.2 and 0.9, and a genetic correlation between host plants of 0.6. These data indicate that plasticity in egg size is capable of evolving in response to selection for greater plasticity. There was no difference in degree of phenotypic plasticity between the two study populations despite a difference in availability of host plants: at one site only paloverde was present, whereas both hosts were present at the other site. However, because gene flow between the two populations is suspected to be relatively high (they are separated by 50 km), this between-population comparison provides only a weak test of local adaptation (Fox et al. 1997, 1999).

Example 4: Plasticity in Male Mating Strategies in Dung Beetles as a Function of Brood Mass

In many species of dung beetles (*Onthophagus*), two male morphs occur in the same population (Emlen 1994, 1997a, 2000; Hunt and Simmons 1998, 2000). Major males have large body sizes and long horns, engage in male–male combat using their horns, mate-guard, and assist females to provision brood masses (Emlen 1997a). Minor males have small body sizes and short horns, engage in "sneaky" mating tactics (they avoid direct combat and dig sideways burrows to mate with females), and do not assist females with the provisioning of brood masses. Emlen (1994) determined that a male's horn size and body size are related to the amount of dung the parent(s) put into the larva's brood mass (see also Hunt and Simmons 1998, 2000; figure 9.4). There is a strong positive relationship between body and horn sizes in major males, but only a weak relationship in minor males. Emlen (1996) found no heritability in horn size, although the position (but not the shape) of the sigmoidal reaction norm for horn size was shifted after seven generations of artificial selection. Hunt and Simmons (2002) also found no heritability for this trait in *O. taurus* based on evidence obtained from a quantitative genetic study using a half-sib breeding design but did find evidence for maternal effects and indirect genetic effects on offspring phenotypes.

Eberhard (1982) suggested that in beetles with male dimorphism in horn size, the suppression of horn growth at small body sizes and the sneaker mating strategy employed by minor males were efforts to "make the best of a bad lot." In dung beetles, males adopt the strategy that takes the greatest advantage of the resources that their parents provide to their larval brood mass (Hunt and Simmons 2001). This argument can be generalized to entire reaction norms for morphological and life history traits (e.g., Stearns and Koella 1986; Bernardo 1993). Thus, an optimized reaction norm will result in each larva producing the phenotype "that makes the best" (i.e., maximizes reproductive success and fitness) of its "lot" (i.e., available resources whether "bad" or "good").

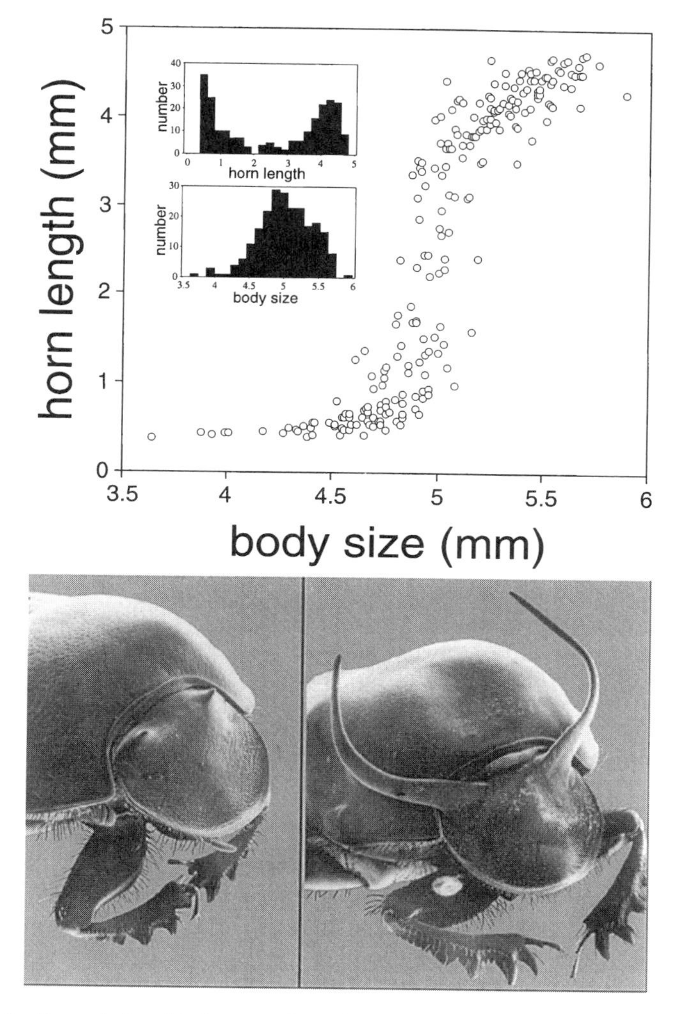

Figure 9.4. Scaling relationship between horn length and body size (prothorax width) in male *Onttophagus taurus*. The relationship produces a bimodal frequency distribution for horn size (A), although body sizes are normally distributed in the population. Lower photographs are of hornless (B) and horned (C) males. [From Stern and Emlen (1999), courtesy of D. Emlen.]

The main data missing from these studies are the fitness values of males in alternative circumstances. Frequency-dependent fitness payoffs for the morphs are likely (e.g., Conner 1989); that is, the reproductive success of one male morph may depend on the rarity of the other male morph (see also Pfennig 1992; Jeschke and Tollrian 2000). The cue in the onthophagine system is the energy available in the brood mass; thus, this is an example of resource-based plasticity. Comparative evidence suggests that the position of the reaction norm varies across species (Emlen 1996). The sigmoidal shape *per se* of the reaction norm, however, can be an adaptive solution to the heterogeneity in resource availability in the developmental environment because it results in a bimodal size distribution of horns, accompanied by a bimodal mating strategy. The passive expression of plasticity (compare with example 5, below) can still be regarded as an adaptation with comparative data if an ancestral species had a nonsigmoidal reaction norm of horn size to body size driven by resource availability. Small males of such an ancestral species would be predicted to not suffer fitness costs by possessing a small horn in mating contests, but might suffer energetic costs if burrowing sneakily was impeded by the possession of a horn (see Moczek and Emlen 2000). The sigmoid shape of the reaction norm is also useful for seeing how a polyphenism can arise from disruptive selection on horn size (a sexually selected continuously varying trait) as a function of body size (Gross 1996; Hunt and Simmons 2001). In an important review, Emlen and Nijhout (2000) present the idea that scaling relationships can be viewed as adaptive reaction norms molded by selection.

This case is somewhat unusual, because morph determination appears to be entirely environmental, with the fitness of alternative morphs possibly being frequency dependent within each dung pat. Therefore, although the position of the threshold can move along the body size axis in response to selection, the shape of reaction norm that results in two morphs will be maintained. This is because disruptive selection for horn size maintains the two strategies and these appear to be evolutionarily stable given the two different ways males obtain matings in this group. Future comparative work may reveal how the position and shape of horn–body size reaction norms among species varies as a function of ecology and behavior (D. Emlen and J. Hunt, personal communication).

Example 5: Plasticity in the Initiation of Maturity of Male Guppies in Response to Food Availability

Age and size at maturation in guppies (*Poecilia reticulata*) vary as a function of food availability and growth rate (Reznick 1983, 1990). Because there is little or no growth after maturation in many animals, and because size can have a large impact on reproductive success, the size at maturation can represent an important component of fitness. Higher food availability and growth rates result in males that are larger and younger when they attain sexual maturity. Being larger appears to enhance mating success, although this relationship is not always evident (reviewed by Reznick 1990).

Reznick (1990) evaluated age and size at maturation in response to food availability in male guppies. He argued that if age and size variation reflect adaptive plasticity, then there should be changes in the timing of maturity as a function of growth rate. An adaptive reaction norm should result in males maturing at ages and sizes that maximize fitness given variation in food availability (e.g., among streams or locations). Such an adaptive reaction norm will produce different age–size combinations than a reaction norm with a fixed size at the initiation of maturity, even though both reaction norms can result in age–size variation. This alternative scenario implies that age–size variation can result from the same pattern of energy allocation being played out in different resource environments. Thus, the adaptation is inferred from flexible development rather than variable phenotypes *per se*. More generally, the goal was to separate patterns of age and size at maturity from the underlying developmental processes, because patterns of age–size variation can result from nonplastic reaction norms in different developmental environments.

Reznick (1990) found that the pattern of development varied with food availability and growth rate (figure 9.5). Males with high growth rates delayed the initiation of maturation and attained larger sizes than they would have had if size at initiation were fixed. Males with low food availability (and, hence, low growth rates) initiated maturation at a small size but at a much earlier age than if they had followed a fixed pattern of development. He argued that this pattern of development (an "L-shaped" reaction norm, Stearns and Koella 1986) was consistent with a hypothesis of adaptive plasticity. Compared with the outcome under a hypothetical fixed pattern of development, fast-growing males matured slightly later but at a much larger size, whereas slow-growing males matured much earlier, but at only a slightly smaller size.

Although this approach makes explicit the nature of the trade-offs associated with plasticity, they have not been evaluated in separate experiments. Genetic variation in plasticity was not evaluated. A comparative study, however, revealed interpopulation variation in plasticity that is consistent with environmental cues (Rodd and Reznick 1997; Rodd et al. 1997), suggesting that plasticity in the development of age and size at maturation can evolve. In these studies, the development of males from environments with low and high mortality was compared. Males from low mortality environments will delay maturation when other mature males are present, whereas males from high mortality environments will not. The trade-offs associated with this interpopulation variation in plasticity would include the costs of delayed maturity on the probability of survival to

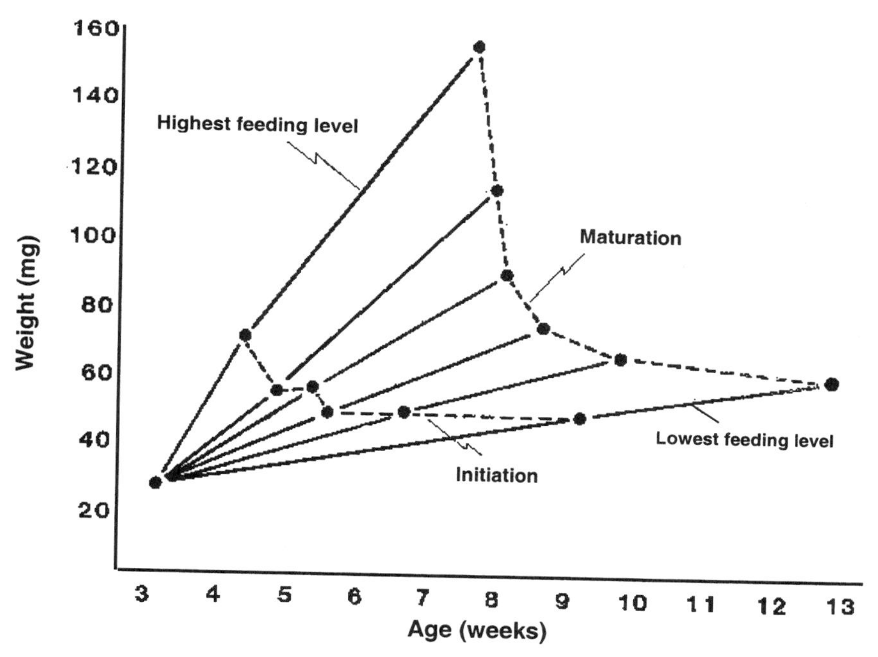

Figure 9.5. Mean age and size of male guppies (*Poecilia reticulata*) at the initiation and completion of maturity in response to variation in food levels during development. [From Reznick (1990).]

maturity versus the gain of larger body size in either environment. These costs and benefits have not been evaluated.

Example 6: Plasticity in the Initiation of Metamorphosis of Spadefoot Toad Tadpoles in Response to Pond Drying

Amphibian larvae often inhabit aquatic environments that can vary in duration, density of conspecifics and heterospecifics, and type and abundance of predators (Newman 1992; Alford 1999). Usually, larvae are deposited in these environments by their parents and have little or no capacity to move to an alternative environment. These circumstances can provide strong selection for plasticity in larval duration, because metamorphosis is often the only means of escape from a body of water. Adaptations of amphibian larvae to their aquatic environments have been studied intensively since the seminal work of Wilbur and Collins (1973). That article reviewed available evidence on patterns of metamorphosis in frog and salamander larvae and presented an adaptive model of development in which individuals were capable of adjusting size and age at metamorphosis.

Spadefoot toad tadpoles (*Scaphiopus* and *Spea*) have been the subjects of a number of studies that evaluated adaptations to ephemeral ponds (Newman 1987, 1989, 1994a; Denver et al. 1998; Morey and Reznick 2000). Here we consider together the results from a number of independent studies on plasticity in spadefoot toad tadpoles, with little description of the experiments. Adult *Scaphiopus couchii* breed in ephemeral ponds formed by unpredictable rains in North American deserts. The duration of water in these ponds can be extremely short (<1 week), and, although tadpoles develop rapidly, high mortality can occur when ponds dry before metamorphosis is complete (Newman 1987, 1989; Morey and Reznick 2000). Newman (1989a,b, 1994a) and Morey and Reznick (2000) quantified duration of water in natural ponds occupied by *S. couchii* in Texas and California, respectively. They also measured several aspects of the life history of *S. couchii* in these same ponds. Pond duration was positively correlated with metamorphic success and the age and size of toads at metamorphosis. Newman (1994a) found that densities of toads increased exponentially as ponds dried. Both investigators found several abiotic changes in ponds as water level decreased, including increased temperature fluctuations and ammonia levels.

Newman (1994a), Denver et al. (1998), and Morey and Reznick (2000) performed controlled laboratory experiments with tadpoles to test for accelerated development in response to an increase in conspecific density and decrease in food, a decrease in swimming volume, and a cut-off of food, respectively. In all studies these environmental factors shortened larval period, provided larvae had surpassed a critical stage in development. This evidence supports the hypothesis that control of metamorphic timing is a result of adaptive phenotypic plasticity. The plastic response in all three studies was a decrease in the larval period in response to cues associated with pond drying. Newman (1994a) found that increases in density mirrored the effects of reduced food availability because density increased as ponds dried, resulting in less food per individual. Denver et al. (1998) found that reduced water volume (or proximity to the surface) alone induced an acceleration of metamorphosis in *S. hammondii* (figure 9.6). That study also showed that the response is reversible because tadpoles slowed development when water volumes were increased. Morey and Reznick (2000) stopped feeding tadpoles of three species of spadefoot toads (*Scaphiopus couchii*, *Spea hammondii*, and *Spea intermontanus*) at specific developmental stages. They found that the risk of pond drying experienced by a species in nature (based on direct measurements of the ponds) was correlated with that species' ability to accelerate metamorphosis and thereby decrease development time providing comparative evidence for adaptive plasticity. Newman (1988a,b, 1994b) presented evidence for genetic variation in plasticity, indicating that the ability to accelerate development can be modified by selection, and Morey and Reznick (2000) presented evidence for interspecific genetic variation and adaptation with their comparative study.

The fitness of tadpoles that do not hasten development in response to loss of water in ponds is clear—they die when ponds dry. Tadpoles that accelerate development and metamorphose at an early age pay a cost in terms of a reduced body size. Evidence from a hylid frog indicated that metamorphic body size is an important determinant of recruitment and reproductive success (Smith 1987). Selection on metamorphic body size may be even stronger in desert species because increased body size likely results in lower rates of evaporative water loss (Newman and Dunham 1994).

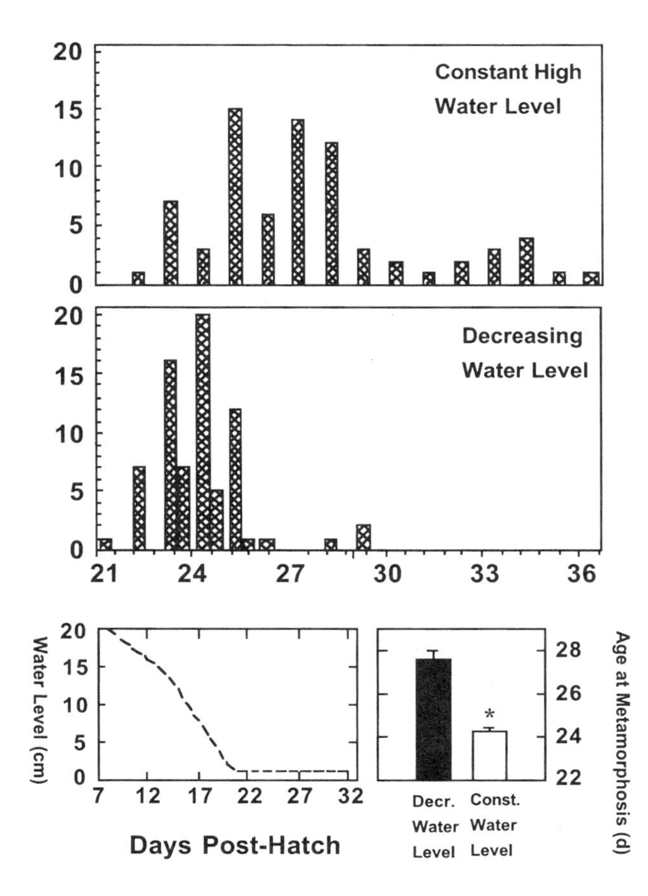

Figure 9.6. Acceleration of metamorphosis of *Spea* (*Scaphiopus*) *hammondii* tadpoles in response to a decrease in water depth. (Top and middle) The number of tadpoles metamorphosing in the constant water depth treatment (top) and the decreasing water treatment (middle). (Bottom on left) Schedule of water depth changes in the decreasing treatment. (Bottom right) Mean age at metamorphosis between the two treatments (*$P < 0.001$). [From Denver et al. (1998), courtesy of R. J. Denver.]

Do Our Examples Demonstrate Adaptive Plasticity?

The study of phenotypic plasticity in animals in the past decade has revealed a diversity of interesting phenomena and provided insight into the evolution and maintenance of adaptive plasticity in animal populations. Yet, most studies do not go beyond a rudimentary description of the plasticity with some comments on the selective factors responsible for its evolution. Meanwhile, theoretical models of plasticity have continued to progress (chapter 6). Future testing of these models requires that more detailed information on the genetics, ecology, and limits on plasticity be collected in well-designed field and laboratory studies. Here we discuss the six case studies described above, then conclude with a brief prospectus of empirical studies of plasticity in animals. How did our six examples fare in evaluating the adaptive significance of plasticity according to our criteria? Below we briefly review the studies with reference to our criteria set out in section 3 (see table 9.2).

What Are the Plastic Responses to the Developmental Environment?

All of the reviewed studies effectively describe environmental influences on the phenotype, and hence all of these cases show putative adaptive phenotypic plasticity. However, most did not consider the developmental pathway to plasticity. The study of maturation in male guppies (Reznick 1990) emphasizes that a pattern of phenotypic responses can be generated from more than one underlying process of development. This is perhaps analogous to Cohen (1978) showing that MacArthur's (1957, 1960) broken stick distribution could be generated by a number of processes. The significance of multiple explanations is that each is open to a different interpretation. There is thus some danger in inferring adaptation from just the final phenotype. For example, variation in body size may result from some individuals eating more food than others and is not necessarily adaptive plasticity. Changing the patterns of allocation of resources to somatic growth, storage, or reproduction as a function of resource availability might be adaptive developmental plasticity (see Emlen and Nijhout 2000; chapter 5). In this regard, the male guppy, seed beetle, and spadefoot toad tadpole examples show that the developmental program is plastic. Few of the remaining studies address a nonadaptive alternative or evaluate the underlying process of development that produces the adult morphology. Ascribing passive responses to developmental environments to adaptation is clearly not desirable, so careful attention must be paid to separating active plasticity caused by a modification of development from passive physicochemical responses. Studies that involve discrete alternatives, such as shell morphology in barnacles (Lively 1986b) or trophic morphology in spadefoot toad tadpoles (Pfennig 1990), carry less of a burden of proof because the alternative morphs were necessarily caused by a change in development.

What Is the Nature and Extent of Environmental Heterogeneity?

Of our six examples, three document natural variation in the environment that generates the observed plasticity (seed beetles, dung beetles, and spadefoot toad tadpoles). It

Table 9.2. Summary of examples of adaptive phenotypic plasticity in animals used in section 4, and how they meet the criteria in section 3 and table 9.1.

Example	Criterion					
	Phenotypic plasticity	Environmental heterogeneity	Fitness reversal	Informative cue	Genetic variation	Comparative evidence
Crab claws	X	—	(X)	—	—	—
Shell morphology	X	X—geographic comparison	(X)	X—water-borne cues	—	1 of 3 studies
Egg size in seed beetles	X	X—geographic	X	—	X	0
Male morphs in dung beetles	X	X—parental provisioning	(X)—frequency-dependent	NA	0	X
Age and size at maturity in guppies	X	(X)—growth rate	—	NA	—	(X)—between populations
Age and size at metamorphosis in tadpoles	X	X—pond	(X)	X—depth, food, and crowding	X	X

Key: X = support; (X) = indirect support; 0 = no support;—= not tested; NA = not applicable is often reasonable to assume that the environmental variation exists, such as variation in a mollusk's exposure to crabs or crayfish, a guppy's access to food, or the type of food available to a crab. Such an inference, however, falls short of demonstrating that the observed plasticity is an adaptation. An important component of meeting this criterion is documenting the temporal and spatial pattern of variation, as scaled against the mobility, generation time, and capacity of an organism to select its environment. Although difficult to collect, we expect such data to be a more prominent feature of studies of plasticity in the future given their importance in recent theoretical treatments of plasticity.

Is There a Reversal in Maximum Fitness of the Alternative Phenotypes in Different Environments?

None of the cited examples presents a complete analysis of the fitness of alternative phenotypes in alternative environments. The best evaluation of fitness was in the seed beetle example, because the larger egg size is required to use one of the alternative hosts. The argument in favor of small eggs on the alternative host is that the beetles can then produce more eggs. A complete argument would be a comparison of the reproductive success of both phenotypes on both hosts. Scenarios could be developed for all of the remaining examples but still require testing. For example, the acceleration of development in spadefoot toad tadpoles is clearly advantageous when a pool is drying, because the alternative to successful metamorphosis is death. The advantage of prolonged development in longer lived pools is presumably associated with metamorphosing at a larger body size. Testing the scenario requires comparing the results of studies on other species of amphibians for which the effects of body size have been evaluated.

For Information-Based Plasticity, Does the Organism Exploit a Reliable Cue That Predicts the Future Selective Environment?

Two of our six examples potentially rely on informative cues (table 9.2, column 4). In these cases, animals responded to cues detected in the developmental environment. It is prudent to consider the time course of the plastic response and whether it will enhance individual fitness. For example, the rate of response seems to be well matched to the time frame of temporal variation for spadefoot toad tadpoles and changes associated with pond drying. Although the remaining examples all reveal a plastic response, it is not possible from the context of the studies to determine how the change in phenotype will influence fitness. Dung beetles and guppies appear to respond to resource availability. The morphological change in marine snails can be induced by water that contains crabs, but low food availability also induces a similar morphological response. Thus, it is possible that the response to crabs is mediated through a change in feeding behavior and hence is partly resource mediated. Spadefoot toad tadpoles respond either to information (water depth) or resources (growth rate).

Is There Genetic Variation for Plasticity?

Only the seed beetle, dung beetle, and one spadefoot toad tadpole example considered heritable variation for plasticity. There was significant within-population variation for seed beetles and tadpoles, but not for dung beetles. In the last case, genetic variation was instead demonstrated as a shift in the position of the reaction norm in response to selection. Although it is not necessary to satisfy this criterion to demonstrate local adaptation, the presence and magnitude of such genetic variation remain of interest for evaluating the potential for evolutionary change and genetic correlations with other traits.

Is There Comparative Evidence That Plasticity Has Coevolved with Environmental Heterogeneity?

Differences in amount or nature of plasticity among local populations or among related species can provide a powerful argument for adaptation if plasticity is correlated with differences in the local environment. Whether or not such differences exist will be a function of the mobility and longevity of the organism and differences in local selection pressures. Local adaptation of plastic responses was evaluated and not found in studies of marine snails and seed beetles. Differences in plasticity were found among populations of guppies and were a function of predation and mortality rate, but the adaptive significance of these differences was not evaluated. Interspecific differences among spadefoot toad tadpoles were correlated with the risk of drying in the larval environment. Only this last example can be argued to be consistent with a hypothesis of adaptive plasticity.

Conclusions

These examples reveal several trends in what is most often missing in investigations of adaptive phenotypic plasticity.

- Arguing for plasticity as an adaptation demands a demonstration of a cause and effect relationship between the agent of selection and the response (e.g., Reeve and Sherman 1993; Reznick and Travis 1996). An effective approach to establishing this link, taken too rarely thus far, is comparative studies among populations that differ in factors that are likely to select for plasticity (e.g., Doughty 1995; Gotthard and Nylin 1995).
- A more consistent effort is needed to couple evaluations of plasticity with theory. Relevant theory emphasizes the spatial and temporal patterns of variation. Evaluations of the natural history of agents that are presumed to select for adaptive plasticity tend to be very rare in animals [e.g., amphibians: Newman (1989), Morey and Reznick (2000); shelled marine invertebrates: Lively (1999a), Trussell (2000)].
- Options for local adaptation versus polymorphism versus plasticity need to be considered. This is a function of the mobility of the animal, the distribution of habitats, the strength of selection, and genetic architecture of the traits involved in the response. For example, the absence of evidence for local adaptation in seed beetles might result from high levels of gene flow. Studies of the interplay among gene flow, selection, and local adaptation are rare in any context, let alone in the context of adaptive plasticity.
- If populations differ in agents of selection and in plasticity, is there a genetic basis to differences in plasticity? Genetic evidence can be revealed by a variety of methods including half-sib breeding designs (e.g., dung beetles) or reciprocal transplants (e.g., Trussell 2000).

- Multivariate response to cues should be examined (e.g., Spitze and Sadler 1996; DeWitt 1998; Tollrian and Harvell 1999). Coupled with studies of selection, investigators should be able to identify what specifically is being selected for in the different environments and what changes occur owing to genetic correlations among traits.

The study of adaptive phenotypic plasticity in animals is in its infancy. Our review is intended to highlight the kinds of evidence necessary to demonstrate adaptive plasticity convincingly and to point out where many studies fall short of an ideal case study. Clearly, much more will be learned about adaptive plasticity in animals in the coming years, and rigorous empirical studies based on sound methodologies will be crucial toward achieving this goal.

Acknowledgments We thank G. Burghardt, T. DeWitt, S. Downes, D. Emlen, C. Fox, T. Frankino, J. Hunt, A. McCollum, and R. Palmer for interesting discussions and helpful comments on the manuscript.

10

The Functional Ecology of Phenotypic Plasticity in Plants

SUSAN A. DUDLEY

Functional ecology is the study of the interactions of individuals, populations, and species with their environments. This broad definition embraces fields ranging from physiological ecology to evolutionary ecology. These fields are unified in their study of the process of adaptation to different environments and how different properties of the organism confer adaptation to the environment. The observation that an organism's traits depend on its environment easily leads to the hypothesis that phenotypic plasticity functions as a way of adapting to a variable environment. The different fields comprising functional ecology have all addressed this hypothesis but have followed very different lines of evidence. In evolutionary biology, phenotypic plasticity is assessed in terms of genetic variation and fitness consequences; in physiological ecology, in terms of the functional consequences to stress tolerance and carbon acquisition; and in developmental biology, in terms of the mechanisms by which the environment can affect development of a trait. In this chapter, I describe some of the areas of concordance from these different fields of plant functional ecology, and how they can inform each other to suggest new areas of research on phenotypic plasticity in plants.

Bradshaw (1965) and many others have suggested that the sessile nature of plants particularly favors phenotypic plasticity as a mechanism for adaptation to a variable environment. There is considerable evidence that plants defend against predators, cope with stress, and compete and forage for resources through cued patterns of growth and environmentally determined development. Acclimation responses demonstrate that a plant's ability to tolerate a particular environmental stress is increased by prior exposure to that stress. Transplant and common garden experiments have demonstrated that the environment rather than the genotype causes much of the character variation within a species.

Phenotypic plasticity is adaptive when the appropriate phenotype in each environment results in higher fitness in that environment compared with the alternative phenotype (Thompson 1991). For several examples of phenotypic plasticity, functional arguments can be made that the optimal phenotype differs between environments and that phenotypic plasticity permits the plants to produce the optimal phenotype in each environment. In a few cases, the fitness consequences of having the appropriate versus the alternate phenotypes have been measured (see below).

A confounding factor in understanding phenotypic plasticity is that some environments favor plant growth and reproduction more than others. Environmental differences in ecosystem primary productivity are explained by the availability of nutrients for leaf production, light energy for photosynthesis, and water needed to replace that lost during stomatal opening. At the level of the individual species, average fitness also frequently differs between environments, although not always in the direction of differences in primary productivity. Such environmental differences can be characterized by the species-specific parameter of *environmental quality*, which is measured for a species by the highest fitness achieved by any individual of that species within that environment (Zhivotovsky et al. 1996).

The sessile nature of plants and differences in environmental quality cause nonadaptive phenotypic plasticity. For example, low nitrogen availability in the environment reduces leaf growth and leaf nitrogen concentrations (Taiz and Zeiger 1998). Lower leaf nitrogen concentrations are correlated with lower net photosynthetic rates (Sage and Pearcy 1987). Smaller leaves and lower photosynthetic rates reduce growth rates. The result is smaller plants that make fewer seeds (Sultan and Bazzaz 1993c). Any potentially adaptive responses to low nitrogen, such as the higher root-to-shoot ratio that increases foraging for nitrogen (Sultan and Bazzaz 1993c), co-occur with responses that have neutral or negative fitness implications.

There are many other reasons why the plasticity we observe may not be adaptive. Some traits may be selectively neutral. Plasticity in other, correlated traits may reflect greater stress and damage in the lower quality environment. As a consequence of adaptive plasticity in one trait, constraints and trade-offs may result in a maladaptive response in another trait. A population may have evolved specialized genotypes for different environments. The population may never have experienced selection for adaptive plasticity in the environments under consideration. In all environments, natural selection will favor increased fitness. Directional selection will also favor increases in components of fitness and performance traits such as fruit number, survivorship, and plant size, although trade-offs among these traits may cause considerable indirect selection on these traits.

Evolutionary theory (chapter 6) considers adaptive phenotypic plasticity as one of the possible outcomes of evolution by natural selection in a heterogeneous environment. The other possible outcomes include local adaptation, adaptive homeostasis, or an intermediate compromise phenotype (Van Tienderen 1997). However, genotypes with the ability to produce the most optimal phenotype in each environment and so have high fitness in multiple environments should be favored over locally adapted genotypes that have high fitness in only one environment (Via and Lande 1985; Van Tienderen 1991), when the environments are both common. Nonetheless, local adaptation is extremely common in plant populations. The early research on plant evolution in heterogeneous environments showed that many species comprise several physiologically distinct ecotypes specialized to differing regions and habitats (Turesson 1922; Hiesey and Milner 1965). In recent years, research has demonstrated local adaptation in response to spatial variation at very small scales of environment heterogeneity (Linhart and Grant 1966; Schmitt and Gamble 1990; Stratton 1994). Hence, one approach to understanding the role of phenotypic plasticity in the functional ecology of plants is to explore instances when local adaptation evolves rather than adaptive phenotypic plasticity.

Local adaptation is demonstrated when genotypes taken from differing environments perform significantly better in their "home" environment than in the "away" environment (figure 10.1A,B) or, more generally, when significant genotype by environment interaction for fitness exists (Stratton 1994; Fry 1996). The aphids growing on clover and alfalfa (Figure 10.1A) show local adaptation with environments of equal quality, so that the populations have similar fitness when in their home environments. The jewelweed grown in the floodplain and hillside environments in figure 10.1B also show local adaptation, but the environments differ in their quality. Even though in each environment, the home population performs better than the away population, both populations perform best in the floodplain environment. Finding local adaptation is expected when gene flow is low relative to the grain of environmental heterogeneity. But finding local adaptation when the gene flow is high and environmental grain size is small (chapter 6) provides two pieces of information. First, that local adaptation has evolved suggests that

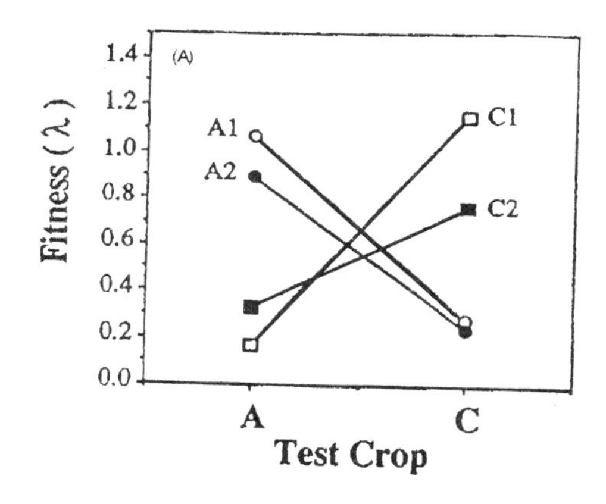

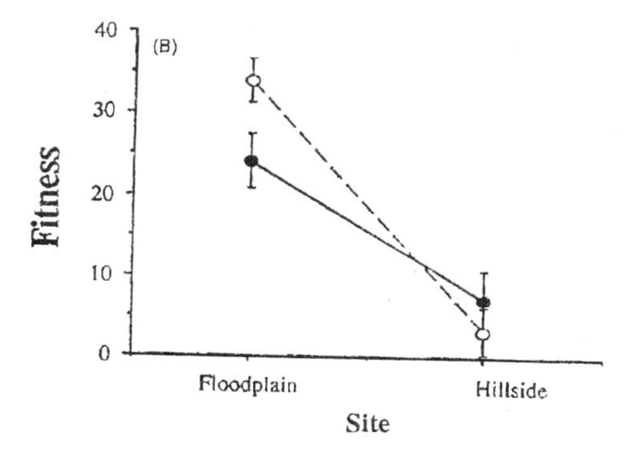

Figure 10.1. Local adaptation in aphids (A) and jewelweed (B). (A) The fitness of two collections from alfalfa fields (Al, A2) and two collections from clover fields (C1, C2) raised on alfalfa (A) and clover (C) (from Via 199l). (B) The fitness of a population from a floodplain (open circles) and a population from a hillside (solid circles) raised in the floodplain and hillside environments. [From Bennington and McGraw (1995).]

the environmental heterogeneity would select for adaptive phenotypic plasticity. Second, that local adaptation has evolved instead of adaptive phenotypic plasticity suggests that genetic constraints and functional costs may be slowing or preventing the evolution of adaptive phenotypic plasticity.

To analyze the functional ecology of phenotypic plasticity in plants, I first present a detailed example of phenotypic plasticity—plant responses to density. Then I discuss patterns that emerge from the breadth of studies of phenotypic plasticity. Which environments evoke plasticity compared with those that select for local adaptation? What patterns are seen in how traits respond to the environment? How can the question of adaptation be approached? What do functional and selection studies tell us about the nature of adaptation through phenotypic plasticity to heterogeneous environments? How does the scale of heterogeneity (fine vs. coarse grained) affect the nature of plant responses? I suggest that traits exhibiting apparent adaptive phenotypic plasticity tend to affect three distinct aspects of plant performance: (1) plasticity in traits can function to reduce damage or stress imposed by an environment, (2) plasticity in traits can function to increase carbon acquisition in an environment, and (3) plasticity in some traits alters reproductive strategy, presumably to maximize fitness in an environment.

An Example of Phenotypic Plasticity—Responses to Density

Density-dependent stem elongation is perhaps the best-studied example of phenotypic plasticity in plants. The presence of other plants provides a complex environment for a plant. Neighboring plants reduce the availability of water, light, and nutrients. Plants alter the microclimate, increasing local atmospheric humidity, lowering wind speed, and reducing soil and leaf temperature. The roots are exposed to the leachates of other plants that contain nutrients and other compounds, including allelopathic substances. Plants are also known to respond to volatile hormones, including ethylene and jasmonates, produced by other plants (Taiz and Zeiger 1998). The presence of other plants is a dominating factor in an individual's environment: although other plants will sometimes benefit an individual (Bertness and Callaway 1994), competition for resources most often decreases growth (Casper and Jackson 1997).

Plants perceive the presence of other plants by the use of spectral light quality, especially the ratio of red to far-red (R:FR; Smith 1982). Because chlorophyll absorbs more red than far-red light, the R:FR ratio of light is lowered by the presence of other plants (Smith 1982). Plants can sense and respond to R:FR through light-stable phytochrome, primarily phytochrome B of *Arabidopsis* (Schmitt and Wulff 1993; Smith 1995; Smith and Whitelam 1990; Von Arnim and Deng 1996), although phytochromes C, D, and E of *Arabidopsis* also have some red-sensing function (Whitelam and Devlin 1997). Plants respond to the light reflected off neighbors before they are large enough to shade one another (Ballaré et al. 1987; Novoplansky et al. 1990; Smith et al. 1990). This ability to "anticipate" future competition potentially allows plants to express more competitive phenotypes only when necessary.

Plants in open, high-light sites have more elongated stems when also at higher densities (Jurik 1991; Weiner 1985). The increased elongation of internodes in high density is phytochrome mediated. In a stand of plants, small differences in height can determine whether a plant has its uppermost leaves in full sun or shaded by other plants. Increased stem elongation is hypothesized to allow the plant to forage for light in the presence of competitors. Although the elongated stems are thin and fragile, the plants in high density support one another (Smith 1995). Conversely, elongating in the absence of neighbors would not affect light reception, and would be costly both in terms of the biomass allocated to stems instead of leaves and roots, and the mechanical fragility of elongated stems (Casal et al. 1994). These functional hypotheses suggest that stem elongation will have fitness consequences through its effects on plant carbon acquisition.

For increased stem elongation at high density, we can test whether phenotypic plasticity is adaptive. In this system, both the light quality cues that plants use to sense density and the mechanisms of signal of perception are known. This knowledge has been exploited in two studies to produce the "right" and "wrong" phenotypes in high- and low-density environments. To test the adaptive value of stem elongation plasticity, Schmitt et al. (1995) used mutant and transgenic genotypes to obtain short and elongated phenotypes in both density environments.

To test the advantage of nonelongation at low density, Schmitt et al. (1995) compared the biomass and flower number of wild-type *Brassica rapa* and a phytochrome B–deficient mutant of *Brassica rapa* at high and low density. The phenotype of the *Brassica rapa ein* mutant is constitutively elongated (i.e., at both high and low density). They found that the biomass of *ein* plants and wild-type plants did not differ at high density. However, at low density, the elongated *ein* plants had half the biomass of nonelongated wild-type plants. Flower number followed a similar pattern: at high density, the *ein* plants had roughly half as many flowers as wild-type plants, but at low density, the *ein* plants had only one quarter of the flowers of wild-type plants. These results support the hypothesis that the elongated phenotype is disadvantageous in low-density stands.

To test the hypothesis that elongation is advantageous at high density, the role of phytochrome A in inhibiting elongation in normal sun was exploited. Normally, this is a transitory role, because phytochrome A levels drop rapidly in high light (Reed et al. 1994). However, inserting a transgenic phytochrome A gene under a constitutive promoter results in continued high levels of phytochrome A. The transgenic plants had reduced elongation at high densities. Performance of two lines of transgenic phytochrome A tobacco (*Nicotiana tabacum* cv. *xanthi*) was compared with wild type in pure stands and in stands with wild-type lines mixed with the transgenic plants. At the time that the canopy closed, no effects of genotype or stand type on plant biomass were seen. But in later measurements, after the plants interacted in competition, both transgenic tobacco lines had reduced relative performance in mixed stands compared with their performance in pure stands. This demonstrates that the wild-type plants that elongated in competition were able to suppress the biomass growth of the transgenic lines. These results support the hypothesis that having the elongated phenotype is advantageous in high-density stands (Schmitt et al. 1995).

In another study, Johanna Schmitt and I used a light quality cue to manipulate the phenotype (Dudley and Schmitt 1996). We manipulated R:FR independently of density to produce short (suppressed) and elongated seedlings of *Impatiens capensis* in the greenhouse, grown at high density, with all other environmental factors held constant. These plants were then planted in high and low density in the field. Fitness and phenotypic characters were followed over time. As predicted, the elongated plants had higher fitness in high density, and the suppressed plants had higher fitness in low density (figure 10.2). This result demonstrates that the suite of responses to R:FR is adaptive but does not prove that the stem elongation response in particular is adaptive.

A feature that plasticity to low R:FR shares with plasticity to other environments is that many traits are affected, not only the trait hypothesized to show adaptive plasticity. Other responses to high density and low R:FR include reduced branching, increased petiole length, reduced leaf area, increased chlorophyll concentration, and changes in nitrate reductase activity (Smith 1982). Also, the low-density environment is a higher quality environment; in our study, both elongated and suppressed plants had higher lifetime fruit production in the low-density environment. Leaf length, which is strongly associated with plant biomass in this species, was also greater in the low-density environment (Dudley and Schmitt 1996). We used measures of phenotypic selection (Lande and Arnold 1983) on height and leaf size to try to determine the fitness consequences of the individual traits. Within both environments, fitness increased with leaf size, but selection favored greater height in high density only. Although the elongated plants were less fit in low density, within light-quality treatments there was no selection against height. Either the high correlation between height and treatment masked selection on height in low density, or the lowered fitness of elongated plants is the consequence of changes in other, unmeasured traits (Dudley and Schmitt 1996).

Elongated plants have lowered root allocation relative to leaf area in both *Impatiens capensis* (Maliakal et al. 1999) and in *Phaseolus vulgaris* (Cipollini and Schultz 1999). Both leaf area and the ability to tolerate the removal of leaf area are reduced in elongated plants (Cipollini 1999). In a study of nonelongated and elongated plants, we found differences in gas-exchange traits (Maliakal et al. 1999), but the association of gas-exchange trait and elongation depended on whether density or light quality alone was manipulated, suggesting that multiple aspects of the high-density environment, not just light quality, affect gas-exchange traits. Reduced root allocation in high density may or may not be adaptive: it is hypothesized that plants allocate more carbon to stems at the cost of reducing below-ground resource acquisition (Cipollini and Schultz 1999). It is also plausible, however, that in high-density stands, mechanical support provided by neighbors reduces the need for anchoring by roots.

Stem elongation responses are cued by both reduced irradiance and low R:FR (Ballaré et al. 1991). Reduced irradiance and low R:FR are also the cues for plant responses to overhead canopy shade (Fitter and Ashmore 1974; Kwesiga and Grace 1986; Mitchell and Woodward 1988; Solangaarachchi and Harper 1987). Like plants growing in high density, plants growing under forest canopies have the problem of reduced light availability. However, the phenotype hypothesized to maximize light capture in canopy shade is very different from the stem elongation response. The optimal shade strategy is to increase the ability to capture low light through such changes as larger, thinner leaves, and increased chlorophyll concentrations, especially chlorophyll b (Taiz and Zeiger 1998), and to exploit transient periods of high light from sunflecks (Chazdon and Pearcy 1991). In canopy shade, stem elongation is arguably maladaptive, because increased height cannot increase light reception and the low mechanical stability and high carbon allocation to stems are costly (Givnish 1982; Morgan and Smith 1979).

Because the optimal traits in shade differ from the optimal traits in high densities, but both environments provide similar cues, it has been hypothesized that plants cannot adapt through phenotypic plasticity to environments heterogeneous in density and over-

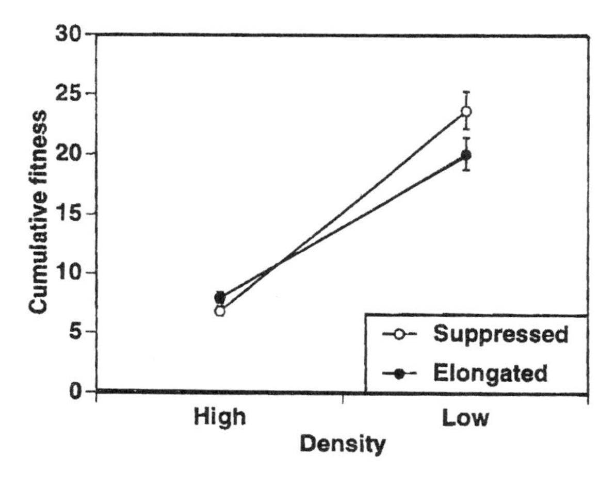

Figure 10.2. Fitness of elongated and suppressed (shorter) plants at high and low density. The interaction between treatment and density ($p < 0.05$) indicates that elongated plants performed better than did suppressed plants at high density but more poorly at low density.

head shade. Instead, populations are expected to show either local adaptation to variation in shade or local adaptation to variation in density (Morgan and Smith 1979). Several studies have supported this hypothesis. At the species level, it has been shown that woodland species are less responsive to R:FR than are open species in stem elongation rate, petiole length, and leaf:stem ratio (Fitter and Ashmore 1974; Morgan and Smith 1979). Moreover, the elongation response to R:FR is greater for genotypes (Bain and Attridge 1988; Skávlová and Krahulec 1992) and populations (Donohue and Schmitt 1999; Dudley and Schmitt 1995) from open sites compared with those from woodland sites. Arguably, the direction of the source of low R:FR is overhead if trees are present and lateral if herbs are present. However, the studies finding greater responses to R:FR in plants from open sites used artificial canopy shade suspended over the plants so that they measured responses in an environment similar to tree canopy shade. Therefore, these results appear to demonstrate that, although plants sense the presence of other plants, they are unable to differentiate between trees and herbs.

An alternate explanation for the apparent inability to differentiate between trees and herbs is that the spatial scale of woodland and open patches could be relatively large to the distance of gene flow, favoring the evolution of local adaptation over adaptive phenotypic plasticity. But competition among similar-sized plants of differing species creates a natural selection pressure to recognize other plants at a very small scale. Local adaptation to different species of competitors (Martin and Harding 1981; Miller 1995; Turkington and Harper 1979) and to different genotypes of a single species of competitors (Aarssen and Turkington 1985) has been demonstrated. There is less evidence for adaptive phenotypic plasticity to competitor species. One study did find that that the plastic response of clover differed between the species of grass competitors, but the differences could explained simply by the effects of the grass species on irradiance and R:FR (Thompson and Harper 1988). Although plants sense the presence of other plants, limitations of plant perceptive abilities appear to constrain the evolution of adaptive phenotypic plasticity to specific plant species or growth forms. Instead, variation among populations in phenotypic plasticity appears to have evolved as part of local adaptation to the presence or absence of tree canopy shade.

Environments That Evoke Phenotypic Plasticity and Affect Natural Selection

Evolutionary theory suggests that adaptive phenotypic plasticity can evolve in response to heterogeneity in any aspect of the environment that affects selection on a trait. But not all plant environments or important environmental factors appear to evoke plasticity. Phenotypic plasticity occurs in response to temperature (Zangerl and Bazzaz 1983), light (Boardman 1977; Sultan and Bazzaz 1993a; Winn and Evans 1991), water (Sultan and Bazzaz 1993b), nutrients (Pigliucci et al. 1995a; Sultan and Bazzaz 1993c), and even CO_2 (Reekie and Hiclenton 1994). The presence of other plants (Smith 1995), the presence of other seeds (Bergelson and Perry 1989), type of soil (DeLucia et al. 1989), damage by herbivores (Karban and Baldwin 1997), and wind and other forms of mechanical perturbation (Cipollini 1999) can effect plant phenotype. Development of flowers, inflorescences, and fruits is frequently regulated by pollination (O'Neill 1997). But despite diversity in the types of environments that induce plasticity, many of the environmental factors known to result in the evolution of genetic differentiation and local adaptation (Linhart and Grant 1966) do not evoke phenotypic plasticity.

Local adaptation has been shown in response to toxic soils, nutrients, herbicides, clipping and grazing, maritime exposure, moisture, temperature, elevation, competitors, pollinators, herbivory, and parasites (Linhart and Grant 1966). Several of these factors, and others hypothesized to be important selection agents on plants, do not induce a plastic response in morphological and physiological traits. Plants do not adapt through phenotypic plasticity to sudden fires, disastrous floods, lethal levels of grazing, and other causes of rapid mortality, even though the frequency of such disasters is thought to be an important selective force (Grime 1977). The number and type of pollinators are important agents of selection on flower characters, but plants are not observed to change such flower traits as color, shape, or stigma and anther positions in response to whether hummingbirds or bees are more abundant or whether pollinators are scarce. Plants sense the presence of other plants (see the examples discussed above) but do not seem able to differentiate between whether these plants are siblings, conspecifics, or other species, another important determinant of selection. The lack of response to these environmental factors is thought to result from inabilities to perceive pollinators and the species of nearby plants. This functional constraint means that no responses to these environments, much less genetic variation in response to these environments, can exist, creating a genetic constraint (Arnold 1992) on the evolution of adaptive plasticity.

It is often difficult to determine which factors are important to plant growth, development, survivorship, and fecundity and which factors act as agents of natural selection, because natural environments vary in several environmental factors simultaneously (Wade and Kalisz 1990). In a study of plasticity in *Mimulus* (Galloway 1995), fitness of experimental populations did not differ between two habitats—sand bank and organic soil—that also differed in proximity to a stream and presence of competitors. However, fitness varied strongly within these habitats, and Galloway found orders of magnitude differences in fitness for plants in apparently identical sites as close as 1.5 m. These differences are presumably due to cryptic variation in the environment, although genotypic variation is an alternative explanation. It has been shown that factors as subtle and cryptic as the diversity of mycorrhizal fungi can affect diversity and productivity of vascular plant communities (van der Heijden et al. 1998).

Because of the desire to determine effects of individual factors, controlled environment experiments are common in studies of phenotypic plasticity. But even single factors have great complexity in their effects on plants. Drought stress is a secondary effect of many types of stress, including heat, cold, freezing, flooding, salinity, and wind (Taiz and Zeiger 1998), leaving open the question of whether a plant response is to the primary stress or to the drought that stress has induced. And controlled studies can omit important aspects of the environmental factors. Studies that manipulate light levels through neutral shade cloth, a common technique, quantify the importance of light intensity as a resource for growth, but miss such ecologically crucial nuances of the natural light environment as the temporal fluctuations caused by natural sunfleck patterns and the changes in light spectral quality. Given that these problems are recognized, controlled environmental studies have been more informative than studies in nature in determining what aspects of the environment these traits respond to.

On the other hand, studies in natural environments have clearly demonstrated that there is spatial heterogeneity in natural selection at the scale of centimeters to several meters. These distances are of the magnitude observed for seed and pollen movement (Schaal 1975). Such coarse-grained variation, where a given individual experiences only one environment but some of its progeny experience very different environments, should favor the evolution of adaptive phenotypic plasticity (Via and Lande 1985; chapter 6). Natural populations show variation in phenotype correlated with environmental parameters (Lechowicz et al. 1988), variation in phenotypic selection (Gross et al. 1998; Kalisz and Teeri 1986), variation in phenotypic selection that was correlated with environmental parameters (Stewart and Schoen 1987), and local adaptation at small scales (Schmitt and Gamble 1990). Temporal variation in selection on germination time has also been found (Kalisz 1986). Miller et al. (1995), using phytomers (i.e., replicate plants of known genotypes planted into different environmental conditions as an assay of environmental variation) and measuring plant phenotype in several characters, found that the environment tended to vary as a mosaic more than as a gradient. Genotype–environment interaction in fitness for phytomers, indicating spatial heterogeneity in the environment resulting in local adaptation, has been shown at the scale of centimeters in several elegant experiments (Stratton 1994, 1995; Stratton and Bennington 1996).

Attempts to determine which factors in the natural environment cause this variation are more rare. Differences in environments significant to plants are expected at this small scale because of factors such as microclimatic variation in light and temperature, the presence of other plants, and patterns of nutrient deposition and water availability (Gross et al. 1995; Alpert and Mooney 1996). Stewart and Schoen (1987) found variation in selection on several traits at the level of a few meters and were able to correlate this variation to environmental variation in light and water. Lechowicz et al. (1988) found that different edaphic factors were correlated with performance for two vegetatively similar annual species of *Impatiens*. For *I. capensis*, growth and reproduction were highest in sites rich in potassium and phosphorus and high in organic matter. For *I. pallida*, performance increased with calcium and magnesium and high light levels.

A broad generalization I would make about the environments that evoke plastic responses is that they either stress and actively damage plants (e.g., high temperatures) or affect photosynthetic carbon acquisition and growth. We find plasticity to the availability of the resources for photosynthesis: light, carbon dioxide, water, mineral nutrients, and oxygen. Stresses that damage plants tend to decrease rates of growth and carbon acquisition in surviving plants. Insect damage, toxicity, and heat and cold result in a reduced capacity for carbon acquisition or in reduced rates of metabolism (Taiz and Zeiger 1998).

More evidence for this argument is that environmental differences that evoke plasticity also affect fitness (Bazzaz and Sultan 1987). Particularly in nature, but also in greenhouse studies, it is common to see that the plants grown in different environments differ in fitness by several orders of magnitude. Differences in fitness between environments tend to be considerably greater than the differences in such characters as leaf size and stem length (Winn and Evans 1991; Sultan and Bazzaz 1993a; Dudley 1996; Dudley and Schmitt 1996). The fitness differences between environments appear to be a consequence of environmental differences that affect carbon acquisition. Size at reproduction is a primary determinant of fitness in annuals (Samson and Werk 1986). Because plants grow by producing new modules (e.g., aboveground modules consisting of an internode, node, leaf, and axillary bud), plants can grow indeterminately, with larger plants having more reproductive modules and a higher reproductive biomass.

A few exceptions can be found where, although differences in natural selection can be inferred from local adaptation, the environments did not differ in overall quality. *Anthoxanthum odoratum* demonstrates the evolution of local adaptation to soil pH, but average fitness did not differ between pH levels (Snaydon 1970). *Polygonum persicaria* demonstrates considerable plasticity in response to flooding, with apparent genotype–environment interaction for total biomass and fruit biomass. However, the average biomass of the plants was very similar in field-capacity and flooded environments (Sultan and Bazzaz 1993b).

Despite these exceptions, the rule is to find tremendous variation in environmental quality for plants. In plants, it is common to see differences in maximal fitness between environments despite the evolution of local adaptation (e.g., figure 10.1B). This variation may be greater than what animals experience because of the differences in resource requirements and motility in these groups. Plants, unlike animals, lack the option of leaving poor environments. Moreover, all plants need the same resources for photosynthesis, whereas animals may specialize on food types that differ in size, handling, or source but not availability or quality (Futuyma and Moreno 1988). Therefore, the pattern found in aphids, where locally adapted populations experienced similar maximal fitness on peas and alfalfa (figure 10.1A), may be much more common in animals than in plants.

That plant environments are very likely to differ in quality has important implications for how we should interpret plasticity. Many phenotypic traits change through development, a phenomenon referred to as ontogenetic drift (Coleman et al. 1994). Many traits can change with size and age. For example, if root:shoot ratio lessens with age, then plants in a wetter environment could have a lower root:shoot ratio than plants in a dryer environment simply because plants in the poorer environment grow more slowly and are at an earlier ontogenetic stage (Gedroc et al. 1996). Such size-generated plasticity may be adaptive, but its genetic basis and implications for the evolution of adaptive plasticity differ from those of plasticity generated by changes in development. Comparing traits at a common size as well as a common age (Coleman et al. 1994) demonstrates that traits can vary both ontogenetic state and because of changes in development.

Environmental quality, as measured by the maximal fitness achieved in that environment, has been shown to be a crucial parameter in modeling the evolution of adaptive plasticity (Zhivotovsky et al. 1996). Natural selection will favor reaction norms that express the optimum phenotype in frequent, high-quality environments, and as a consequence the traits expressed in environments that are infrequent and of low quality will tend to deviate more from their optimal value (Via and Lande 1985; Zhivotovsky et al. 1996).

Models of evolution in environments of heterogeneous quality predict that populations will be slow to adapt to low fitness environments either by adaptive plasticity or by local adaptation (e.g., Holt and Gaines 1992; Kawecki 1995a). A very interesting question that does not appear to have been addressed by such models is whether asymmetry in environmental quality predisposes evolution either toward adaptive plasticity or local adaptation.

The Complexity of Plant Responses to Differing Environments

A major problem for understanding phenotypic plasticity in plants is that the simplicity of most models, with one trait expressed in two environments, is far surpassed by the complexity of what actually occurs in plants. The responses of four *Arabidopsis* populations to variation in water, light, and nutrients (Pigliucci et al. 1995a) provide an excellent example of this complexity. For any one genotype responding to any one environmental factor, it is observed that some traits will change and others remain constant; for example, population K in figure 10.3 responded to low light (compared with the optimal conditions) with an increase in the number of leaves, a decrease in total height, and a constant life span. There is commonly genetic variation for the responses to a single environmental factor; for example, the response of population T to low light differed from that of K with a slight increase in number of leaves, a slight increase in total height, and a marked increase in life span (figure 10.3). The responses are specific to the environment; for example, population T responded to low nutrients with a very large decrease in leaf number, a decrease in height, and an increase in life span (figure 10.3). Of the nine traits measured in this study, all responded to at least one environmental change, and all but three varied among populations in how they responded to the environmental changes. No environment caused a response in only one trait. The patterns found in this data, high genetic variation, specific responses to each environment, and responses in multiple traits, are common in genetic studies of phenotypic plasticity in plants.

Variation among populations in their responses to the environment has been found by many other researchers (Macdonald and Chinnappa 1989; Schlichting and Levin 1990; Schmitt and Wulff 1993). Variation in responses to the environment is also commonly observed between species (Zangerl and Bazzaz 1983). Within populations, the variation among genotypes in responses to the environment may be as great as that observed among populations and species (e.g., Sultan and Bazzaz 1993a,b,c; Zangerl and Bazzaz 1983).

High genetic variation for plasticity implies that phenotypic plasticity can evolve readily. Certainly, plant responses to R:FR appear to evolve on a short time scale. Populations (Bain and Attridge 1988; Dudley and Schmitt 1995) and genotypes (Skálvová and Krahulec 1992) differ in responsiveness of stem elongation to R:FR. Species within the genus *Acer* differ greatly in their growth, morphological, and physiological responses

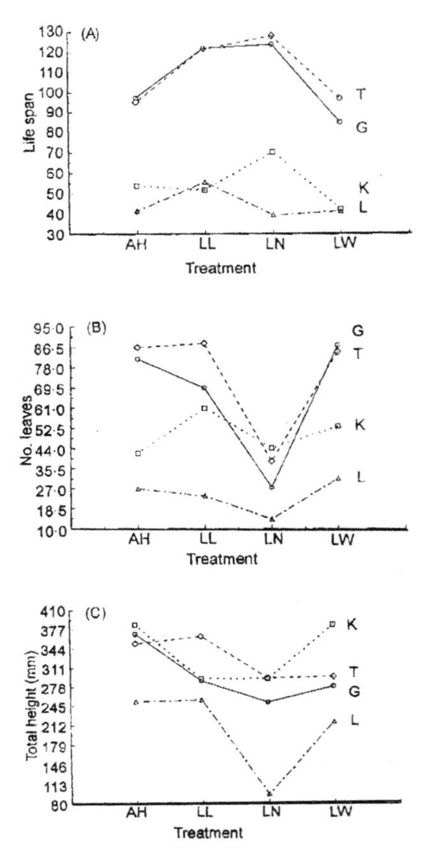

Figure 10.3. Norms of reaction for *Arabidopsis* grown in four environments: high resource (AH), low nutrients (LN), low water (LW) and low light (LL). (A) Life span. (B) Number of leaves. (C) Total height. Each line indicates a different population (inbred line). [From Pigliucci et al. (1995a).]

to the transition from low R:FR shade to full sun; a natural transition is likely to be cued both by R:FR and irradiance (M. J. Lechowicz, unpublished manuscript). For a variety of dicot species, internode elongation in response to R:FR depends on the species habitat (Morgan and Smith 1979). Within the angiosperms, the traits that are cued by phytochrome vary from internode length (Smith 1995), to tiller number and height in grasses (Skávlová and Krahulec 1992), to flowering time in *Arabidopsis* (Bagnall et al. 1995), to leaf width and angle in *Plantago* (Van Hinsberg 1997).

But although responses to R:FR evolve readily, the signal perception molecule for R:FR is of ancient origin and evolves slowly. Phytochrome has been found in algae, moss, ferns, and conifers (Quail 1991). The individuation of the phytochrome gene family into forms specialized to different functions, a key event in its evolution (Smith 1990), predates the separation of the monocots and dicots (Sharrock and Quail 1989).

Figure 10.3 demonstrates a second common phenomenon of plasticity, the specificity of responses to each environmental factor. Studies that have measured plants of the same genotypes subjected to reduced availability of the most common resources—light, water, and nutrients—have found that the changes in phenotype are very specific to the resource (Pigliucci et al. 1995a,b; Schlichting and Levin 1984, 1990; Sultan and Bazzaz

1993a,b,c). Chapin et al. (1993) suggest plant responses to a stress include both potentially adaptive responses that are specific to that stress, and a more generalized set of responses they term the stress resistance syndrome, characterized by reduced relative growth rate, low rates of resource acquisition, low rates of tissue turnover, and high concentrations of plant secondary metabolites. The stress resistance syndrome is hypothesized to be an adaptive strategy that increases survival in stressful environments.

Pigliucci et al. (1995a,b) also demonstrate a third common phenomenon of plasticity, that environments evoke responses in multiple traits. It becomes very difficult to pose hypotheses about the adaptive value of responses in multiple traits. If the environment directly affects an individual trait, the direct response can result either from adaptive changes in development cued by the environment or from the nonadaptive impacts of environmental resource levels and stresses on the trait expression. An individual trait can respond indirectly to the environment as a consequence of its correlation with a trait directly affected by the environment, and these indirect responses may be either neutral or costly.

Some researchers have hypothesized that multiple traits respond to the environment because of a primary response in an underlying or causal trait with other traits constrained to respond as well (e.g., plants able to acquire more carbon will have more leaves, more branches, greater height, and greater root biomass as a consequence). Some candidate causal traits include size (Coleman et al. 1994), timing of reproduction and size at reproduction (Kudoh et al. 1996), meristem limitation (Geber 1989; Schmitt 1993), and seed size (Chapin et al. 1993).

Another hypothesis is that changes in hormonal level cued by the environment cause plasticity in multiple traits (Chapin et al. 1993). For example, genetic differentiation between sun and shade populations and plasticity to R:FR in *Plantago lanceolata* appear largely explained by changes in the level of gibberellins (Van Hinsberg 1997). Several traits are tightly correlated, and both plasticity and genetic differentiation can be mimicked by application of either gibberellins or gibberellin inhibitor to give a range between erect rosettes with few long, narrow leaves or prostrate rosettes with many short, broad leaves (Van Hinsberg 1997). This result is most consistent with the idea that adaptive plasticity responses in multiple traits may comprise a suite of coevolved functional responses. The traits cued by gibberellins in this example appear to result from functional interrelationships rather than from changes in one variable causing a change in another.

Particularly in the study of plants, the effect of a potentially adaptive character on fitness can frequently be seen as acting through intermediary traits. These traits may be a component of fitness such as survivorship (Arnold and Wade 1984b) or an ecologically relevant performance trait such as stress tolerance (Arnold 1983). For phenotypic plasticity, plant size is the most important performance trait because of its overall importance in predicting fitness in plants (e.g., Samson and Werk 1986), and because environmental factors that evoke plasticity also tend to affect carbon acquisition. Bigger plants have more carbon to be used to make structures to acquire nutrients and water, defend against predators and disease, compete, grow, and reproduce (Ehleringer 1985), and fecundity is observed to be strongly correlated with size in annuals (Samson and Werk 1986).

A problem in the empirical study of plant phenotypic plasticity is that performance traits tend not to be recognized as such. Instead, they may be treated as characters that could potentially show adaptive plasticity, or as measures of fitness. The fact that they comprise aspects of both kinds of trait makes them more complicated than either, and the expectations for their evolution are quite complex. Indirect measures of plant size include characters highly correlated with size such as branch number, leaf number, total leaf area, and height. These indirect measures may also encompass potentially adaptive changes in addition; for example, plant height increases with both size and elongation. The relationship between height and plant size changes in response to the density environment (Dudley and Schmitt 1996). Allometric changes such as this may be common in plastic responses (Schlichting and Pigliucci 1998).

The relationships among multiple intermediary traits determine fitness. For example, life history theory suggests that the components of fitness, age- or stage-specific survivorship and fecundity, are always selected to increase, but there may be trade-offs (negative correlations) between components of fitness (Lande and Arnold 1983). It has also been hypothesized (Arnold and Wade 1984a) that trade-offs will be found between performance traits, such that both are under positive selection but are negatively correlated.

Because intermediary traits are always selected to increase, it could be concluded that plasticity in performance traits and components of fitness, like plasticity in fitness itself, could not be adaptive. But it has been suggested that the optimal combination of life history traits depends on the environment (Stearns 1976). It has been theorized that trade-offs in allocation may determine fitness in environments where acquisition rates are almost uniformly high. However, in environments where acquisition rates are low, the rate of acquisition may be the best predictor of fitness (Van Noordwijk and De Jong 1986). For *Amphibromus scabrivalvis*, a perennial corm-forming grass, trade-offs between clonal reproduction and storage and between clonal and sexual reproduction occur in fertilized treatments but not in unfertilized treatments (Cheplick 1995). For *Impatiens capensis* plants grown in high density, branch number and flower number had a positive genetic correlation, but for plants grown in low density the genetic correlation was negative, indicating a trade-off (Donohue and Schmitt 1999). For *Arabidopsis* grown in greenhouse conditions, the environment affects how several measures of plant size (branch number, number and weight of basal leaves, height of plant and first flower, and inflorescence weight) and days to flowering contribute to fitness (Pigliucci et al. 1995a,b).

Distinguishing between potentially adaptive characters, performance traits, and fitness is key to the most common analysis of plasticity, the measurement of genotype–environment interaction (*G*×*E*). This statistic, the *G*×*E* term from an analysis of variance, is generally used to test whether genotypes differ in how they respond to the environment and so is a measure of genetic variation in plasticity. Another important model measures the genetic constraints on the evolution of plasticity as the genetic correlations between trait states in the alternate environments (Via 1987). The theories about whether selection acts on the trait in each environment (character state model) or on plasticity itself (reaction norm model) (Schlichting and Pigliucci 1998) are intensely debated. But the continued popularity of *G*×*E* is most likely for statistical reasons. For *G*×*E*, genetic constraints are lowest when the *G*×*E* is highest, that is, when the null hypothesis is rejected. For the genetic correlation model, genetic constraints are lowest when the genetic correlation is zero, that is, when the null hypothesis is accepted. But low sample size or high variation within families versus between families can also lead to acceptance of the null hypothesis. Therefore, for statistical rather than theoretical reasons, *G*×*E* will remain a key measure in studies of plasticity.

There are well-established interpretations of plasticity and *G*×*E* for potentially adaptive traits. Characters such as physiological, morphological, biochemical, behavioral, and phenological traits can evolve adaptive plasticity if they are under differing natural selection in different environments. Figure 10.4A demonstrates conditions selecting for adaptive plasticity, with the optimal value for the traits differing between the environments. Plasticity in such traits may be adaptive but cannot be assumed to be so. In figure 10.4B, the genotypes indicated by solid symbols show adaptive plasticity because plasticity increases fitness compared with a lack of plasticity (e.g., in high density, greater stem elongation is favored, and plants do produce longer stems in high density). The genotype indicated by the open squares shows maladaptive plasticity that decreases the fitness of the genotype compared with a lack of plasticity. The differences among genotypes in their norms of reaction (significant *G*×*E*) are interpreted as genetic variation for plasticity, which will allow the population to evolve toward optimal phenotype in each environment. An absence of *G*×*E* in adaptive traits indicates genetic constraint that may explain why genotypes are specialized to each environment.

When the trait measured is fitness, the presence of *G*×*E* suggests the presence of genotypes specialized to each environment (Fry 1996; Stratton 1994). Higher fitness is of course favored in all environments (figure 10.4C). The pattern of fitness differences between environments typical of genotypes showing local adaptation (figure 10.4D) is very typical of *G*×*E* for fitness with crossing norms (changes in rank). Fry (1996) argues that any *G*×*E* for fitness implies the potential for local adaptation, even without crossing norms. Thus, although *G*×*E* in a character trait indicates a genetic architecture that favors the evolution of phenotypic plasticity in that trait, *G*×*E* in fitness suggests genetic constraints on the evolution of phenotypic plasticity.

G×*E* in performance traits or component of fitness, such as biomass, mating success, or longevity, are often interpreted similarly to *G*×*E* in fitness. The evolutionary expectation for these intermediary traits is similar to that for fitness: intermediary traits are directly selected to increase in all environments such that, all else being equal, larger plants will produce more seeds (figure 10.4C). Variation in performance variables can help us understand variation in fitness. For example, to understand why fitness differs between environments, we may look for environmental differences in growth or survival or in mating success. Similarly, *G*×*E* for performance traits may result in *G*×*E* in fitness; for example, genotypes may be specialized to different environments because genotypes may show high mating success in one environment but not in the other. The evolution of performance traits is more complex, however, because fitness is the consequence of the interactions among performance traits. If performance traits are correlated, then trade-offs among these traits complicate the interpretation of their plasticity. Potentially, the relative importance of performance traits in determining fitness may differ between environments; for example, variation in size may most strongly predict fitness in a poor-quality environment, but mating success may most strongly predict fitness in a high-quality environment.

Some traits are plastic simply because plants experience stress, damage, or low resource availability to a greater degree in one environment. The expectation for traits that indicate damage is for adaptive homeostasis: that the optimal value for the trait is

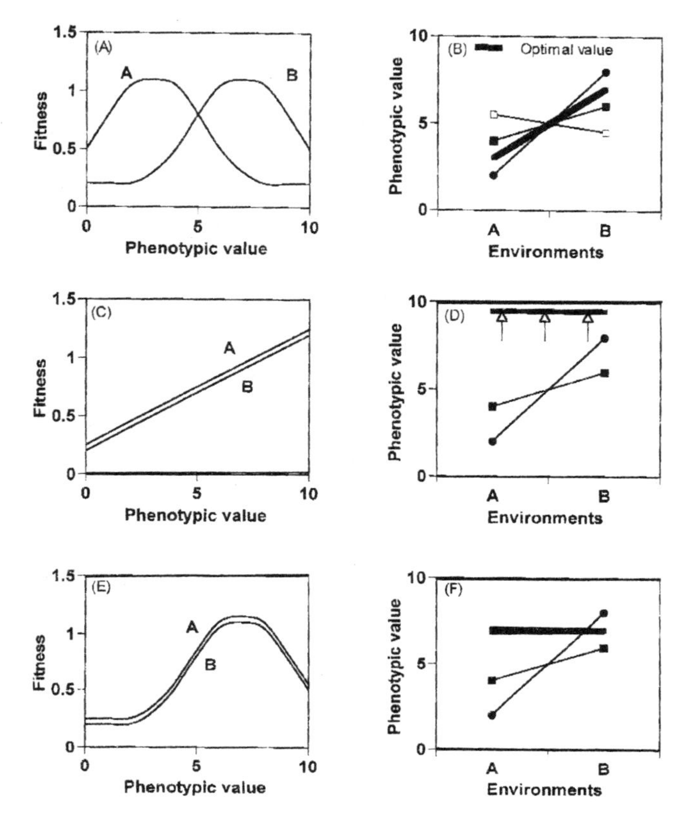

Figure 10.4. Plasticity in different types of traits in a low quality environment (labled A) and a high quality environment (labled B). (A) Selection for adaptive plasticity: the optimal trait value differs between environments. (B) Norms of reaction for adaptive plasticity showing the optimum norm (indicated by the thick line) and three genotypes (indicated by thin lines with different symbols). (C) Selection on fitness, components of fitness, and performance traits. (D) Norms of reaction for a fitness or performance trait showing two genotypes. For these traits there is no optimum norm because a higher value is always favored. (E) Selection for adaptive homeostasis, on a trait that indicates damage or stress. (F) Norms of reaction for a stress or damage trait showing the optimum norm and two genotypes.

the same in both environments. Figure 10.4E shows natural selection for stress damage to a plant. The optimal value is obtained in the more favorable environment, but the less favorable environment results in the expression of a suboptimal value. Leaf turgor potential may show this pattern in environments that differ in water availability. In mesic conditions, the plants can maintain the optimal high turgor. In more xeric conditions, although high turgor is still optimal, plants may lose turgor and even wilt because of the low water availability. In this example (figure 10.4F), the genotype represented by a square maintained more adaptive homeostasis, and the genotype indicated by a circle was stressed in the low-quality environment. *G*×*E* for stress damage suggests that plants differ in their ability to maintain homeostasis and may explain why genotypes are specialized to different environments.

It can be surprisingly difficult to even determine whether a trait is an adaptation or whether it indicates damage. In my own work, I found that leaf size was an adaptive trait with different optima in environments with different light availability in *Cakile edentula* (Dudley 1996), but a performance trait closely correlated with plant size across densities in *Impatiens capensis* (Dudley and Schmitt 1996). Necrosis of leaf tips suffering nickel deficiency has a straightforward interpretation as damage induced by the environment. However, necrosis is also a part of the adaptive hypersensitive response to pathogens. Programmed cell death surrounds and isolates the pathogen with dead tissue that prevents a pathogen from infecting the plant (Taiz and Zeiger 1998). These examples demonstrate how profoundly the fitness implications of traits depend on the environment.

The Consequences of Phenotypic Plasticity for Performance and Fitness

Determining the fitness implications of a trait in different environments, and of plasticity in a given trait, requires understanding how traits function and how traits are selected in different environments. Plant physiological ecology and developmental biology provide *a priori* hypotheses about costs and benefits of traits in different environments, the causes of plasticity in certain traits, and the functional interrelationships among traits. Measuring selection in multiple environments and measuring the fitness implications of plasticity are still emerging areas in evolutionary biology (Schmitt et al. 1999; Scheiner and Callahan 1999). Each method of studying the adaptive value of phenotypic plasticity has strengths and weaknesses; each provides supporting evidence rather than conclusive proof of adaptation. Given these qualifications, we can observe that apparent adaptations fall into three functional categories: (1) to reduce damage or stress imposed by an environment, (2) to increase carbon acquisition in an environment, and (3) to alter reproductive strategy, presumably to maximize fitness in an environment.

The first line of evidence for adaptation is knowledge of the biology of the organism. Knowing how a trait functions is still the most common method for assessing its adaptive value (Sultan and Bazzaz 1993a). However, determining how the phenotype affects performance and fitness is made difficult by the interrelationships among traits. For example, changes in leaf size affect the area of the photosynthetic surface but also can affect leaf temperature, which in turn affects the rates of photosynthetic acquisition and water loss (Ehleringer and Clark 1988). Optimality models provide a methodology for analyzing interactions among traits and predicting the best trait value in each environment (Givnish 1982, 1986).

Smith (1990) suggests that if a response to the environment is the consequence of a signal perception-transduction-response chain, it is almost certainly an adaptation, for example, the stem elongation response. However, not all adaptations result from signal transduction. Increases in root/shoot allocation in response to low nutrients, although thought to be adaptive, are thought to be a simple consequence of the sink strength of roots being greater than that of shoots because the roots have more access to nutrients (Clarkson 1985). Another indication of adaptation is if the response requires energy and resources to produce, particularly if the environment is associated with lowered growth, for example, production of specific proteins in response to a stress that reduces overall protein synthesis. The functional argument is that the plant would not waste resources to change its phenotype unless the trait was adaptive.

Examining genetic differences in traits between populations from contrasting environments provides another line of evidence for adaptation. This method is also confounded by differences in multiple traits. Genetic differences between populations in multiple traits make it difficult to assess the importance of individual traits. Differences in individual traits may increase fitness or may arise from trade-offs between traits (Hanson and Hitz 1982). Van Tienderen (1990) used this method to assess the adaptive value of plasticity in *Plantago lanceolata*. Hayfield and pasture plants were locally adapted. Differences in growth phenotype resulting from plastic responses paralleled genetic differences, suggesting that plasticity was adaptive in direction, but the response from plasticity was not as great as the genetic differences between populations.

Evolutionary methods for directly measuring the fitness consequences of phenotypic plasticity are discussed by Schmitt et al. (1999) in detail. It is possible to show that the overall change in plant phenotype evoked by an environment is adaptive. The existence of acclimation responses to various environmental stresses, such that exposure to prior stress increases tolerance of that stress compared with a naive plant, indicates the adaptive value of the stress-induced phenotype in a stressful environment (Taiz and Zeiger 1998). More recently, phenotypic manipulation has been used to explore the costs of the stress-induced phenotypes induced by mechanical hardening (Cipollini 1999) and elongation (Cipollini and Schultz 1999) in a benign environment. Assessing the fitness consequences of the alternate phenotypes in both environments has been done by Dudley and Schmitt (1996) and Schmitt et al. (1995) for plant responses to density (discussed above), and for induced plant responses to herbivores by Baldwin (1998). These approaches, however, do not permit the researcher to dissect out the fitness consequences of individual traits.

A line of evidence that addresses the fitness consequences for a given trait is the estimation of natural selection in alternate environments from the relationship between traits and fitness. Measurements of selection on traits in differing environments (Wade and Kalisz 1990) may be expected to indicate the adaptive value of individual traits in differing environments. Problems with this approach include the problem of unmeasured, correlated traits (Mitchell-Olds and Shaw 1987) and phenotypic plasticity itself, which alters the correlations among traits (Schlichting 1989) and also changes the range of trait values (Donovan and Ehleringer 1994; Dudley 1996). Phenotypic manipulation in combination with measurements of selection does permit comparison of natural selection over the same range of trait values (Dudley and Schmitt 1996). Because of the difficulty of measuring selection, relatively few studies have been done, although I have found that hypotheses from physiological ecology can predict selection in natural environments (Dudley 1996; Dudley and Schmitt 1996). In contrast, Winn (1999) found that selection on leaf traits did not differ between summer and winter conditions, despite compelling functional arguments that the observed plasticity was adaptive.

Plasticity in traits that function to reduce stress is very common. Examples of responses that apparently reduce damage by an environment include heat shock responses and ultraviolet light responses. Plants subjected to heat stress produce heat-shock proteins. The small chloroplast heat-shock protein has been shown to protect photosystem II during heat stress (Heckathorn et al. 1998). Ultraviolet light responses include the induction of ultraviolet-absorbing pigments and reduction in stem and leaf elongation. The pigments almost certainly function to reduce damage by ultraviolet light, but the implications for fitness of the changes in morphology have not been established (Barnes et al. 1996).

Plasticity in traits that affect carbon acquisition is also common. For example, the responses to low light—thinner, larger leaves with increased chlorophyll concentration and increased ratio of chlorophyll b to chlorophyll a—provide an increased surface with more antennae to maximize capture of light. The responses to density, particularly increased stem elongation, are thought to allow plants to increase light capture by placing their leaves higher in the canopy.

It is undoubtedly common for plasticity both to reduce the risk of stress and increase carbon acquisition. The reduction in water loss in dry environments, accomplished through reduced leaf area and reduced stomatal conductance, will decrease the likelihood of declining water potential, which in the extreme results in wilting, damage to the plants, and death. These responses do decrease net assimilation compared with a plant in a mesic habitat. However, because small declines in turgor potential substantially reduce photosynthetic rates, reduced water loss may allow plants to maintain higher net assimilation in the long run (Ehleringer 1985; Fitter and Hay 1987).

Changes in reproductive strategy in response to the environment can also be identified. Earlier flowering time is seen in response to some low-productivity environments; for example, earlier flowering was found in low-fecundity environments (Galloway 1995), in high density in *Impatiens* (Schmitt et al. 1987), and in response to low R:FR in *Arabidopsis* (Halliday et al. 1994). However, four *Arabidopsis* populations responded to low light intensity with later flowering, and three of the four flowered later in response to low nutrients as well (Pigliucci et al. 1995a). In monoecious and andromonoecious plants, allocation to male flowers is affected by plant density (Lundholm and Aarssen 1994), plant size (Ackerly and Jasienski 1990), and resource availability (Solomon 1985; D'Antonio and Vitousek 1992; Diggle 1993). The fitness consequences of these changes are not yet known.

Nonadaptive responses in physiological, morphological, and other potentially adaptive characters could potentially be generated by genetic and phenotypic constraints and trade-offs. In a study of water use efficiency and leaf size in plants grown in environments differing in water availability (Dudley 1996), I found that leaf size was greater in the wet environment, as predicted by functional arguments and measured selection. However, water use efficiency was lower in the dry environments, contrary to the predictions of functional arguments and measured selection. This discrepancy may be explained by the positive correlation between leaf size and water use efficiency I observed in these populations (Dudley 1996).

Functional Ecology of Fine-Grained Responses

Evolutionary biologists recognize that the scale at which environmental heterogeneity occurs is important. Evolutionary models (chapter 6) and empirical studies have mostly considered responses to coarse-grained variation, where an individual experiences only one environment during its lifetime (but see Winn 1996a,b, 1999). Conversely, developmental studies most often consider responses to fine-grained temporal changes in environment. The evolution of responses to fine-grained heterogeneity, spatial and temporal variation in the environment experienced by a single plant, has not been well studied

(Schlichting and Pigliucci 1995). Both fine- and coarse-grained variation are hypothesized to select for adaptive phenotypic plasticity, but fine-grained environmental variation selects directly on the reaction norm within an individual, whereas coarse-grained environmental variation can only select on the expressed character state in a generation (Via et al. 1995). This difference may affect the evolution of adaptive plasticity quantitatively, with a slower approach to the adaptive optimum when environmental variation occurs at a coarse grain. An interesting question is whether fine- and coarse-grained variation can cause qualitative differences in the nature of the selection on the traits, and what response to the environment evolves.

Sources of fine-grained temporal variation are common. Temperature and rainfall may occur unpredictably depending on weather, although with a predictable seasonal trend. Forest canopy closure results in a seasonally predictable change in light level. Factors such as nutrients, light, and water that show coarse-grained spatial variation may also vary spatially in a fine-grained manner. The plant itself determines at what scale it experiences the environment: between-year variation is coarse grained for an annual and fine grained for a perennial.

Plants grow by adding new modules such as leaves, internodes, and branches. A functional distinction can be made between responses to temporal change that occur at the level of the modules and those responses that occur over the whole plant. Stomatal conductance and root:shoot ratio are examples of whole-plant responses. Another example is a chemical defense against herbivory; foliar herbivory in *Nicotiana* stimulates increased nicotine concentrations in the remaining leaves resulting from xylem transport of nicotine synthesized in the roots (Karban and Baldwin 1997). Whole-plant responses vary in their impact and reversibility: stomatal conductance is extremely labile, but root:shoot ratio can only change slowly (Gedroc et al. 1996).

Quite often changes in the environment cause changes in the morphological characters of newly produced modules; for example, changes in light quality affect the internode length only for the internodes still growing and expanding (Dudley and Schmitt 1995, 1996). Older modules are already fixed for this character. Modules can differ in their responsiveness to the environment (Dudley and Schmitt 1995; Winn 1996b), and this individuation has been shown to play a role in local adaptation (Dudley and Schmitt 1995). For modular traits, coarse-grained environmental variation could be considered the special case where all the modules are produced in the same environment. Even if selection acts only on the fine-grained responses, adaptive plasticity to coarse-grained environmental variation could evolve as a correlated response in modular traits; that is, the ability to produce the optimal module phenotype in a given environment conveys adaptation to both fine- and coarse-grained variation in the environment.

The phenotype of previously produced modules is not directly reversible, but their effect on the whole plant's phenotype can be modified. One mechanism is the loss of modules, such as the drought-induced deciduousness of leaves (Taiz and Zeiger 1998). Another mechanism is a dependence of later modules on the state of previous modules. We found that later-produced internodes were shorter in plants that had previously elongated internodes (Dudley and Schmitt 1996). Similarly, the effects of position within an inflorescence (architecture) on flower morphology can be modified by previous developmental events such as nutrient storage and fruit set (Diggle 1997).

Some costs of phenotypic plasticity will be incurred in response to both fine- and coarse-grained variation. For the induced response to herbivory described above, the costs include the investment of energy and biomass in the nicotine defense. The reliability of the cue also has fitness consequences for responses to both fine- and coarse-grained variation; for example, both not producing a defense when herbivores are common, and investing in defense when herbivores are absent have been shown to decrease fitness (Baldwin 1998). But other costs of plasticity will only be incurred in response to fine-grained variation. For the induced response to herbivory, there is the cost of damage during the time between the initial damage of a leaf and the induction of a nicotine concentration to defend against herbivory. Such time-dependent costs of remaining in the wrong phenotype after the environment has changed depend on the lability and reversibility of traits. Changing traits can also have an energetic cost, for example, the lost biomass in the discarded leaves for drought-induced leaf senescence.

Temporal variation has both predictable and unpredictable components. Responses to temporal variation in the environment often function to aid plants to adapt to unpredictable changes in the environment, for example, heat-shock responses. But other responses are cues to a date or a season rather than to the immediate environment. These seasonality cues share common mechanisms for environmental perception with responses to the unpredictable component; for example, phytochrome is involved in photoperiodic responses where day length is used by plants as an indicator of date (Thomas 1991). Other seasonal cues include vernalization and stratification requirements, where a long period of cold is necessary before a plant flowers or a seed germinates. Plant responses can also be cued to season through internal clocks or developmental states, such as ontogenetically determined changes in modular characters such as leaves (Winn 1996b) or in root and shoot allocation (Gedroc et al. 1996).

The agent of selection on these seasonality cues is the predictable component of temporal heterogeneity in the environment, making the evolution of these cues a process of local adaptation. When the environment is seasonally predictable, these cues may allow plants to anticipate changes in the environment and show a suitable response (e.g., Winn 1996b). This anticipation may allow plants to avoid the cost of the lag between perception of the cue and the response to that cue. They may also be important when the costs of making the wrong decision are high; responses such as germination are irrevocable, and the cost of germinating during a brief period of warm weather in mid winter is mortality. In some cases, there is a main and a backup seasonality cue. For example, in many species flowering requires either vernalization (several days of cold temperature) or a short-day photoperiod requirement (Taiz and Zeiger 1998). Both cold temperatures and short days are characteristic of fall, but if the weather stays unseasonably warm, the plants will still flower before winter.

Tolerance of limited resources is typical of plant responses to coarse-grained heterogeneity, for example, sun–shade responses. In contrast, active foraging for resources is typical of plant responses to fine-grained heterogeneity, for example, phototropism, a response involving active growth toward light (Taiz and Zeiger 1998). Plants respond to sunflecks, temporal variation in light, by opening stomata and induction of higher photosynthetic rates that permit them to maximize exploitation of the brief periods of high light levels (Chazdon and Pearcy 1991). *Portulaca*, a prostrate species, uses R:FR as a cue to place shoots in areas free of other plants (Novoplansky et al. 1990).

Roots search for nutrients and water by increasing root proliferation in patches of high resources (Gross et al. 1993). Root biomass in conditions with varying resource availability has been demonstrated to follow models of ideal foraging behavior, and this strategy appears to increase plant growth (Gersani et al. 1998). Therefore, although plants respond to coarse-grained differences in nutrients by increased root allocation in low nutrient patches (Sultan and Bazzaz 1993c), for a given plant in low nutrients, root growth will be greatest in high-nutrient patches.

Foraging for resources is most strongly demonstrated in the study of integrated clonal plants. These plants experience greater spatial heterogeneity than does a unitary plant and may share resources between ramets (Salzman and Parker 1985; Alpert 1996). A few studies have demonstrated that phenotypic plasticity in clonal morphology (i.e., stolon or rhizome internode length, branching intensity, or branching direction) plays a role in how clonal plants find and exploit resources (Evans 1995; Humphrey and Pyke 1997). Evans (1995) explicitly demonstrated differences in clonal morphology between heterogeneous and homogeneous environments. In contrast, a review by de Kroon and Hutchings (1994) suggests that plasticity in positioning of ramets is relatively rare and not very effective in allowing genets to exploit patchy resources. They suggest that more regular, unresponsive positioning of ramets may most efficiently search for resources, with phenotypic plasticity of the ramets themselves being most important in exploiting patches of resources.

Conclusions

Phenotypic plasticity is a common theme in the study of plants. Integration of many disciplines is important in studies of phenotypic plasticity, because the complexity of plant responses makes it difficult to discern patterns and to make predictions from norms of reaction and selection gradients on a suite of traits chosen at random. Research on the genetics of plasticity has benefited from the incorporation of molecular and developmental biology. What fields should be incorporated into the study of the functional ecology of phenotypic plasticity? Developmental biology, for one, because of the importance of determining how the environment affects the trait. Another is the small but growing field of plant evolutionary physiology. The study of the fitness consequences of plasticity in different environments is largely the study of the consequences for survival and especially for carbon acquisition of mass and energy flow traits in different environments, that is, plant physiological ecology.

A major advance in the study of phenotypic plasticity is the ability to manipulate the phenotype and determine the fitness consequences of adaptive phenotypic plasticity. However, the study of nonadaptive plasticity, particularly in performance characters, also deserves more attention. These are more difficult characters to study because they in part reflect aspects of adaptation but largely function as major determinants of fitness. Recognizing the intermediary nature of these traits and approaching their plasticity as a problem in life history may reveal what strategies plants use to adapt to good- and poor-quality environments.

Acknowledgments I thank S. Scheiner, J. Schmitt, J. Sleeman, A. Fast, M. DeGuzman, C. Jankowski, and an anonymous reviewer for helpful comments on this chapter.

11

The Genotype–Environment Interaction and Evolution When the Environment Contains Genes

JASON B. WOLF
EDMUND D. BRODIE III
MICHAEL J. WADE

Genetic Context

The context in which genes are expressed is often a major determinant of the phenotype (including fitness) associated with a given genotype (Schlichting and Pigliucci 1998; Wolf et al. 2000; chapter 1). Contexts that influence genetic effects span a hierarchy that begins within the cell and extends beyond the individual as far as the ecological community. Below the level of the individual, such context dependence is called *epistasis*, wherein the genetic background provided by other loci influences the effect that a given locus has on the phenotype. Epistasis may also arise when the cytoplasmic environment, provided by organelles such as mitochondria, influences the expression of nuclear genes and vice versa. Beyond the individual, we often think of genotype–environment interaction ($G \times E$) as a result of interactions between the genotype and the context deriving from the abiotic ecological environment. But other contexts include the environment provided by conspecifics in social interactions (e.g., Wade 1980; Moore et al. 1997; Brodie 2000; Wolf 2000) and by other species in community interactions (e.g., Wade 1990; Goodnight 1991). Although the tools and perspective of plasticity and $G \times E$ research can help us understand the general phenomenon of context-dependent expression, they often ignore the fact that many, if not most, of the ecologically relevant contexts are themselves genetically determined.

Contexts with a genetic component represent a unique situation wherein the *context itself can evolve*. Contexts containing genes may be of particular evolutionary importance because they are often associated with large fitness effects providing a large opportunity for selection (Crow 1958; Arnold and Wade 1984a). In addition, when fitness effects are reciprocal, wherein individuals simultaneously experience contexts and provide contexts to others, the opportunity for coselection (the driving force for coevolution or coadaptation) exists. For example, food sources of essentially all animals contain genes, as do their predators and their parasites, with obvious fitness effects to both the animals and their contexts. The reciprocal coevolution of organism and context that can result from these interactions is a critical element of coadaptation between species and in the adaptive integration of ecological communities (Wilson 1980; Thompson 1982; S. Swenson et al. 2000). Social interactions between individuals such as those between mates and those between parents and offspring are also major determinants of fitness in many plants and animals (Frank 1998; chapter 10). When coselection occurs within a species (i.e., genes and their contexts are contained in the same genome), coadaptation (e.g., Huey and Bennett 1987) or counteradaptation (e.g., Rice and Holland 1997) can occur.

Theoretical investigations of context-specific effects have used two basic frameworks. When the context is ecological and implicitly assumed to have no genetic component, models of the genotype–environment interaction and the evolution of phenotypic plasticity have been developed (see Schlichting and Pigliucci 1998; chapter 6). In these cases, contexts may be variable, but they cannot respond to changes in the evolving species. Therefore, although context and genotype interact with fitness consequences, only the genotype can respond evolutionarily. Adaptive plasticity evolves because it maximizes fitness across the variable array of nongenetic contexts by improving the fit between phenotype and context. In some evolutionary genetic models, however, context harbors an explicit genetic component, which may be identified as either epistasis or genotype–genotype interactions (see Wolf et al. 1998; Wolf 2000). In these cases, both genotype and context can respond evolutionarily to the interaction between them. In this chapter, we combine these two approaches under a single conceptual framework to examine evolution at various levels in the hierarchy of context. This approach allows us to investigate general patterns and processes that are important whenever context-specific genetic effects are present.

The Evolutionary Importance of Context Dependence

The evolutionary importance of context dependence becomes clear when we consider how interactions influence evolutionary processes. In general, context-specific fitness effects make the evolutionary fates of alleles in populations (the critical process in evolution) dependent upon the context experienced by those alleles. When the context has a genetic component, the ultimate evolutionary fate of an allele depends not only on the context it experiences but also on the evolutionary trajectory of the context itself. For example, if the context of a particular allele consists of other loci within the same genome, then the allele will evolve differently on different genetic backgrounds (Goodnight 2000). Context-specific fitness effects can also lead to the reciprocal coevolution of genotype and context, via either the evolution of the covariances between these factors or the evolution of the interaction itself. For example, interactions between genes in the same genome can lead to the development of linkage disequilibria (a measure of genic covariance) when epistatic selection occurs (Lande 1984; Brodie 1992). Evolution of linkage disequilibria may lead to "coadapted gene complexes" when sets of interacting loci evolve in concert.

The interaction between genotype and context can evolve whenever the phenotype produced by a genotype in a given context can evolve (i.e., the genotype–phenotype relationship can change). Because the phenotype mediates the interaction between genotype and context, we can consider the phenotype part of the context experienced by the genotype and, thus, the evolution of the genotype–phenotype relationship can be considered part of the evolution of the genotype–context interaction. Viewed as a hierarchy, the environment provides a context for the phenotype, and the interaction of the phenotype and environment provides a context for the genotype. For example, under some conditions, selection might favor a disassociation of genotype and phenotype when selection favors a single phenotype, leading to "canalization." Genetic canalization can evolve to diminish the degree of sensitivity that a locus shows to genetic context, and environmental canalization can evolve to diminish sensitivity to environmental context (Wagner et al. 1997). Thus, with canalization, a gene's phenotypic expression is independent of context.

In contrast to canalization, the evolution of phenotypic plasticity may increase the degree of context specificity of genetic effects, such that different adaptive phenotypes are produced in different contexts by the same genotype. Thus, plasticity and canalization can be viewed as alternative sides to the same basic process, lying at opposing ends of a spectrum of sensitivity to context. Determining where on the spectrum a population will appear, or where selection will push a population, will depend on the form of selection (i.e., how contexts affect fitness), the heritability of context-specific gene expression (Scheiner and Lyman 1989; S. H. Rice 1998), and the predictability of the context experienced by a genotype (Gavrilets and Scheiner 1993a; Scheiner 1993a).

In the following discussion, we use the concept of the genotype–context interaction ($G \times C$) to cover all forms of interaction including gene–gene ($g \times g$ or epistasis), genotype–genotype ($G \times G$), genotype–environment ($G \times E$), and genotype–species ($G \times S$) interactions. Using this framework, we attempt to identify some unify concepts that link evolutionary processes in the face of interactions with contexts, regardless of the origin of the context. We start with some basic evolutionary concepts that do not depend on the form of context being considered. We then discuss processes that are specific to situations where the environment contains genes and evolves. We will focus primarily on the latter, drawing on a few simple examples as illustrations and relating the processes to evolution of plasticity throughout.

Modes of Effect of G×C on Evolution

The $G \times C$ interaction affects the evolutionary process through three basic modes. The first mode is the most fundamental and captures the essence of context specificity: (1) the strength and direction of selection on a genotype are dependent upon the mean value of the context. Both of the other modes of influence involve the covariance between the genotype and context: (2) the covariance between the genotype and context affects net selection on the genotype, and (3) whenever change in the covariance between genotype and context affects mean fitness, all factors causing this covariance and influencing its "heritability" are subject to evolution (i.e., when the covariance has a positive effect on mean fitness, factors enforcing the covariance are favored by selection, and when the effect is negative, factors diminishing the covariance are favored). We describe and examine these effects in more detail below. Note that we focus on the $G\times C$ interaction, although this interaction will obviously be mediated by the phenotype in most cases. As a result, when we discuss $G\times C$ interaction, we include situations where the phenotype interacts with context, and all processes that we describe can be applied to questions about phenotypic distributions. When considering cases where phenotypes interact with context and genotypes build phenotypes, the context of the genotype will include both the external environment and the phenotype itself.

To examine the three modes of effect, we start by examining a simple model where fitness is determined by the interaction of two factors. In this case, we can imagine that the two factors are the genotype of an individual (X) and the value of the context (Y). For illustrative purposes, in this section we discuss a single example (assigning X and Y characteristics), although we will discuss many other examples in the sections that follow. Imagine a case where the context is some abiotic environmental factor, such as temperature, and the genotype determines the thermal tolerance of the individual. We choose a case where the context is not genetic for this example because of its simplicity and because we examine cases where the context is explicitly genetic in the sections that follow.

We use the term interaction to indicate a nonadditive influence of a pair of factors (X, Y) on the third factor (fitness). It is important to keep in mind that, whereas an individual might "interact" with their environment (in the physical sense), the resulting influence of the environment on their fitness can be independent of their genotype, resulting in no interaction between these two factors with regard to their effect on fitness (see figure 11.1A). Interactions can take on two basic forms: (1) the absolute effect of factor X changes as a function of factor Y, but the rank order or direction of effect of X does not change (figure 11.1B); and (2) the relative effect of factor X changes rank order or direction of effect as a function of the value of factor Y (referred to as "crossing-type" interaction; figure 11.1C). The impact that the interaction has on evolution will often depend on which type of interaction is occurring. In general, crossing-type interactions have a much more important impact on evolution, because they can result in changes in the *qualitative* relationship between a trait and fitness. Because of this, we will assume a crossing-type relationship in most examples we discuss. In the example of thermal tolerance, we would find that, under crossing-type interactions, the genotype favored by selection would change depending on the thermal environment (figure 11.1C), whereas in the non-crossing-type interaction, the same genotype would be favored by selection in all thermal environments (figure 11.1B).

We use a mechanistic model to define how each of the factors influences fitness. Note that this model *defines* how these factors affect fitness, and thus it is not a statistical model. As a result, the statistical partitioning of effects on fitness does not necessarily directly reflect the underlying mechanistic relationship of factors and fitness. This distinction is analogous to the distinction between epistasis at the level of gene action and epistasis at the level of statistical/quantitative genetic parameters, such as additive and epistatic variances (Wade 1992; Cheverud and Routman 1995). As in the case of epistasis, purely nonadditive effects in the definition of fitness can produce either additive or nonadditive effects measured at the population level (Falconer and Mackay 1996, p. 131; Goodnight 2000).

We assume that an individual's fitness (w) is influenced solely by the interaction between the values of its genotype and the context:

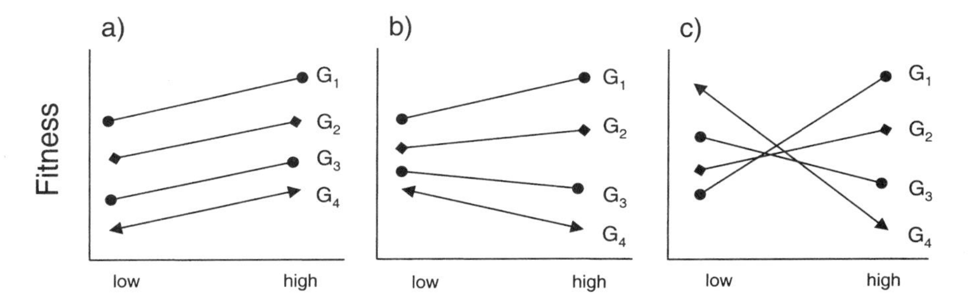

Figure 11.1. Three possible relationships between two factors (genotype and temperature) and fitness. Each line represents the fitness of a particular genotype across the gradient of temperatures (that genotype's norm of reaction). These norms of reaction are linear in this example but could be curved in a real system. (A) No interactions between genotype and context, resulting in parallel lines. The relative effect of the genotype on fitness is the same in all environments. (B) Genotype interacts with thermal environment, resulting in nonparallel lines, but the rank order of genotypes remains constant across the gradient. (C) Genotype interacts with thermal environment, resulting in a change in rank order of genotypes across the gradient of temperatures.

$$w(X) = a_x + b_{xy}XY \qquad (11.1)$$

where a_X represents the components of fitness uncorrelated with X or Y, and b_{XY} represents the effect that the interaction has on individual fitness. Again, note that b_{XY} is not a regression coefficient because this linear equation does not represent a regression equation. [Equation 1 from Gavrilets and Scheiner (1993a) presents an analogous more general equation where the interaction takes on any form.] Equation 11.1 could be expanded to include terms reflecting the independent influences of X or Y, but here we are interested primarily in interactions and will not include such terms. Fitness is assigned to the genotype here, but the results apply equally well to situations where fitness is assigned to the phenotype, or where the phenotypic value is influenced by the $G{\times}C$ interaction. Such an analysis would simply require the inclusion of a term that describes the relationship between the phenotype and the genotype.

It is also important to note that equation (11.1) is defined for the individual possessing genotype X. An analogous equation could be created with respect to individuals "possessing" Y (e.g., when Y is a genotype of a different individual or species; see below). However, there is no reason to expect *a priori* that the equation defining fitness of an individual with characteristic Y experiencing context X will have the same form as equation (11.1). For example, the equation for fitness associated with Y may or may not contain an interaction component and will almost certainly contain a value for b_{YX} that differs from b_{XY}. These sorts of reciprocal effects, where X is a context for Y and vice versa, are likely in social interactions, where individuals are usually both focals and contexts.

In our example of context being thermal environment, the Y value would be temperature, the X value would be the genotypic value of an individual for thermal tolerance, and the b_{XY} term would represent the strength of the interaction between the factors. It is important to keep in mind that the trait we are calling thermal tolerance is completely dependent on the thermal environment (with a crossing-type relationship). Figure 11.2 shows a plot of fitness as a function of genotype and temperature. Both factors are expressed as deviations from the mean. Figure 11.2 shows that genotypes with positive thermal tolerances do well in higher than average temperatures but very poorly in low temperatures and vice versa (b_{XY} determines the degree of curvature on this surface).

To understand how selection will act on genotype X, we take the covariance between X and fitness (Price 1970), which yields the genetic selection differential (S_X) (following Bohrenstedt and Goldberger 1969):

$$S_X = \text{cov}(X,w) = b_{XY}\bar{Y}V_X + b_{XY}\bar{X}C_{XY} \tag{11.2}$$

where the V_i represents the variance of variable i and C_{ij} represents the covariance of variables i and j. The two terms in this equation capture the first two modes of influence that context specific fitness has on evolution. The first term, $b_{XY}\bar{Y}V_X$ illustrates that *the strength and direction of selection on* X *depend on the mean context.* Selection may change depending on the average abiotic environment experienced, average value of the allelic state at an interacting locus, the average character value expressed in an interacting species like a parasite, or the average trait value of social partners. In our example, we would find that selection on thermal tolerance would change as mean temperature changes. If temperatures were to increase, directional selection would favor genotypes with higher thermal tolerances, and if temperature were to decrease, selection would favor lower thermal tolerances. This example illustrates that, although equation (11.1) shows that fitness is determined by an interaction between thermal tolerance and temperature, we would find only directional selection acting on temperature in a population, although the sign of the selection coefficient would change depending on the mean temperature.

The second term in equation (11.2) captures the first of the two modes that involve the genotype–context covariance; the interaction between X and Y only results in this additional component of selection on X when the two factors covary. That is, when there is a nonrandom distribution of genotypes across contexts, the $G{\times}C$ interaction will have an additional additive contribution to selection. This occurs because covariation between the interacting units makes its effect on fitness appear to be additive. Thus, we see the second mode of influence: *the covariance between the genotypic value (or phenotypic value) and the context determines, in part, net selection on the genotype.* Depending on the nature of the context, this covariance could result from a variety of processes, such as nonrandom distribution of genotypes among host plants due to maternal oviposition site preference, habitat preference of particular genotypes, nonrandom assortment of social partners due to relatedness or preferential assortment, population subdivision and differentiation, linkage disequilibrium between loci, or phenotypic fit to context owing to phenotypic plasticity. In our example, this covariance would represent a nonrandom distribution of thermal tolerance genotypes across different local thermal environments, perhaps due to an associated habitat preference.

Lastly, it is important to understand that the covariance, C_{XY}, can evolve. Whenever selection is nonadditive, the covariance between X and Y affects population mean fitness and can be subject to selection. Taking the expectation of equation (11.1) and assuming that the mean of a_X is zero, population mean fitness is, by definition,

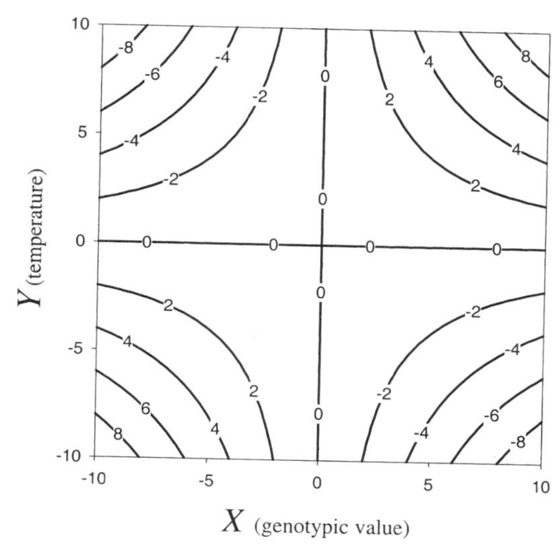

Figure 11.2. Fitness as a function of genotype X and context Y. The coefficient b_{XY} has the value of 0.1 for this surface (see equation 11.1). Contours represent fitness isoclines and are labeled with values. The axes are labeled for the example where X is the genotypic value of thermal tolerance and Y is the thermal environment.

$$\bar{w} = b_{XY}\,(\overline{X}\overline{Y} C_{XY}) \tag{11.3}$$

Note that the first term in parentheses is the product of the means of X and Y, not the mean of the product XY. The mean of the product XY is given by the entire term in parentheses. When both b_{XY} and C_{XY} are both positive or both negative, then population mean fitness is an increasing function of the association between the genotype and context, so that increased values of C_{XY}, are favored by selection. Conversely, when b_{XY} and C_{XY} are of opposite sign, then population mean fitness is a decreasing function of the association between the genotype and context, such that decreased values of C_{XY}, are favored by selection. In both cases, we can consider the covariance "adaptive" because it fits the population to the topography of the selection surface (as defined by b_{XY}), such that genotypes are associated with contexts to which they are best adapted. In our example, it is easy to see that, if genotypes can assort into the thermal environments to which they are best adapted, population mean fitness would be increased because there would be fewer "bad" combinations and more "good" combinations. As a result, we expect that selection would favor any processes that would allow genotypes to be associated with the thermal environment to which they are adapted.

In order to predict how selection will modify the covariance (C_{XY}), we can examine an equation that approximates the change in the covariance between X and Y due to selection ($\Delta C_{XY(t)}$) within generation t (based on Tallis and Leppard 1988). This simplified equation does not give the exact change in the covariance but can be used to illustrate how selection acts to change the covariance:

$$\Delta C_{XY(t)} = C^{*}_{XY(t)} - C_{XY(t)} = b_{XY} C^{2}_{XY(t)} + b_{XY} V_X V_Y \tag{11.4}$$

In this expression, the asterisk (*) indicates the value of the parameter C_{XY} after selection has occurred. Equation (11.4) shows that, as a result of the interaction between the genotypic value and the contextual value for fitness (as measured by b_{XY}), selection alters the covariance between these two factors and can build the covariance even if they are initially randomly distributed. This process is analogous to the process by which epistatic selection builds nonrandom associations, or linkage disequilibrium, between alleles at different loci. Equations (11.3) and (11.4) show the third mode by which interactions affect evolution: *factors that contribute to the adaptive covariance of genotype and context, and its cross-generational stability (i.e., the "heritability" of* C_{XY}*), are favored by selection*. When the sign of the covariance matches the sign of the interaction, selection acts to build a positive covariance because it provides an adaptive fit between genotype and context, but when the sign of covariance opposes the sign of the interaction, selection builds a negative covariance.

Changes in the covariance between X and Y wrought by selection persist across generations in proportion to the heritability of C_{XY}. The covariance between X and Y in the next generation (i.e., after a generation of re-assortment) can be expressed as

$$C_{XY(t+1)} = C^*_{XY(t)} - rC^*_{XY(t)} \quad (11.5)$$

where t represents the current generation (where selection occurs) and $t + 1$ represents the next generation. The parameter r represents the proportional loss in the covariance due to forces that impose random assortment of genes and context. These randomizing forces differ depending upon the level of the hierarchy of context. For example, when X and Y represent loci, then equations (11.4) and (11.5) can be used to model changes across generations in linkage disequilibrium. In the two-locus case, r represents the rate of recombination between loci, and b_{XY} represents a measure of epistasis for fitness between these factors. When X and Y represent heritable characters, then these equations can be used to model changes across generations in the genetic covariance between traits. This expression shows that, in every generation, selection builds the covariance and forces akin to recombination (e.g., random dispersal in the environment or migration among demes) destroy the covariance. Whether selection favors an increase in the covariance between genes and context or an increase in the randomizing forces that diminish it depends upon the sign of the product of b_{XY} and C_{XY}.

Obviously, the relative efficiency of various processes to build, maintain, or destroy associations between genotypes and context will depend largely on the form of context being considered. For example, evolution of linkage disequilibrium for loci located close to one another on the same chromosome is relatively easy because recombination between them is unlikely, whereas nonrandom association between genotypes in different species may be difficult to maintain because vertical transmission of the association may not be possible. In our example, equation (11.4) would tell us how selection, by eliminating low fitness combinations of thermal tolerances and thermal environments, would modify the covariance of genotype and environment (figure 11.3). Equation (11.5) would tell us what the association between genotype and environment would be in the next generation. If individuals disperse and randomly settle into thermal environments, then r would be 1 and, every generation, dispersal would remove the covariance built by selection. However, if individuals remained in the area where their parents were and thermal environments are somewhat stable, then the covariance may be maintained across generations.

In our discussion of the implications of the above equations, we do not focus on plasticity *per se*, but rather on general evolutionary processes that involve the evolution of

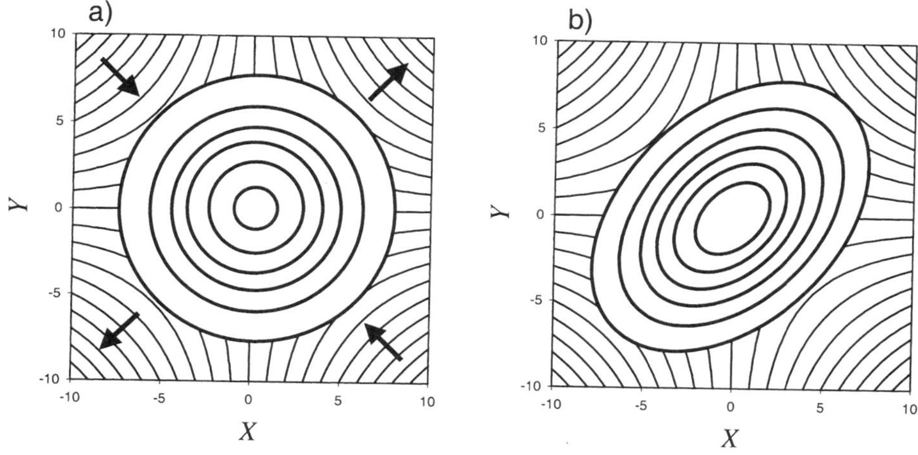

Figure 11.3. Populations lying on the fitness surface from figure 11.1 before and after selection. Contours of the surface represent fitness isoclines. The joint distribution of *X* and *Y* values for all individuals in the population are represented by the concentric rings or ovals, where each ring represents an isocline of probability density for *X–Y* combinations. (a) The population with no covariance between *X* and *Y* before selection has occurred. The arrows indicate how selection is expected to deform the distribution. The parts of the distribution that lie on the "downhill" regions of the saddle are diminished by selection, whereas the parts lying on the "uphill" regions are favored by selection. (b) The population after selection, showing the positive covariance between *X* and *Y* that was built by selection.

the genotype–context covariance. However, in keeping with the theme of this volume, we discuss the potential role that plasticity (or canalization when considered as part of the plasticity continuum) can play in each of these cases. In all cases, adaptive plasticity can be viewed as a phenomenon that, by creating a particular context-dependent relationship between the genotype and phenotype, leads to adaptive phenotype–context covariances. In our example, plasticity would allow a genotype to express a thermal tolerance that matches the thermal environment, leading to an adaptive covariance between thermal tolerance and thermal environment.

Context and Coadaptation: Evolution When the Context Is Conspecific

The affect of interaction effects between loci within the same genome of a species (epistasis) is well established and will not be discussed in any detail here (see Wolf et al. 2000). Rather, in this section, we focus specifically on situations where the context consists of conspecifics, that is, where the interaction between genes and context occurs via a social interaction (Wade 1980, 1998; Wolf and Brodie 1998; Wolf 2000, 2001). Further, we will focus on the particular case where the context itself is influenced by genetic effects and indirect genetic effects occur (Moore et al. 1997; Wolf et al. 1998). Indirect genetic effects (IGEs) are influences on the phenotype of one individual that are due to genes expressed in another individual. These effects contrast with direct genetic effects (DGEs), which occur when genes possessed by an individual influence that individual's phenotype. Models and studies that focus on IGEs provide a useful framework because they explicitly address the situation where the phenotype of one individual is influenced by an environment that contains genes. Most studies of IGEs focus on the situation where effects of IGEs and DGEs on the phenotype are strictly additive (Moore et al. 1997; Wolf 2000). This additivity implies no genotype–context interaction (i.e., no DGE×IGE) such that the influence of the social environment on the phenotype of an individual does not depend on the genotype of that individual. However, it appears likely that nonadditivity occurs, and genotype–context interaction is an important component of IGEs (Wolf 2000).

Whenever the phenotypic effect of a gene in one individual depends upon the genes possessed by other individuals, genotype–genotype (i.e., among genotype/genome or $G \times G$) epistasis can be said to occur (Wade 1998; Wolf and Brodie 1998; Wolf 2000). $G \times G$ epistasis is likely to be an extremely common and potentially important type of genotype–context interaction given that genetically based environments provided by conspecifics are ubiquitous and often have major effects on fitness. Unlike intragenomic epistasis, which occurs because of physiological interactions between loci within individuals, $G \times G$ epistasis results from interactions among loci composing the genotypes of different individuals. As a result, this sort of interaction exists at the level of a population of individuals, not at the level of the individual itself (Wolf 2000). $G \times G$ epistasis has been seen as important in the evolution of competitive and cannibalistic interactions (Griffing 1977; Goodnight 1991; Wade 1988, 1998, 2000; Wolf 2000), genetic correlations between maternal and offspring characters (Wolf and Brodie 1998), and traits mediating sexual interactions (Johnson 1982; Meffert 1995, 2000; Clark et al. 1999).

To illustrate the effects that these sorts of interaction effects have on evolution, consider a general situation in which individuals possess genes that determine their competitive ability, but the fitness of an individual with a particular competitive ability is determined by the competitive phenotype of the individual's social partners (also genetically based). This situation has been described in a number of plant species (e.g., Gustafson 1953; Roy 1960; Wiebe et al. 1963; Griffing 1989) and insect species (e.g., Park 1936; Lewontin 1955; Lewontin and Matsuo 1963; Weisbrot 1966; Wade 1988, 2000; Miranda et al. 1991), where the fitness of a genotype is largely dependent upon the genotype of the individuals with whom the individual interaction (i.e., the social context). This situation could be described by equation (11.1), where the coefficient b_{XY} can be viewed as a coefficient describing how the competitive environment influences fitness. Using this simple example we can now ask how this sort of context dependence influences the evolution of competitive ability (or any phenotype showing similar context dependence), and what factors determine the evolutionary outcome.

We can start by examining the implications of equation (11.2), which says that the way in which selection acts on a particular competitive phenotype is determined by the mean competitive environment experienced in a population, and that the covariance between genes and the competitive environment will influence how selection acts. What does this mean in biological terms? First, it means that the evolutionary fate of a gene influencing competitive ability will depend largely on the competitive context impacting a gene. This context dependence can be particularly important in a metapopulation, where drift might create variance in competitive contexts across demes, and thus selection within each deme can favor a different competitive phenotype (Wade 1988, 2000). In this case, the selection coefficient defined by equation (11.2) would not be a single value, but would be a distribution of values across demes, and the values of $\bar{X}$ and $\bar{Y}$ would be expected values, sampled from the metapopulation distribution. The among-deme variance in selection coefficients would thus depend on the forces that contribute to the variation in the mean (i.e., standard error of the mean) genetic and context values across demes, such as drift and selection. The probability that different demes would evolve in different directions, leading to population differentiation, could be calculated using equation (11.2) under the assumptions that (1) the means were sampled from a normal probability density function, with (2) a mean at the grand mean of the metapopulation and (3) a standard deviation of the sampling distribution equal to the standard error of the mean. If demes were able to diverge as a result of this selection, the social incompatibility of individuals from one deme with the competitive environment in another deme could lead to further genetic differentiation of demes.

Because the mean context (e.g., competitive environment) can also evolve, the competitive phenotypes or genotypes that are favored by selection will evolve as the mean context evolves, resulting in concerted co- or counteradaptation of genes and their environments. In this situation, a population may evolve adaptations to simply keep up with the degrading social competitive environment, in the manner of the so-called intraspecific red queen (Rice and Holland 1997) or a treadmill environment (Dickerson 1955). The evolution of the mean biotic environment has been examined empirically using artificial selection experiments in experimental populations by S. Swenson et al. (2000) and W. Swenson et al. (2000).

The other major factor to consider in the coadaptation of genes and social context is the genetic covariance between genes and context. This covariance can arise from many factors, each of which influences selection. This covariance can arise if individuals assort and interact based on their phenotype (cf. Wilson 1980), when related individuals interact (cf. Wade 1980), or when demes in a metapopulation have diverged in allele frequencies or inbreeding coefficients (which can create a covariance between genes and context at the level of the metapopulation). Selection will also alter the genetic covariance because it favors individuals that have combinations of genes and contexts that perform well together (i.e., intermix well). Evolution may also lead to some form of plasticity that allows a genotype to create a phenotype that does well in a particular social context, leading to an adaptive covariance of phenotype and competitive environment (thereby leading to a positive covariance of the genotype and fitness). Plastic responses to the social competitive environment are well known in plants, where the shade avoidance response to conspecifics shows that the expression of characters related to size and shape of plant structures are strongly modulated by the light environment provided by conspecifics (chapter 10). In animals, plastic responses to the social environment have been seen for many behaviors such as cases where males modify their mating displays to adapt to the preferences expressed by females (e.g., Aragaki and Meffert 1998; see below).

Consider the example of individuals that develop in groups, wherein developmental success is influenced by the combination of an individual's genes and its social context [as it is in *Drosophila* (e.g., Lewontin 1955; Miranda et al. 1991)]. Selection will establish a covariance between genes and context because groups that have gene–context combinations favored by selection will have higher fitness (i.e., if we look across groups we will find that there is a net positive covariance). If individuals develop with unrelated individuals and remix as adults that randomly deposit eggs to produce the next generation, then this covariance will not persist across generations [i.e., the parameter r in equation (11.5) equals 1, meaning that all covariance is broken down by reassortment]. On the other hand, if individuals remain in a local area to produce the next generation, then the covariance can build up, and local populations can show coadaptation between genes and context (i.e., $r < 1$ because individuals do not assort completely randomly). Relatedness between interactants also provides a mechanism by which the selected covariance between genes and context can persist across generations because individuals share genes with their context (Wolf and Brodie 1998).

The coadaptation (or counteradaptation) of genes and context is seen in experimental studies of male–female interactions mediated by seminal proteins in *Drosophila*. Rice (1996) and Clark et al. (1999) have demonstrated that male success in sperm competition is dependent upon the genotype of his mate (the context) and that female fitness is determined by her response to male seminal proteins that can be toxic to females. In this case, males and females provide contexts for one another, and fitness of each is strongly influenced by the genotype combination of mates. Rice (1996; see also Rice and Holland 1997) has demonstrated that this context dependence leads to constant reciprocal evolution of male and female characters, leading to the evolution of counteradapted gene complexes in populations. By experimentally arresting the evolution of the context (in this case, the female adaptation to seminal proteins), thereby holding the mean context constant, the male seminal characters showed rapid directional selection, as would be expected. This sort of process could contribute to speciation as genes and context coevolve to produce population differentiation (W. R. Rice 1998). In effect, the evolution of the context (female adaptations to male seminal proteins) results in a situation where reproductive isolation evolves when males from one population are not adapted to the environment provided by females from a different population, and vice versa. Similar results have been demonstrated for houseflies, where the outcome of male–female courtship interactions shows a dependence upon the genotype combination of mates (Meffert 1995; Meffert et al. 1999). Selection can also lead to plasticity in the male phenotype, such that male genotypes create phenotypes that are adapted to the female provided environment. For example, Aragaki and Meffert (1998) have demonstrated that male houseflies appear to modify their mating behavior to match the specific preferences of different females. This sort of plasticity in mating display is expected to reduce the opportunity for mate choice and prezygotic reproductive isolation. The conditions under which selection leads to plasticity for social behavior should match those conditions predicted to favor the evolution of plasticity under other conditions (Gavrilets and Scheiner 1993a), although the nature and distribution of environments will differ in important ways.

Wolf and Brodie (1998) modeled the evolution of the covariance between genes and context by explicitly addressing interactions between early expressed genes and genes that influence the maternally provided environment (see also Wolf 2001). Using a model that is analogous to the one presented here, we examined how selection on offspring could lead to the buildup of adaptive covariance between offspring genes and maternal genes that produce the maternal environment (e.g., uterine environment, lactate quality, or egg/seed provisioning), which result in maternal effects on progeny phenotypes or fitness. This analysis was motivated by the observation that genetic covariances between maternal and offspring characters are common (see table 7.5 in Roff 1997), and as with "ordinary" genetic covariances, these covariances could reveal functional relationships (i.e., gene–context interactions). A number of experimental analyses have demonstrated a strong interaction between offspring and maternal genotypes, further supporting the interpretation of functional relationships (Wolf and Brodie 1998; Wolf 2000). For example, Cowley et al. (1989; Cowley 1991) have demonstrated that the development rate of mice of a given genotype is largely dependent upon the genotype of the maternal uterine environment. In the Wolf and Brodie model, the genetic covariance between genes and context evolves because selection on offspring favors individuals that have genotypes that function best in the maternally provided environment. Because of the association between the individual's genotype and their mother's genotype, owing to relatedness, this selection is able to act on combinations of direct effect genes and context genes simultaneously. Selection leads to a buildup of linkage disequilibrium and also favors genes that have pleiotropic effects on the maternal and offspring characters, which leads to the establishment of genetic covariance between genes and context because both are the result of the same loci. The evolution of pleiotropy can be a particularly powerful means of achieving coadaptation between genes and contexts because recombinatorial forces cannot erode the covariance.

The coadaptation of early expressed genes and maternal context is clear when one considers an example such as the coadaptation of oviposition and larval performance genes (where performance is a measure of developmental success in the oviposition environment). In many phytophagous insects, the mother provides a developmental environment for her offspring when she chooses a suitable oviposition site (Futuyma 1983; Thompson and Pellmyr 1991). Characteristics of the oviposition environment, such as the content of plant secondary defensive compounds, can be a major determinant of early developmental success (e.g., Leather 1985). The ability of larvae to develop on a given host plant can be determined by the individual's genotype at loci that influence the ability to detoxify the plant's defensive compounds. Selection favors the coinheritance of genes that influence the oviposition choices of females and larval performance, leading to coadapted complexes that are often associated with strong genetic covariances between performance and preference within populations (Via 1986; Craig et al. 1989; Jaenike 1990; Fox 1993; Thomas and Singer 1998; but see Wade 1994). These sorts of associations between preference and performance have been considered as factors that can help maintain genetic variation for host-plant use within species (Mitter and Futuyma 1983; Rausher 1983; Fox 1993; Thomas and Singer 1998). This process has also seen as a major mode through which populations can differentiate, which can then lead to speciation (Rausher 1983; Bush 1975; Feder and Filchak 1999; Filchak et al. 1999, 2000) when postzygotic isolation of populations evolves and offspring are not adapted to the maternally provided environment.

An alternative to the evolution of genetic coordination (i.e., adaptive genetic correlation) between maternal and offspring genotypes is the evolution of offspring or maternal plasticity. In this case, incompatibilities between maternal environments and offspring genotypes could be diminished or avoided if progeny show plastic responses to the maternally provided environment or mothers show plasticity in the maternally provided environment. In the example of oviposition sites, progeny could show norms of reaction to oviposition environments, which might allow their phenotype to match the environmental requirements, leading to the evolution of plasticity. The degree to which individuals show plastic responses to maternally provided environments is not clear, although a related phenomenon has been reported by Olvido and Mousseau (1995). However, given the significant fitness effects associated with maternal environments, it appears to be a worthwhile avenue of investigation. In contrast, there is ample evidence that mothers show plasticity in the expression of the maternally provided environment, such that mothers can create environments that increase the fitness of their progeny in certain environments (reviewed by Fox and Mousseau 1998). For example, Fox et al. (1997, 1999) have demonstrated that female seed beetles (*Stator limbatus*) adjust egg size in response to host species (where hosts are seeds from different species). They referred to this maternal plasticity as an adaptive maternal effect because egg size appeared to be adjusted to match larval requirements on the two different hosts examined. In this system, females lay large eggs on the host that produces seeds with a coat that is difficult to penetrate, presumably because a large egg is required to produce larvae that can penetrate the seed, whereas they lay small eggs on the host with the seeds that are easy to penetrate. Thus, mothers can produce a positive covariance between progeny phenotype and host context via plasticity, leading to positive selection on this form of cross-generational plasticity.

Context and Coevolution: Evolution When the Context Is Heterospecific

When the context of interest is a separate species, another level of complexity must be considered. As with the case of contexts that are conspecifics, the contexts in this case can have a genetic basis and therefore the potential to evolve themselves. In reciprocal interactions each species can be considered both a context and a focal genotype. However, the context for each focal genotype is a separate genome, so different rules apply. Most important, each species has its own individual fitness function, which may be context dependent although not necessarily in the same way. That is, a given combination of variables in two species may have opposing fitness consequences for the two players, X and Y. Also, the covariances of importance to fitness are no longer within the genome of a single population, so covariances cannot be built through within genome phenomena such as linkage disequilibrium, and are not degraded by recombination (Wade 1988; Goodnight 1991). We must look to other analogous forces to understand how covariances between genotype and context evolve and influence fitness.

Coevolutionary studies tend to emphasize the comparative approach to species interactions (Ehrlich and Raven 1964; Wilson 1980; Thompson 1994). Matching topologies of victims and exploiters are usually taken as evidence of coevolution, and modern applications use rigorous phylogenetic methodology to evaluate phenotypic evolution based on character states and the topology of relationships of the taxa of interest. Patterns of parallel adaptive radiation that are expected to result from interactions are evaluated by examining complementary character state evolution such as that exemplified by the form of gastropod shells and the crushing chelae of molluscivorous crabs in Lake Tanganyika (West et al. 1991) and other gastropod taxa studied over extended periods of geological time (e.g., Vermeij 1994). Because this approach focuses on species (or clade) level characteristics, it implicitly emphasizes the importance of the context mean rather than the covariance between genotype and context. In many ways, this is a fair and classic view of coevolution, because it focuses attention on the average features of the selective environment that are likely to drive the lockstep phenotypic matching in interacting species. The context mean is also going to be important in species interactions because it determines the direction and pace of evolution, and differences in mean contexts across populations can produce differentiation of populations (e.g., Goodnight 1991) resulting in speciation or, in some cases, cospeciation (Thompson 1997).

Although invaluable, the comparative approach fails to consider the interactions among individuals that drive the process of coevolution, and, in doing so, misses some features of the $G{\times}C$ that can influence evolutionary dynamics. Thompson's Geographic Mosaic Theory (GMT) recaptures some of this complexity by emphasizing the importance of finer scale interactions (Thompson 1994, 1999a,b,c). The GMT is based on the observation that species are typically divided into genetically differentiated subpopulations. These subpopulations may not experience the same kinds of selection pressures and therefore may obtain different phenotypic values, but populations are not completely isolated, and so traits "remix." The end result is a complicated and specific process that causes species-level traits to be geographically variable, and therefore results in variation in the coevolutionary "context" among populations.

Under the GMT, then, the mean context will vary among populations, so in most systems we might not expect to see broad evidence of coevolution if we focus on the scale of species-level traits. Similarly, if we concentrate on the mean context within subpopulations, we again miss much of the interesting dynamic. As with context dependence at lower levels of hierarchical organization, covariances among focal genotype and context play a crucial role in the dynamics of coevolution.

What constitutes the covariances between genotype and context in interspecific interactions? The answer becomes clearer if we consider genotypes of one species, say, a herbivore, as the genotype (G) and genotypes of the other, say, a host plant, as the context (C). In most interspecific interactions, we could, of course, reverse this designation because of the reciprocal (although not necessarily equivalent) nature of context dependence—this is why we need two sets of equations analogous to equations (11.1–11.5). To derive this second set of equations, describing selection in the second species (assuming that the first set of equations apply to the first species), we would need to simply swap the Xs and Ys in the first set of equations. However, note that we have no *a priori* reason to believe that the values of the coefficients will be similar in the two equations [or that the form of the equations will even be the same, because one species can experience linear selection and the other experience nonlinear (e.g., stabilizing) selection, or possibly, one species experiences selection whereas the other does not]. However, both equations will contain the same covariance because C_{XY} is equal to C_{YX}. Thus, any covariance between genotypes of interacting species can potentially influence the evolution of both species, even if it is generated simply through the behavior of or selection on one of the species alone. Such associations might exist within subpopulations, or among subpopulations as predicted by the GMT. The way in which selection acts on either of the two species will then be determined by the sign of the coefficients, the sign of the covariance, and the sign of the means.

Although the same covariance applies to both species, selection may favor a positive covariance for one species and a negative or no covariance for the other species. This antagonism between species would result in selection favoring factors that develop or maintain a covariance, whereas in the other species, selection favors factors that reduce or eliminate the covariance. For example, in a mutualism that requires some matching of characteristics of the two species in order for the combination of traits in the two species to function together, selection favors mechanisms that either allow the two species to assort in order to establish the adaptive association, or that prevents the loss of the covariance built by selection. The latter would include coinheritance, as might occur in a system such as insects and their bacterial symbionts (Moran 1988; Clark et al. 2000), or possibly reduced migration, allowing for easy reestablishment of the coadapted combination. Either species (or both) might also show adaptive plasticity that leads to the matching of phenotypes of the two species. On the other hand, if the interaction is not a mutualism but a parasite–host relationship where particular genotype combinations benefit the parasite but hurt the host, selection will favor mechanisms that allow the host population to break the covariance of X and Y (e.g., genetic recombination and sexual reproduction; Lively 1996, 1999b), whereas it favors mechanisms that allows Y to maintain the adaptive covariance (e.g., vertical transmission). Of course, plasticity could also play a role in antagonistic interactions by allowing one or both species to produce phenotypes that counter or mediate traits expressed in the other species.

Using this framework we can examine the conditions that favor the coadaptation of species versus those that result in counteradaptation. Although we will assume that genotypes lead to a single phenotype, the conclusions of this discussion can apply equally well to cases where either species shows plasticity. When plasticity can occur, we might expect that selection favoring positive or negative covariances would favor plasticity in either species to achieve the covariance. First consider the situation where both species experience directional selection provided by the other species. Assuming that both means are of the same sign and we start with no covariance, selection will act in the same direction in both species whenever b_{XY} and b_{YX} are of the same sign, meaning that combinations of genotypes that favor one species will also favor the other. Assuming heritability of both traits, this selection will drive both means to increasing values, which will result in a process of self-reinforcing directional selection. This condition can be viewed as a process of coadaptation between the species because each provides reciprocal selection favoring elaboration of coadapted characters in the two species. If b_{XY} and b_{YX} are of opposite sign, then selection acts in opposite directions in the two species and, as the two means evolve, they may enter an evolutionary cycle where they each swing from positive to negative. This latter scenario can be viewed as a process of counteradaptation.

When selection acts on both species, it will also act on the covariance between the species. Equation (11.5) describes the change in the covariance when selection acts on X, but we now need to consider selection on X and Y, because both alter the covariance. The net change in the covariance of X and Y will be the sum of the changes in the covariance produced by selection on each of the two species and the resulting covariance will reflect balance of the strength and direction of selection in the two species. We can approximate the change in the covariance as

$$\Delta C_{XY(t)} = [C^2_{XY(t)} b_{XY} + V_X V_Y b_{XY}] + [C^2_{XY(t)} b_{YX} + V_X V_Y b_{YX}] \quad (11.6)$$

When the signs of b_{XY} and b_{YX} are the same (and thus combinations that favor one species favor the other), selection will act to build positive covariances and the two species will show coadaptation due to selection. However, when the signs of these coefficients are opposite, the net change will be closer to zero (depending on the net balance), and selection will not build coadapted combinations of characters between the two species.

The opportunity for covariances among genotype or phenotype and context is present in all species interactions and has been measured in the laboratory (Goodnight 1991). Unfortunately, evidence of such covariances in natural populations is still indirect (e.g., Benkman 1999; Brodie and Brodie 1999; Lively 1999b), owing to a combination of logistical difficulties and an underappreciation of their evolutionary importance. Within-population covariances can arise through any type of behavioral sorting of genotype with context. For example, phenotypic and genetic variation for resistance to tetrodotoxin exists within populations of garter snakes that feed on the toxic newt *Taricha granulosa*. Preliminary laboratory studies indicate that individual snakes with low resistance reject individual prey with high toxin levels, whereas highly resistant individuals readily consume the same prey (E. D. Brodie, Jr., unpublished data). The result of this type of interaction in the wild would be a covariance between predator genotype and prey phenotype (and presumably genotype). Likewise, insects often sample a variety of host plants before selecting an individual upon which to oviposit. Genetic variation for host plant preference is present among populations and species of many insects, including western swallowtails of the genus *Papilio* (Thompson and Pellmyr 1991; Thompson 1994). Variation of this sort within populations will lead to nonrandom interactions of larval herbivore genotype and plant (i.e., context) genotype. A covariance between a species and its host might also arise from plasticity in either of the interacting species. The resulting covariance can affect evolutionary processes in ways that are similar to a genetic covariance because the origin of the covariance is not important when considering selection that results (equation 11.2). However, the evolutionary consequences (i.e., resulting responses to that selection) are likely to depend largely on the genetic architectures of the covarying traits in the two species. Also note that, although such covariances may arise from the actions of only one of the two interacting species, the covariance has evolutionary consequences for both species regardless of the cause.

Even in cases where covariances are weak or absent, we expect the fitness consequences of interactions to generate covariance among genotype and context, especially among populations. Selection can create covariances within populations as long as genetic variation is maintained, but such covariances can be relatively short-lived because directional selection in a population can lead to fixation of selected alleles. When different subpopulations experience different contexts, selection might act in different directions in different subpopulations. This could lead to fixation of different alleles in different populations. Thus, the covariances generated by within population processes will be translated into among population covariances. Similar processes might exist in any interacting system, and this is the crux of the GMT. Because any number of ecological factors could cause different genotypes to spread in different subpopulations, we expect to see a geographic mosaic of character states or genotypic values across the range of coevolving species (Thompson 1994, 1999a,b,c).

Conclusions

In our analysis, we have examined the connection between evolutionary consequences of context-dependent fitness of a genotype across a hierarchy of levels of organization in biological systems. We have demonstrated that there is a set of modes through which the interaction affects evolutionary processes, which have analogous effects at each level of the hierarchy. The impact of the interaction on selection and evolution will depend on the average context, factors that cause changes in the mean context, associations of the genotype and context, and forces that maintain or dissipate the associations of genotypes and contexts. When the context is genetic, these processes take on special meaning because the genotype and its genetic context can coevolve. Within species, this genotype–context coevolution can influence the evolution of linkage disequilibrium, pleiotropy, plasticity and canalization. Between species, genotype–context coevolution determines whether coadaptation or counteradaptation occurs. When evolution leads to the concerted evolution of multi-locus systems, it can lead to the evolution of genetic correlations and evolution of population differentiation. When the context is a different species, we must consider evolution in each of the genomes, although the same basic processes will determine how genes and their context coevolve and covary.

12

The Role of Phenotypic Plasticity in Diversification

CARL D. SCHLICHTING

Although phenotypic plasticity has gained acceptance as an important means of adaptation to heterogeneous environments, its role in population differentiation, speciation, and other macroevolutionary events is poorly understood. My goal here is to persuade you, first, that phenotypic plasticity actually plays *any* part in the diversification of taxa and, further, that its role may be greatly underestimated. I discuss three specific areas where the possession of plasticity may facilitate evolutionary change. Its greatest contribution appears to be in promoting the occupation of new niches by means of the production of phenotypic alternatives, paving the way for genetic differentiation. Another key role may be to shield genetic diversity from the discerning eye of selection. Finally, it may enhance the long-term survival of taxa, via species selection.

Genetic Assimilation

The process now referred to as genetic assimilation (GA) has been something of a phoenix. The basic premise of GA is that a phenotypic change, initially strictly plastic (i.e., environmentally induced), over time can be incorporated into the genetically determined phenotypic repertoire. It provides a mechanism whereby seemingly Lamarckian shifts can be explained with Darwinian principles. Originally hailed as a means of promoting evolutionary change, it was later considered to be an equally potent buffer against diversification. Resurrected as an evolutionary catalyst independently in Russia and England, it was subsequently dismissed as being of minor importance. Now a markedly diverse group of scientists has called for its revival again as a key factor in evolutionary change (e.g., Ho and Saunders 1979; Rachootin and Thomson 1981; Hinton and Nowlan

1987; Thompson 1991; Wolpert 1994; Rollo 1994; Jablonka and Szathmary 1995; Sarà 1996a,b).

The theory of GA has been the major vehicle for incorporating plasticity as an evolutionary catalyst (Waddington 1942, 1961; Schmalhausen 1949). There are several excellent sources for detailed historical perspectives on GA (e.g., Wcislo 1989; Gottlieb 1992; Robinson and Dukas 1999), so the background I provide here is short. First introduced as "A New Factor in Evolution" by Baldwin (1896), it was quickly joined by similar concepts developed by Lloyd Morgan (1896) and Osborn (1897b). Their ideas never really caught on: a mechanism for microevolutionary change was of minor interest at a time when the major battle over the form of genetic control of phenotypes was brewing between Mendelians and biometricians (Provine 1971). The role of plasticity in evolution was further limited by Wright (1931), who simultaneously praised plasticity as a major object of microevolutionary change and damned its role in macroevolution, by pointing out that individual adaptability "is not only of the greatest significance as a factor of evolution in damping the effects of selection . . . but is itself perhaps the chief object of selection" (p. 147). The view of plasticity as an inhibitor of evolutionary change has been reiterated over the years (Stearns 1982; Schaal and Leverich 1987; Levin 1988).

The Heretics

A decade after Wright's proclamation, both Schmalhausen (1949) and Waddington (1942) were independently developing new and more robust conceptual versions of GA (chapter 2). For each, GA played a central role in processes of evolutionary change. Figure 12.1 depicts the basic scenario they envisioned. A new environmental factor induces a plastic response that has some advantage under those new conditions (e.g., relative to a lack of response). This can occur in the absence of genetic variation for that trait in the population (Rollo 1994; Schlichting and Pigliucci 1998). This purely plastic response may enable the population to persist, allowing subsequent alterations in the reaction norm toward better adaptations, because of new mutations or recombination. Both Waddington and Schmalhausen suggested that, if this new environmental state persists, selection would fix the character state—that is, the plasticity would be eliminated. This step is the culmination of GA: an initially facultative phenotypic change subsequently becomes constitutively expressed by means of conventional natural selection. This could occur for several reasons: (1) there is a cost to maintaining plasticity in the absence of the original environment, (2) random mutation may disable the now hidden plastic response, or (3) there may be selection for canalization such that the original environmental conditions would no longer elicit that phenotype.

Waddington and colleagues experimentally demonstrated GA for a variety of traits in *Drosophila* (presence or absence of crossveins in the wings, dumpy wings, enlarged anal papillae in the larvae). For example, after induction with a high temperature treatment of the pupa, the frequency of individuals with broken crossveins (*crossveinless*) after heat shock was greater than 90% after about 15 generations, and after selection *crossveinless* was expressed even without heat shock at 100% in some lines (Waddington 1953). However, the magnitude of change that could be incorporated in this way was demonstrated in spectacular fashion by Waddington's experimental GA of *bithorax*

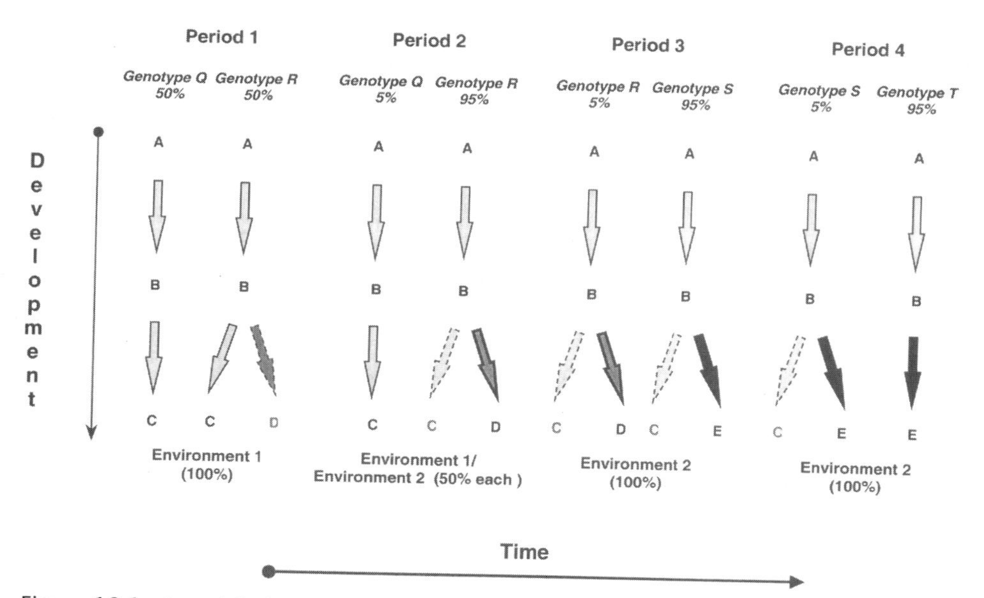

Figure 12.1. A model of the evolutionary progression of GA. The development of genotypes is shown at four periods in evolutionary time. The optimal phenotype in environment 1 is C; in environment 2 it is E. In environment 1, development proceeds from phenotype A, through phenotype B to phenotype C. Although it is only necessary that there be one genotype with the potential for phenotype D, I have represented two genotypes, the nonplastic *Q* and the plastic *R*, that both produce phenotype C, but only *R* has an unexpressed capacity to produce D. In time period 2, the environment fluctuates between conditions 1 and 2. Because environment 2 is equally frequent, the ability of genotype *R* to express D is favored because it is closer to the optimum, phenotype E. In time period 3, environment 2 is constant and mutations arise producing a genotype *S* that expresses phenotype E. Note that *S* still retains the capacity to produce C in environment 1. A final stage (period 4) may be attained, when a nonplastic genotype *T* replaces the plastic genotype *S*. Dashed arrows represent an unexpressed, or relatively unfit phenotype. Darkness of arrows indicates progressive development of the phenotype which is being genetically assimilated.

phenotypes (Waddington 1956). In these experiments, Waddington found that a slight enlargement of the halteres could be magnified into full expression of two pairs of wings and eventually almost fully genetically assimilated—more than 80% in the absence of the original environmental stimulus (Waddington 1956). Although the stimulus in that experiment (ether) was extreme and not likely to be a natural phenomenon, clearly Waddington believed that GA could produce macroevolutionary change (Waddington 1975). However, despite the persistent efforts of him and his students to promote the importance of GA (Waddington 1961), it ultimately was perceived as playing no more than a minor role (Simpson 1953; Williams 1966).

Revival

The early 1980s saw the beginnings of the current renaissance in several outlying camps. The descendants of Waddington continued to work on GA (Ho et al. 1983a,b; te Velde et al. 1987; te Velde and Scharloo 1988). Rachootin and Thomson (1981), recognizing the failure of the framers of the synthesis to adequately incorporate development, explicitly argued for the importance of understanding epigenetic processes. In this context, they considered that changes in development via GA, even if rare, might promote evolutionary change by opening new adaptive zones.

Matsuda (1982) constructed a scenario whereby an environmental response, such as the facultative neoteny of *Ambystoma tigrinum* in response to cold temperature, could be genetically assimilated by means of mutations affecting thyroid hormone secretion or sensitivity, as in the constitutive neoteny of *Ambystoma mexicanum*. Matsuda argued that GA could be the basis for the evolution of neoteny in other salamanders, as well as in other examples of morphological change mediated by hormones (e.g., ecdysteroids and juvenile hormone in insects). Such changes, for example, have been investigated in several species of wing dimorphic insects (Fairbairn and Yadlowski 1997; Zera and Denno 1997; Zera et al. 1998a,b; Nijhout 1999a).

An in-depth study of the nymphalid butterfly *Precis coenia* examined both environmental and genetic influences on seasonal color polyphenism. Rountree and Nijhout (1995a,b) found that the shift from a beige color in the spring and summer to a dark red brown in the autumn is mediated by ecdysteroid hormones. The effectiveness of these hormones, however, is confined to a period between 28 and 48 hours after pupation, and produces beige butterflies. Under short-day conditions, such as in autumn, ecdysteroid levels rise only after about 60 hours after pupation, leading to production of reddish brown adults. They characterized a recessive mutant that constitutively produces reddish brown adults, due not to an alteration of the ecdysteroids but to a change in physiology after the hormone-sensitive period (Rountree and Nijhout 1995b). As these examples show, changes in timing of production or in sensitivity of cells or tissues to morphogenetic hormones might result in manifold changes in phenotypes (Harvell 1994), some having macroevolutionary implications. Nijhout (1999a) reviews the details of the hormonal control of polyphenism in insects.

West-Eberhard provided detailed arguments for the importance of polyphenism or behavioral shifts as precursors of evolutionary change (West-Eberhard 1986, 1989, 1992, 2003). Although none of her publications specifically mentions GA, from the descriptions it is clearly the primary process of evolutionary change. Here, as in Matsuda (1982), the spotlight is on plasticity via developmental conversion: discrete phenotypes produced by a single genotype in response to environmental alternatives. Many polyphenisms in insects represent examples of this, and West-Eberhard argues that such polyphenism facilitates speciation. For example, a species with two environmentally induced states could be easily converted to separate forms via loss of plasticity in either of the two environments, for example, the Laysan finch, *Ranunculus*, *Wyeomyia*, or *Pieris* (West-Eberhard 1989). Gottlieb (1992) explicitly invokes GA in the process of speciation.

We can create a hypothetical scenario of GA using the geometrid moth *Nemoria arizonaria* (Greene 1989a,b). In this species, caterpillars that hatch early feed upon the inflorescences (catkins) of oak trees and develop into catkin mimics; caterpillars hatching after catkins are gone feed upon the newly emerging leaves, and the higher diet tannin levels shift development toward a twig morphology. (These two morphs were initially identified as separate genera.) This developmental conversion is, reasonably, an evolved, plastic response that minimizes predation. GA could subsequently result in the fixation of either catkin or leaf specialists in isolated populations whose hatch dates were cued earlier or later by temperature or photoperiod. Populations in secondary contact would be partially reproductively isolated based on phenology, and speciation could follow. Many shifts from generalized to specialized feeding behavior may have followed this route.

In her 1989 review, West-Eberhard also promotes the case for a macroevolutionary (speciation and higher) role for behavioral plasticity. First, she notes that behavior is more labile than morphology. Adaptive behavioral plasticity (chapter 8) is more likely to evolve because of availability of cues to stimulate different behaviors (and probably the direct feedback between behavior and reward). Finally, changes in behavior are known in some cases to result in suites of correlated changes in morphology (e.g., Smits et al. 1996b; Thompson 1999d).

Wimberger (1994) examined the consequences of behavioral plasticity for diversification in fishes. His focus is on trophic polymorphisms and he posits the following sequence of events: (1) behavioral plasticity leads to foraging specialization; (2) the morphological differences arising from plasticity tend to reinforce this specialization by altering foraging efficiencies; and (3) this promotes partial reproductive isolation, (4) which can be solidified by philopatry (offspring remain in natal territory), (5) leading ultimately to speciation. Plasticity has been especially prominent in discussions of diversification of cichlid fishes, being invoked, either implicitly or explicitly, numerous times (Rachootin and Thomson 1981; Meyer 1987; Witte et al. 1990; Wimberger 1991, 1992; Sturmbauer and Dallinger 1992; Rollo 1994; Galis and Drucker 1996; Smits et al. 1996a; Galis and Metz 1998).

Adding a new host plant or food type, developing facultative resistance to predators, and markedly altering physiology or morphology to optimize energy capture all are important components of adaptation. However, they do not necessarily lead to diversification in the strict sense of an increase in taxon diversity. It might be as easily argued that simply the gains of distinct morphological or behavioral phenotypes during the origins of polyphenism are events of significant evolutionary importance themselves. For example, the development of castes in insects appears largely to be due to plastic responses to cues early in ontogeny.

There have been several recent discussions in which GA plays a prominent role. Ancel (1999) examined the conditions that would favor the assimilation of a plastic response in changing environments and showed that selection will initially favor broad norms of reaction, with subsequent selection favoring reduction of reaction norm breadth (costs of plasticity are an assumption of the model). The probability of such canalization is strongly influenced by the likelihood of environmental change—a shift in the probability of an environmental transition from 1% to 5% per generation drastically reduced the degree of canalization. Budd (1999) reinterpreted the view of the role of *Hox* gene evolution as a stimulus of morphological evolution, arguing that the reverse pattern is more likely. He suggests a pattern of gradual GA of new *Hox* alleles after the evolution of functionally integrated morphologies.

We have suggested, from first principles, that GA is likely to have played a major role in evolutionary change in general (Schlichting and Pigliucci 1998). We note the following conditions: (1) the low probability of having appropriate mutations available to deal with any particular environmental change (low mutation rate × low probability of advantageous mutation × probability of particular environmental change), (2) the overwhelming likelihood of some form of environmental change, and (3) the pervasiveness of plastic responses (high likelihood of plasticity × low to moderate likelihood that plasticity is in appropriate direction—up vs. down). Given these, the evolution of phenotypes via phenotypic plasticity and GA may be more likely than by means of allelic substitution, the mode of phenotypic evolution that we are taught (or teach!) exclusively. Thus, plasticity in the "right" direction can permit persistence of a genotype during new environmental conditions. If those conditions alternate with the former ones, then plasticity itself will be selectively favored. If the new conditions become predominant, then GA can continue until the new phenotype matches the optimum. We argued (Schlichting and Pigliucci 1998) that appropriate mutations are unlikely to be available. Why should it be any more likely that appropriate plasticity is? This is where the buffering role of plasticity may be important.

Plasticity as a Buffer and Hidden Reaction Norms

Wright (1931) was not wrong about the buffering role of phenotypic plasticity—selection is indeed effectively blind to anything but the phenotype(s) expressed in that particular environmental condition, but with two important caveats. First, a genotype produces a single phenotype only in constant conditions (a coarse-grained environment). Fine-grained environments may elicit several phenotypes during the course of an organism's lifetime, and that individual's total fitness will reflect the summation of effects of selection on those various states. Second, an organism's fitness extends necessarily into its evolutionary future. Thus, the extended fitness of a particular genotype will incorporate the fitness of its offspring as well, illuminating the importance of demic structure (Scheiner 1993a, 1998). In this context, even coarse-grained environmental variation gets averaged in, based on the relative frequencies with which offspring inhabit alternative environments, and whether genotype fitness is enhanced by a plastic reaction norm that adjusts appropriately.

The ability of plasticity to buffer natural selection on a trait will depend on the extent to which there is overlap between the genes that determine the trait and those that determine its plasticity: plasticity of a trait only counteracts direct selection on that trait if there is considerable overlap in their genetic control. A number of studies have examined the relationship between trait evolution and trait plasticity and have found evidence ranging from some to substantial independence of the two (Stearns 1983; Scheiner and Goodnight 1984; Schlichting and Levin 1986; Jinks and Pooni 1988; Scheiner and Lyman 1991; Schlichting and Pigliucci 1998). For example, Andersson (1989) compared differentiation of means and plasticities of traits in populations of the plant *Crepis tectorum* and found, contrary to expectation, that more plastic traits appear to evolve faster.

The extent to which plasticity promotes evolutionary stasis is directly proportional to the strength of selection favoring a specific phenotype under particular conditions. If different genotypes can be selected to converge on that phenotype (in essence, canalization of the reaction norm), then the other facets of their genetic differences will remain hidden (Gupta and Lewontin 1982; Sultan 1987; Rollo 1994). The evolutionary process of canalization for performance in a particular range of environmental conditions creates hidden reaction norms. If selection favors different genotypes for their ability to plastically converge on specific phenotypes, then the genetic and phenotypic variability will be expressed only outside of the zone of canalization, i.e., in environments that have not imposed significant selection on the reaction norm.

A study on *Drosophila* by Robertson (1964) demonstrates these ideas. He found that wild-type flies were well canalized relative to mutants when reared on standard *Drosophila* food. However, abnormal food conditions (deficiencies in various constituents such as protein, RNA, or choline) revealed substantial genetic variability within the wild type. In some cases, similar overall changes in body size were attained by altering growth rate either in the early exponential phase or in the later time-independent phase, and it was discovered that the two growth phases were substantially genetically independent. Thus, plastic responses of body weight followed different genetic pathways but achieved similar final phenotypes.

How much variability is being suppressed by selection for convergent reaction norms? The concept of the hidden reaction norm has received little attention, but there are several lines of evidence suggesting that the storehouse may be substantial. Studies of yeast and plants exposed to different media, temperatures, or pathogens have found differential and on/off gene expression in those environments (yeast: Smith et al. 1996; Wodicka et al. 1997; Jelinsky and Samson 1999; *Arabidopsis*: Ruan et al. 1998). Both Oliver (1996) and Moxon and Higgins (1997) speculate that a large fraction of the putative genes of currently unknown function may be useful in different environments. The production of phenocopies of known mutants by means of changing environmental conditions is also evidence for latent genetic potentials (Goldschmidt 1938; Gibson and Hogness 1996; Ren et al. 1996; Serna and Fenoll 1997).

The strongest evidence for hidden reaction norms comes from studies that expose organisms to multiple or new environments. Gregor (1956) pointed out the strikingly higher variability of plant species in common gardens compared with their natural habitats. Subsequently, a large body of experiments on populations of animals and plants confirmed that genetic variability (V_g) does indeed change under different environmental conditions (e.g., Yampolsky and Ebert 1994; Pigliucci et al. 1995a; Bennington and McGraw 1996; Imasheva et al. 1997, 1998; Hoffmann and Schiffer 1998; Sgrò and Hoffmann 1998a; Suvanto et al. 1999; see also references in Schlichting and Pigliucci 1998, their chapter 3; and in Hoffmann and Merila 1999). This change in variability is an important source of $G{\times}E$ interaction. One detailed example is the work of Sgrò and Hoffmann (1998b) on *Drosophila melanogaster* grown at 14°C, 25°C, and 28°C. They found contrasting results for two traits: heritabilities and evolvabilities for fecundity were highest when parents were exposed to 14°C, whereas those for development time were higher at 28°C. Clausen et al. (e.g., 1940, 1948) did many studies with clonal plant material transplanted to new habitats and observed many instances of the expression of hidden variability (figure 12.2).

Other studies have focused on environments that are novel for the organism. Several of these have found that V_g is higher in novel conditions (Service and Rose 1985; Guntrip et al. 1997; Holloway et al. 1997; but see Kawecki 1995b). Pigliucci et al. (1995a) proposed that the changes in genetic variability can be directly linked to previous selection history and the differential expression of genes in environments that have seldom been

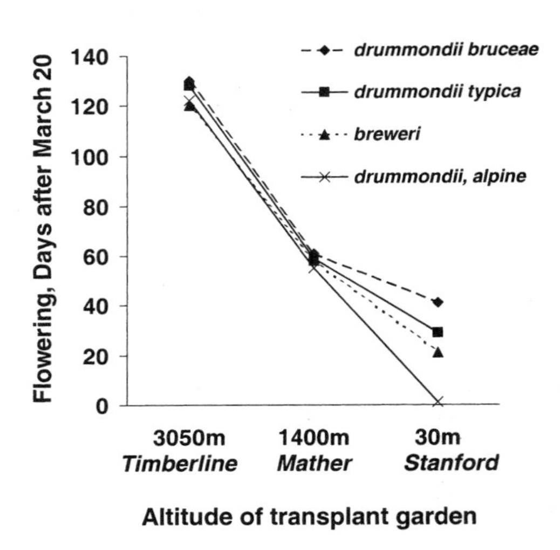

Figure 12.2. An example of hidden reaction norms. Graphed are the mean date of flowering of transplants of four types of the species *Potentilla drummondii* at three stations along the transect from the coast to the Sierra Nevada. These subspecies/varieties all occur at elevations above about 1,500 m and up to 3,650 m. Although they are generally indistinguishable at the elevations they normally inhabit, there is significant differentiation under novel environmental conditions. [Redrawn from Clausen et al. (1940).]

encountered. Hoffmann and Merila (1999) review this and other hypotheses for why genetic variability changes with environment.

The intriguing finding of Rutherford and Lindquist (1998) that Hsp90, a chaperonin that assists in proper protein folding, is capable of masking differences in a wide variety of structural mutations of proteins, reveals another facet of the hidden reaction norm. Mutations in Hsp90, or environmental stress that reduces the available pool of the chaperonins, increase the expression of the mutant phenotypes. Rutherford and Lindquist's results indicate that the expression of some of these mutants can be readily stabilized, again raising the possibility of trait evolution via GA (McLaren 1999).

Plasticity and Peak Shifts

The movement of populations on adaptive landscapes has been of interest since Wright first proposed the concept (Wright 1932). Although most attention has been paid to the roles of mutation and drift, the effects of selection and environmental change on the likelihood of peak shifts have been considered more recently (Price et al. 1993; Whitlock 1996). Pál and Miklos (1999) have also suggested that plasticity and subsequent GA could produce peak shifts.

The developmental conversions discussed above represent environmentally induced peak shifts without genetic change. In addition to the evolution of such putative adaptive changes, the multivariate nature of correlated plastic responses may play a part in generating distinctive phenotypes. Schlichting and Pigliucci (1998) proposed that the environmentally induced changes in correlations produced through correlated plastic responses of multiple traits "represent a prime route for either (1) moving between peaks without traversing valleys, or (2) for altering the landscape itself" (p. 321). Thus, they envisioned that changes in trait correlations could produce novel combinations of characters (peak shifts or new peaks) or could alter the relationship between traits and fitness (change the heights of peaks and valleys). The likelihood that environmental change would result in coordinated changes in characters is increased by the modular architecture of organisms.

Plasticity and Lineage Selection

The role of plasticity in lineage (species) selection has received little attention. Species selection occurs by means of either differential survival or differential extinction of taxonomic lineages. Factors that promote diversification or survival, such as limited or widespread dispersal, respectively, have been the foci of most studies (Stanley 1975; Jablonski 1987; Norris 1991; Grantham 1995; Ridley 1997). Here I present a few instances where phenotypic plasticity might be seen to be an object of lineage selection.

Bürger and Lynch (1995) developed a model to examine the likelihood of extinction under different conditions. He discovered that for a population under directional selection, the mean phenotype will parallel, but lag behind, the optimum favored by environmental change. The magnitude of this lag is directly related to the susceptibility to extinction. Additionally, he found that in finite populations, stochastic variation in genetic variance can contribute to increased likelihood of extinction (see also Lande and Shannon 1996). Plastic responses could clearly ameliorate both problems. A plastic response in the direction of the optimum would reduce the strength of selection on the population, and also perhaps reduce the lag between environmental change and attainment of a fully adapted phenotype. This would appear to provide ideal conditions for species selection favoring plasticity as a safeguard against extinction. In the case of stochastic variation in V_g, the more that plasticity damps the effects of selection, the lower the probability of extinction via random genetic drift.

Eshel and Matessi (1998) have proposed that a breakdown in canalization per se can have similar repercussions. They present a model that indicates that a canalization system (i.e., those mechanisms enforcing canalization) that is inactivated when exposed to extreme environments is more advantageous than one that is resistant. This result is achieved in their model because the inactivated system generates enough random phenotypic plasticity to produce some individuals that are better adapted to the new environment than is the canalized phenotype. Additionally, they argue that such a system of adaptive breakdown of canalization can "postadapt" a population to subsequent environmental changes resembling historical environmental extremes; that is, the genotype contains hidden genetic variability for response to such environments.

Björklund (1994) examined selection on the genetic variance/covariance structure at the population level and found that a higher degree of integration can respond to selection faster: species selection favoring higher integration. Because adaptive plastic responses often require a genetic framework for integrating the responses of a variety of traits (Schlichting and Pigliucci 1998), there could be a component of higher level selection operating here as well. The most common problem associated with detecting a signal that species selection has operated is that the signal will be obscured if selection at the individual level is operating in the same direction. Several of the examples above share this difficulty, in that the plastic responses could easily be envisioned as adaptive for the individual first.

Robinson and Dukas (1999) point out that the hypothesis of plasticity as a hindrance or stimulus for evolutionary change can be tested comparing either the diversity of taxa in more or less plastic lineages, or the evolutionary divergences among species using the genetic versus the total phenotypic trait variances within species (caveat: there are many assumptions in such comparisons). Data in Schluter (1996) did not distinguish the two, suggesting that at least environmental variation was not an impediment. Stearns (1983) proposed another approach—comparison of the evolutionary rates of taxa that are more or less plastic.

Summary

I have presented a number of lines of evidence implicating phenotypic plasticity in the process of evolutionary change. A primary role of phenotypic plasticity may be to just "buy time": the appropriate mutational variation does not have to be lying concealed in the gene pool for a population to persist through an environmental challenge. This may also be an essential part that plasticity plays in the process of GA, suggested as a primary mode for the evolution of phenotypes (Schlichting and Pigliucci 1998). GA has been of special interest to workers in behavioral evolution (Hinton and Nowlan 1987; Wcislo 1989; West-Eberhard 1989; Gottlieb 1992) and can easily be envisioned to be a major mode of change (e.g., in changes in host range of phytophagous insects). Another area where GA can clearly be important is in the evolution of the effects of hormones on morphogenesis. Both the timing of hormone production and the sensitivity of tissues are known to be environmentally altered, and both have been seen to have significant effects on morphology and behavior. GA provides a prime route by which initially environmentally induced phenotypic alternatives can be incorporated as constitutive components.

West-Eberhard in particular has championed the role of plastic changes through developmental conversion as an important springboard for evolutionary change. She has argued that the existence of separate adaptive, and well integrated, developmental pathways is the perfect prologue to divergence. These developmental pathways have themselves been molded as a result of selection for divergent phenotypes under different environmental conditions, and they are expressed as plastic responses (condition-dependent polymorphism or polyphenism). All told, there appears to be a substantial amount of variation in the genomes of most organisms in the form of hidden reaction norms (Schlichting and Pigliucci 1998). The buffering role of plasticity and the phenotypic convergence of the reaction norms of different genotypes on the same phenotype (canalization) lead to hidden reaction norms and the possibility of a subsequent explosion of phenotypic diversity when the typical environmental range is exceeded.

Several authors have independently proposed that the process of development is itself a form of reaction norm to internal (and external) environmental conditions (Wolpert 1994; Sarà 1996b; Schlichting and Pigliucci 1998). Viewed in this light, they postulate a strong role for GA in the evolution of development itself, whereby it results in the substitution of a constitutive expression of a phenotype for a previously facultative phenotypic response to the internal environment.

Acknowledgments I thank Courtney Murren, Jan Conn, Sam Scheiner, and Juan Núñez Farfán for comments on earlier drafts.

13

Future Research Directions

SAMUEL M. SCHEINER
THOMAS J. DEWITT

Where Have We Been?

Over the past 20 years the study of phenotypic plasticity has grown from a small cottage industry investigating an odd phenomenon, to a major research agenda (figure 13.1). Although its roots go back to the very beginnings of modern evolutionary biology (chapter 2), only recently has phenotypic plasticity been studied in a widespread and systematic fashion. This book is testimony to the substantial progress that has been made, especially in the past decade. So, where do we go from here?

The empirical focus of plasticity studies has shifted considerably in the past two decades, from simply demonstrating the existence of genetic variation for plasticity and the response of plasticity to artificial selection, to whether plasticity in nature is adaptive and what types of environmental heterogeneity will select for plasticity. Perhaps the most important advance has been the general recognition that phenotypic plasticity is a property of organisms that can evolve. Still contentious is the form of that evolution. Is phenotypic plasticity the direct target of selection or a does it evolve as a correlated response to selection on other aspects of the phenotype? How important is phenotypic plasticity in overall evolutionary dynamics? What constrains the evolution of plasticity? And, that enduring evolutionary mystery, does phenotypic plasticity contribute to speciation?

Challenges of Studying Plasticity

The challenges of studying phenotypic plasticity come from its very nature, which makes experiments technically difficult. Plasticity is often the property of a genotype, not a

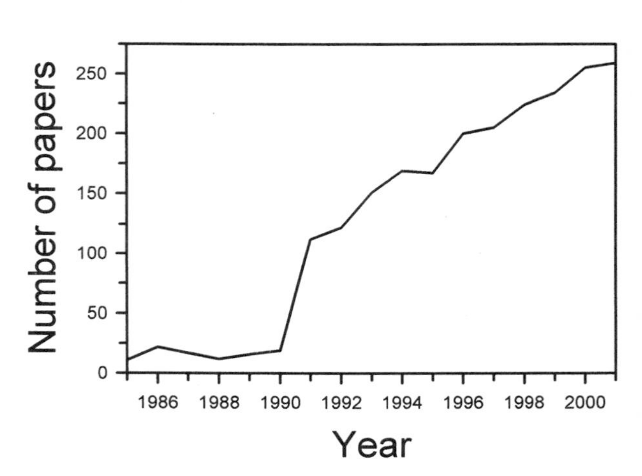

Figure 13.1. Numbers of papers published each year from 1985 to 2001 on phenotypic plasticity. In 1991 the number of papers increased fivefold, and then doubled over the next decade. In the first eight months of 2002, another 202 papers were published, suggesting that the numbers will continue to climb. Data are from a search of the Institute for Scientific Information Science Citation Index using the key words "phenotypic plasticity," "norm of reaction," "*G*×*E*," and "genotype-environment interaction."

single individual. That is, to "see" the plasticity one needs to raise separate individuals in multiple environments. The number of individuals raised and measured for such studies increases multiplicatively with the number of environments. Even when plasticity is expressed within a single individual that alters its phenotype during its lifetime, multiple measures are required.

The genetic nature of plasticity makes it even more challenging. It is easy to study the genetics of a trait that has a fixed phenotypic expression. Plasticity, in contrast, is context dependent. By its very nature, plasticity must always be defined relative to a set of environments. One can never simply speak of the plasticity genes for limb development, for example. Instead, we have genes for limb development that are plastic with respect to temperature. These may be different from the genes for limb development that are plastic with respect to food level. This context dependence has been the source of an enduring dispute over what is meant by plasticity genes. So, the challenge comes from the fact that the target of our investigations is much wider and more difficult to pin down.

One of the grand challenges of evolution is understanding why there is myriad diversity of organisms. G. Evelyn Hutchinson (1959) framed this question in ecological terms more than 40 years ago with his famous essay "Homage to Santa Rosalia, or, Why Are There So Many Animals?" Plasticity adds another layer of complexity on this issue. Why are there so many species, rather than a few highly plastic species? The question that Hutchinson actually addressed in his essay is why there are so many different kinds of forms. If multiple forms can be equally successful, why is it that more species are not plastic? Why is there not, for example, a single plant species that is able to assume the forms of a forb, a grass, a vine, a shrub, an understory tree, and a canopy tree? (One species, *Rhus radicans* or poison ivy, is able to take on three of those forms, but for some reason it has not been well studied in this regard.) Obviously, the answer involves a diversity of potential constraints, yet our effort to date has largely remained focused on demonstrating cases where plasticity has solved adaptive problems. It is much harder to demonstrate why something does not evolve than to demonstrate why it has evolved.

Perspectives of Trait Evolution

How should we study the evolution of phenotypic plasticity and trait evolution? A trait can be studied from four perspectives: functional, developmental, historical, and adaptational. These perspectives are alternative explanations for why a trait exists in the form that it does and define the four classes of constraints for trait evolution. In studies of phenotypic plasticity these perspectives are intertwined.

The functional perspective asks, given the limitations on how an organism is put together and operates, what the constraints are on plasticity. Dudley (chapter 10) classified plant traits along functional lines and concluded that environments that create stress or affect carbon acquisition tend to invoke plastic responses, whereas other types of environmental factors (e.g., pollinators) do not. Sih (chapter 8) detailed how behavioral plasticity and morphological plasticity can often face similar sorts of functional constraints and therefore can be studied with a common framework. More such analyses are needed before we can understand why adaptive plasticity is found less often than we would expect based on theoretical predictions.

The developmental perspective asks how the developmental system generates different forms from a single genotype, or how the same phenotype can be achieved through entirely different developmental pathways. Until now, studies of the evolution of development have focused almost exclusively on extremely large-scale differences (e.g., comparisons among fruit flies, round worms, and humans). Studies of sister taxa and within-species variation are few and far between. But as laid out by Frankino and Raff (chapter 5), such approaches can expand our views of the developmental process. How is it, for example, that some species of ants can develop an amazing array of forms among genetically identical individuals? The study of the evolution of developmental mechanisms is entering a period of explosive growth that will soon provide a deep understanding, especially about the genetic architecture underlying phenotypic plasticity. Because details of genetic architecture matter (chapters 3, 4, and 6), this knowledge will provide important insights into the evolutionary dynamics of phenotypic plasticity.

We have advanced significantly beyond the old arguments over plasticity genes (Via et al. 1995). We now understand that plasticity genes are context dependent. Some genes may always be considered plasticity genes, whereas others are plasticity genes with regard to some environments but not others. But as Windig et al. (chapter 3) make abundantly clear, we have yet to assemble a complete picture of the genetic architecture of plasticity for even a single system.

A developmental perspective is also necessary to answer questions concerning the cost of plasticity. Although theoretical models indicate that costs can substantially alter whether plasticity is favored (chapters 6 and 7), currently there is little evidence of costs—few studies have demonstrated their existence, and the costs are based on the actions of one or a few genes (e.g., Krebs et al. 1998). Where entire life histories or body plans are altered, costs appear to be negligible (e.g., DeWitt 1998; Krebs and Feder 1998; Scheiner and Berrigan 1998; Winn 1999; Donohue et al. 2000; Relyea 2002). None of these studies has examined the possible mechanistic basis of costs. More work on the existence of costs may be warranted, but we think limits (*sensu* DeWitt et al. 1998) should begin to receive more attention, especially in a developmental context.

The historical perspective asks how phylogeny determines current trait values and constrain future evolution. Are certain lineages more plastic because they happen to arise from an ancestor that was highly plastic (Doughty 1995; Schlichting and Pigliucci 1998)? Very little has been done to look at plasticity in a phylogenetic perspective (but see Diggle 1993; Pigliucci et al. 1999).

The adaptational perspective asks how the ecology of the organism molds its traits. This perspective has been the primary focus of work over the past decade and represents a significant advance in the field. Much of the work in the 1980s was devoted to simply demonstrating that traits were plastic and measuring genetic variation for plasticity. Now the focus has shifted to ascertaining whether and how this plasticity is adaptive, as exemplified by the pioneering work of Schmitt and colleagues (e.g., Schmitt 1993; Schmitt et al. 1995, 1999; Dudley and Schmitt 1996; also see chapters 7–10). Progress in this area will be accelerated with detailed attention to (ecological) limits of plasticity, such as documenting geographic variation in reliability of induction cues.

Advantages of a Plasticity Perspective

Why should we study evolution from a plasticity perspective? Phenotypic plasticity is the intersection of development, genetics, and natural selection (chapter 1). Most studies of adaptation either concentrate entirely on natural selection (optimality approaches and many quantitative genetic studies) or have focused exclusively on constraints of genetic architecture. A plasticity perspective often brings development issues to the forefront. In this regard, plasticity studies are the microevolutionary component of the study of the evolution of development. This perspective also highlights physiological and behavioral accommodations and so requires a dynamic view of organisms interacting with an environment. Rather than simply performing, living, or dying as passive responses to environmental conditions, organisms actually respond and adapt to their environment within a generation. Thus, an organism's *reaction to* the environment determines its performance—the environment simply sets a context.

A plasticity perspective emphasizes different aspects of scale than are often considered. Scale is a popular buzzword in ecology (e.g., Allen and Starr 1982; Palmer and White 1994; Lyons and Willig 1999; Scheiner et al. 2000). Spatial scale has played an important role in the development of evolutionary theory with regard to population structuring and the roles of migration and isolation. Less attention has been paid to temporal scale. Previous work on temporal structure has focused on generalist versus specialist adaptation as a function of long-term environmental autocorrelation. Although long-term temporal patterns are important for plasticity, equally important are short-term patterns, particularly the relative temporal scales of environmental heterogeneity and plastic (developmental) response.

The ratio of these temporal scales defines the domain of adaptational models of the evolution of phenotypic plasticity: development under uncertainty (chapter 6). If developmental responses are fast and reversible relative to environmental change, then functional and developmental constraints will control trait evolution. If the response is slow or nonreversible, then adaptational constraints will predominate. Thus, a plasticity perspective widens our consideration of temporal patterns.

A plasticity perspective provides a new answer to the question, Why are there missing forms? The traditional answers were based on adaptational or functional constraints. During the 1970s and 1980s, historical and developmental constraints received renewed interest within evolutionary biology, with the latter forming part of the basis of the current field of the evolution of development. The focus of that field, however, is on macroevolutionary patterns. Little consideration has been given to microevolutionary patterns or the intersection of developmental and adaptational processes. This intersection, divergent natural selection within a species, is explicitly the focus in studies of phenotypic plasticity.

A plasticity perspective changes how we conceptualize evolution. The old paradigm (e.g., Hutchinson 1959) envisaged two evolutionary strategies: specialist and generalist. Each strategy consisted of a fixed phenotype. It was generally assumed that the generalist had a lower fitness in a particular environment relative to the specialist for that environment. Models then explored what environmental conditions (e.g., degree of spatial heterogeneity, migration rate) and genetic architectures (e.g., haploid vs. diploid vs. multigenic) favored one strategy over the other. With phenotypic plasticity, there are now three strategies—specialist, jack-of-all-trades generalist, and plastic generalist—and models must be expanded to consider additional issues (e.g., costs and limits of plasticity and multitrait reaction norms; see chapters 6 and 7).

Where Are We Going?

One way to parse the process of evolution by natural selection is phenotypic variation, phenotypic selection, heritability, and evolutionary response (Endler 1986). Each of these represents an open frontier in the study of phenotypic plasticity. We have already commented on the first, the deepening of considerations of developmental and functional constraints on the range of phenotypic variation (see also chapter 4). This is an area of current and growing interest, particularly with regard to within-individual plasticity (e.g., Winn 1996a,b, 1999).

Much less has been done to examine patterns of phenotypic selection. The vast majority of studies of variation among natural populations have been pairwise comparisons between a constant and a variable environment. We need to expand those comparisons to account for environmental predictability, which theory tells us is a critical variable (chapter 6). Of particular interest is the use of genetic engineering to expand the range of phenotypic expression so as to better map phenotypic selection (Schmitt et al. 1999).

Almost all tests of hypotheses about trait plasticity consist of univariate comparisons. We needed to integrate selection on plasticity with selection on other traits. How does plasticity fit into coadaptive solutions to complex and variable natural environments? Assisting this shift are the development of new conceptual and statistical tools for measuring plasticity and its selection (Gibert et al. 1998a; Scheiner and Callahan 1999) and the study of selection on trait interactions (Arnold 1983; Phillips and Arnold 1989; Scheiner et al. 2000b, 2002). Natural selection should favor maximal fitness everywhere. But fitness is made up of several components (e.g., survivorship and fecundity) that are in turn the product of trait interactions. Depending on circumstances, plasticity in some traits and lack of plasticity in others could yield high fitness across environments (chapters 8–10).

Research should be expanded beyond single species to ask how plasticity affects community ecology through "trait-mediated indirect effects" (Abrams et al. 1996b), which is the current buzzword for what was previously called interaction modification

(Wootton 1994). For example, nonpredatory species, because of their relation to predatory species, induce shell shape changes in snails. This (unnecessary) induced response makes the snails more vulnerable to alternative predators such as crayfish (Langerhans and DeWitt 2002). Thus, the nonpredatory fish can have demographic impacts on snails, but only indirectly by affecting snail traits.

The genetic architecture of plasticity is discussed throughout this book (see especially chapters 3 and 4). Noticeably rare, however, is discussion of the coming genomics revolution. We are on the verge of being able to examine expression patterns of thousands of genes simultaneously through the use of genetic arrays (gene chips; e.g., Chuang et al. 1993). This technology, especially when coupled with advances in quantitative trait loci (QTL) techniques (e.g., Paterson et al. 1991), has great potential to rapidly advance our knowledge of the genetic basis of traits and trait plasticity. The challenge is to separate signal from random variation and noise and then to associate that signal with quantitative trait variation. One of the most exciting lines for future research involves the application of the genomic and genetic technology revolution to the ecology and evolution of plasticity. Currently, projects looking at plasticity on a genomic scale are underway with *Arabidopsis* (Weinig et al. 2003) and *Drosophila* (Pletcher et al. 2002). These species are obvious starting points because of the wealth of available genetic information. In bacteria and fungi, the application of gene cloning techniques and evolutionary experiments are particularly feasible (e.g., Bennett et al. 1992; Bennett and Lenski 1993; Leroi et al. 1994). We expect such studies to widen to many more species in the coming decade.

Predicting and understanding evolutionary response is the realm of theory, including both formal mathematical models and verbal or conceptual models. The pace of theory development for plasticity evolution has slowed considerably from that of 1985–1995 (chapter 6). Part of this change in pace may be the feeling that theory has vastly outpaced empirical data. Rather, the problem has been that the theories are too generic, lacking sufficient complexity for testing. Considerations of more complex environmental patterns married to more realistic genetics are needed. Few models, for example, have considered the effects of metapopulation structure (Scheiner 1998; De Jong and Behera 2002; Sultan and Spencer 2002). Wolf et al. (chapter 11) point to important new avenues of theory development and ways to unite theories of plasticity evolution with other models of the evolution of structured populations. DeWitt and Langerhans (chapter 7) make some specific predictions, such as the likelihood that greater developmental variance should be more common in noninducing environments.

Remaining is that enduring mystery, speciation. Schlichting (chapter 12) makes the case that a plasticity perspective has been largely missing from the study of speciation. As far as we are aware, none of the mathematical theories of speciation has incorporated phenotypic plasticity. Similarly, hunts for speciation genes have not examined the results with plasticity in mind.

All of this promises exciting times to come for the study of phenotypic plasticity. The authors in this book point to a vast array of future research on phenotypic plasticity. We expect the next 20 years to be as exciting as the last.

References

Aarssen, L., and L. Turkington. 1985. Biotic specialization between neighbouring genotypes in *Lolium perenne* and *Trifolium repens* from a permanent pasture. Journal of Ecology 73:605–614.

Abrahamson, W. G., and A. E. Weis. 1999. Evolutionary Ecology across Three Trophic Levels. Princeton University Press, Princeton, NJ.

Abrams, P. A., O. Leimar, S. Nylin, and C. Wiklund. 1996a. The effect of flexible growth rates on optimal sizes and development times in a seasonal environment. American Naturalist 147:381–395.

Abrams, P. A., B. A. Menge, G. G. Mittelbach, D. Spiller, and P. Yodzis. 1996b. The role of indirect effects in food webs. Pages 371–395 in K. Winemiller, editor, Food Webs: Integration of Patterns and Dynamics. Chapman and Hall, New York.

Ackerly, D. D., and M. Jasienski. 1990. Size-dependent variation of gender in high density stands of the monoecious annual, *Ambrosia artemisiifolia* (Asteraceae). Oecologia 82:474–477.

Adams, M. B. 1980. Sergei Chetverikov, the Kol'tsov Institute, and the evolutionary synthesis. Pages 242–278 in E. Mayr and W. B. Provine, editors, The Evolutionary Synthesis: Perspectives on the Unification of Biology. Harvard University Press, Cambridge, MA.

Adler, F. R., and R. Karban. 1994. Defended fortresses or moving targets? Another model of inducible defenses inspired by military metaphors. American Naturalist 144:813–832.

Agrawal, A. A. 2001. Phenotypic plasticity in the interactions and evolution of species. Science 294:321–326.

Agrawal, A. A., C. Laforsch, and R. Tollrian. 1999. Transgenerational induction of defences in animals and plants. Nature 401:60–63.

Alekseev, V., and W. Lampert. 2001. Maternal control of resting-egg production in *Daphnia*. Nature 414:899–901.

Alford, R. A. 1999. Ecology: resource use, competition, and predation. Pages 240–278 in R. W. McDiarmid and R. Altig, editors, Tadpoles: The Biology of Anuran Larvae. University of Chicago Press, Chicago.

Allen, T. F. H., and T. B. Starr. 1982. Hierarchy: Perspectives for Ecological Complexity. University of Chicago Press, Chicago.

Alpert, P. 1996. Nutrient sharing in natural clonal fragments of *Frageria chiloensis*. Journal of Ecology 84:395–406.

Alpert, P., and H. A. Mooney. 1996. Resource heterogeneity generated by shrubs and topography on coastal sand dunes. Vegetatio 122:83–93.

Ancel, L. W. 1999. A quantitative model of the Simpson-Baldwin effect. Journal of Theoretical Biology 196:197–209.

Andersson, S. 1989. Phenotypic plasticity in *Crepis tectorum* (Asteraceae). Plant Systematics and Evolution 168:19–38.

Andersson, S., and R. G. Shaw. 1994. Phenotypic plasticity in *Crepis tectorum* (Asteraceae): genetic correlations across light regimes. Heredity 72:113–125.

Anholt, B. R., and E. E. Werner. 1999. Density-dependent consequences of induced behavior. Pages 218–230 in R. Tollrian and C. D. Harvell, editors, The Ecology and Evolution of Inducible Defenses. Princeton University Press, Princeton, NJ.

Antebi, A., J. G. Culotti, and E. M. Hedgecock. 1998. *daf*-12 regulates developmental age and the dauer alternative in *Caenorhabditis elegans*. Development 125:1191–1205.

Appleton, R. D., and A. R. Palmer. 1988. Water-borne stimuli released by predatory crabs and damaged prey induce more predator-resistant shells in a marine gastropod. Proceedings of the National Academy of Sciences of the United States of America 85:4387–4391.

Aragaki, D. L. R., and L. M. Meffert. 1998. A test of how well the repeatability of courtship predicts its heritability. Animal Behavior 55:1141–1150.

Arnold, S. J. 1983. Morphology, performance and fitness. American Zoologist 23:347–361.

Arnold, S. J. 1992. Constraints on phenotypic evolution. American Naturalist 140:S85–S107.

Arnold, S. J., and M. J. Wade. 1984a. On the measurement of natural and sexual selection: theory. Evolution 38:709–719.

Arnold, S. J., and M. J. Wade. 1984b. On the measurement of natural selection: applications. Evolution 38:720–734.

Atkinson, D. 1994. Temperature and organism size—a biological law for ectotherms? Advances in Ecological Research 25:1–58.

Atkinson, D., and R. M. Sibly. 1997. Why are organisms usually bigger in colder environments? Making sense of a life history puzzle. Trends in Ecology and Evolution 12:235–239.

Bagnall, D. J., R. W. King, G. C. Whitelam, M. T. Boylan, D. Wagner, and P. H. Quail. 1995. Flowering responses to altered expression of phytochrome in mutants and transgenic lines of *Arabidopsis thaliana* (L.) Heynh. Plant Physiology 108:1495–1503.

Bailey-Serres, J., B. Kloeckener-Gruissem, and M. Freeling. 1988. Genetic and molecular approaches to the study of the anaerobic response and tissue specific gene expression in maize. Plant, Cell and Environment 11:351–357.

Bain, A. B., and T. H. Attridge. 1988. Shade-light mediated responses in field and hedgerow populations of *Galium aparine* L. Journal of Experimental Botany 39:1759–1764.

Baldwin, I. T. 1998. Jasmonate-induced responses are costly but benefit plants under attack in native populations. Proceedings of the National Academy of Sciences of the United States of America 95:8113–8118.

Baldwin, I. T. 1999. Inducible nicotine production in native *Nicotiana* as an example adaptive phenotypic plasticity. Journal of Chemical Ecology 25:3–30.

Baldwin, J. M. 1896. A new factor in evolution. American Naturalist 30:441–451.

Baldwin, J. M. 1902. Development and Evolution. MacMillan and Co., New York.

Ballaré, C. L., R. A. Sánchez, A. L. Scopel, J. J. Casal, and C. M. Ghersa. 1987. Early detection of neighbour plants by phytochrome perception of spectral changes in reflected sunlight. Plant, Cell and Environment 10:551–557.

Ballaré, C. L., A. L. Scopel, and R. A. Sánchez. 1991. Photocontrol of stem elongation in plant neighborhoods: effects of photon fluence rate under natural conditions of radiation. Plant, Cell and Environment 14:57–65.

Barker, J. S. F., and R. A. Krebs. 1995. Genetic variation and plasticity of thorax length and wing length in *Drosophila aldrichi* and *D. buzzatii*. Journal of Evolutionary Biology 8:689–709.

Barnes, P. W., C. L. Ballaré, and M. M. Caldwell. 1996. Photomorphogenic effects of UV-B radiation on plants: consequences for light competition. Journal of Plant Physiology 148:15–20.

Barton, N. H., and M. Turelli. 1989. Evolutionary quantitative genetics: how little do we know? Annual Review of Genetics 23:337–370.

Bateman, K. G. 1959. Genetic assimilation of four venation phenocopies. Journal of Genetics 56:443–474.

Baur, E. 1922. Einführung in die experimentelle Vererbungslehre. Gebrüder Borntraeger, Berlin.

Bazzaz, F. A., and S. E. Sultan. 1987. Ecological variation and the maintenance of plant diversity. Pages 69–93 in K. M. Urbanska, editor, Differentiation Patterns in Higher Plants. Academic Press, London.

Bell, A. E. 1977. Heritability in retrospect. Journal of Heredity 68:297–300.

Bell, G., and X. Reboud. 1997. Experimental evolution in *Chlamydomonas* II. Genetic variation in strongly contrasted environments. Heredity 78:498–506.

Benkman, C. W. 1999. The selection mosaic and diversifying coevolution between crossbills and lodgepole pine. American Naturalist 153:S75–S91.

Bennett, A. F., and R. E. Lenski. 1993. Evolutionary adaptation to temperature II. Thermal niches of experimental lines of *Escherichia coli*. Evolution 46:1–12.

Bennett, A. F., R. E. Lenski, and J. E. Mittler. 1992. Evolutionary adaptation to temperature. I. Fitness responses of *Escherichia coli* to change in its thermal environment. Evolution 46:16–30.

Bennington, C. C., and J. B. McGraw. 1995. Natural selection and ecotypic differentiation in *Impatiens pallida*. Ecological Monographs 65:302–323.

Bennington, C. C., and J. B. McGraw. 1996. Environment-dependence of quantitative genetic parameters in *Impatiens pallida*. Evolution 50:1083–1097.

Bergelson, J., and R. Perry. 1989. Interspecific competition between seeds: relative planting date and density affect seedling emergence. Ecology 70:1639–1644.

Bernardo, J. 1993. Determinants of maturation in animals. Trends in Ecology and Evolution 8:166–173.

Bernardo, J. 1994. Experimental analysis of allocation in two divergent, natural salamander populations. American Naturalist 143:14–38.

Bernardo, J. 1996. The particular maternal effect of propagule size, especially egg size: patterns, models, and quality of evidence and interpretations. American Zoologist 36:216–236.

Berrigan, D., and J. C. Koella. 1994. The evolution of reaction norms: simple models for age and size at maturity. Journal of Evolutionary Biology 7:549–566.

Bertness, M. D., and R. Callaway. 1994. Positive interactions in communities. Trends in Ecology and Evolution 9:191–193.

Bertness, M. D., and C. Cunningham. 1981. Crab shell-crushing predation and gastropod architectural defense. Journal of Experimental Marine Biology and Ecology 50:213–230.

Berven, K. A. 1982a. The genetic basis of altitudinal variation in the wood frog *Rana sylvatica* I. An experimental analysis of life history traits. Evolution 36:962–983.

Berven, K. A. 1982b. The genetic basis of altitudinal variation in the wood frog *Rana sylvatica*. II. An experimental analysis of larval development. Oecologia 52:360–369.

Björklund, M. 1994. Species selection on organismal integration. Journal of Theoretical Biology 171:427–430.

Blacher, L. I. 1982. The Problem of the Inheritance of Acquired Characteristics. Amerind Publishing Co., New Delhi.

Boake, C. R. B., editor. 1994. Quantitative Genetic Studies of Behavioral Evolution. University of Chicago Press, Chicago.

Boardman, N. K. 1977. Comparative photosynthesis of sun and shade plants. Annual Review of Plant Physiology 28:355–377.

Bohrenstedt, G. W., and A. S. Goldberger. 1969. On the exact covariance of products of random variables. Journal of the American Statistical Association 64:1439–1442.

Bolker, J. A. 1995. The choice and consequences of model systems in developmental biology. BioEssays 17:451–455.

Bonnet, X., G. Naulleau, R. Shine, and O. Lourdais. 2001. Short-term versus long-term effects of food intake on reproductive output in a viviparous snake, *Vipera aspis*. Oikos 92:297–308.

Bookstein, F. J. 1991. Morphometric Tools for Landmark Data: Geometry and Biology. Cambridge University Press, New York.

Bowler, C. 1997. The transduction of light signals by phytochrome. Pages 137–152 in P. Aducci, editor, Signal Transduction in Plants. Birkhauser Verlag, Basel.

Bradshaw, A. D. 1965. Evolutionary significance of phenotypic plasticity in plants. Advances in Genetics 13:115–155.

Bradshaw, W. E. 1973. Homeostasis and polymorphism in vernal development of *Chaoborus americanus*. Ecology 54:1247–1259.

Bradshaw, W. E., and C. M. Holzapfel. 2000. Epistasis and the evolution of genetic architectures in natural populations. Pages 245–263 in J. B. Wolf, E. D. Brodie III, and M. J. Wade, editors, Epistasis and the Evolutionary Process. Oxford University Press, Oxford.

Bradshaw, W. E., and L. P. Lounibos. 1977. Evolution of dormancy and its photoperiodic control in pitcher-plant mosquitoes. Evolution 31:546–567.

Brakefield, P. M., and T. B. Larsen. 1984. The evolutionary significance of dry and wet season forms in some tropical butterflies. Biological Journal of the Linnean Society 22:1–12.

Brakefield, P. M., and N. Reitsma. 1991. Phenotypic plasticity, seasonal climate and the population biology of *Bicyclus butterflies* (Satyridae) in Malawi. Ecological Entomology 16:291–303.

Brakefield, P. M., J. Gates, D. Keys, F. Kesbeke, P. J. Wijngaarden, A. Monteiro, V. French, and S. B. Carrol. 1996. The development, plasticity and evolution of butterfly eyespot patterns. Nature 384:236–242.

Brodie, E. D., III. 1992. Correlational selection for color pattern and antipredator behavior in the garter snake *Thamnophis ordinoides*. Evolution 46:1284–1298.

Brodie, E. D., III. 2000. Why evolutionary genetics does not always add up. Pages 3–19 in J. B. Wolf, E. D. Brodie III, and M. J. Wade, editors, Epistasis and the Evolutionary Process. Oxford University Press, New York.

Brodie, E. D., III, and E. D. Brodie Jr. 1999. Predator-prey arms races. Bioscience 49:557–568.

Brodie, E. D., III, A. J. Moore, and F. J. Janzen. 1995. Visualizing and quantifying natural selection. Trends in Ecology and Evolution 10:313–318.

Brönmark, C., L. B. Pettersson, and P. Anders Nilsson. 1999. Predator-induced defense in crucian carp. Pages 203–217 in R. Tollrian and C. D. Harvell, editors, The Ecology and Evolution of Inducible Defenses. Princeton University Press, Princeton, NJ.

Brown, S., J. P. Mahaffey, M. Lorenzen, R. Denell, and J. W. Mahaffey. 1999. Using RNAi to investigate orthologous gene function during development of distantly related insects. Evolution and Development 1:11–15.

Buchholz, D. R., and T. B. Hayes. 2000. Larval period comparison for the spadefoot toads *Scaphiopus couchii* and *Spea multiplicata* (Pelobatidae:Anura). Herpetologica 56:455–468.

Buchholz, D. R., and T. B. Hayes. 2002. Evolutionary patterns of diversity in spadefoot toad metamorphosis (Anura:Pelobatidae). Copeia 2002:180–189.

Budd, G. E. 1999. Does evolution in body patterning genes drive morphological change—or vice versa? BioEssays 21:326–332.

Bull, J. J. 1987. Evolution of phenotypic variance. Evolution 41:303–315.

Bürger, R., and M. Lynch. 1995. Evolution and extinction in a changing environment. Evolution 49:151–163.

Bush, G. L. 1975. Sympatric speciation in phytophagous parasitic insects. Pages 187–206 in P. W. Pice, editor, Evolutionary Strategies of Parasitic Insects and Mites. Plenum, New York.

Byers, D. L., and T. R. Meagher. 1997. A comparison of demographic characteristics in a rare and a common species of *Eupatorium*. Ecological Applications 7:519–530.

Callahan, H. S., C. L. Wells, and M. Pigliucci. 1999. Light-sensitive plasticity genes in *Arabidopsis thaliana*: mutant analysis and ecological genetics. Evolutionary Ecology Research 1:731–751.

Capy, P., E. Pla, and J. R. David. 1993. Phenotypic and genetic variability of morphometrical traits in natural populations of *Drosophila melanogaster* and *D. simulans*. I. Geographic variations. Genetics, Selection, Evolution 25:517–536.

Capy, P., E. Pla, and J. R. David. 1994. Phenotypic and genetic variability of morphometrical traits in natural populations of *Drosophila melanogaster* and *D. simulans*. II. Within-population variability. Genetics, Selection, Evolution 26:15–28.

Carrière, Y., and B. D. Roitberg. 1995. Evolution of host-selection behaviour in insect herbivores: genetic variation and covariation in host acceptance within and between populations of *Choristoneura rosaceana* (Family: Tortricidae) the obliquebanded leadfoller. Heredity 74:357–368.

Carroll, S. B., J. Gates, D. N. Keys, S. W. Paddock, G. E. F. Panganiban, J. E. Selegue, and J. A. Williams. 1994. Pattern formation and eyespot determination in butterfly wings. Science 265:109–114.

Carroll, S. P., and C. Boyd. 1992. Host race radiation in the soapberry bug: natural history with the history. Evolution 46:1052–1069.

Casal, J. J., C. L. Ballaré, M. Tourn, and R. A. Sánchez. 1994. Anatomy, growth and survival of a long-hypocotyl mutant of *Cucumis sativus* deficient in phytochrome B. Annals of Botany 73:569–575.

Casper, B. B., and R. B. Jackson. 1997. Plant competition underground. Annual Review of Ecological and Systematics 28:545–570.

Cassida, R. C., and R. L. Russel. 1975. The dauer-larva: a post-embryonic developmental variant of the nematode *C. elegans*. Developmental Biology 46:326–342.

Castillo-Chavez, C., S. A. Levin, and F. Gould. 1988. Physiological and behavioral adaptation to varying environments: a mathematical model. Evolution 42:986–994.

Caswell, H. 1983. Phenotypic plasticity in life-history traits: demographic effects and evolutionary consequences. American Zoologist 23:35–46.

Caswell, H. 2001. Matrix Population Models, 2nd edition. Sinauer Associates, Inc., Sunderland, MA.

Chailakhyan, M. K. 1968. Internal factors of plant flowering. Annual Review of Plant Physiology 19:1–36.

Chakir, M., A. Chafik, B. Moreteau, P. Gibert, and J. R. David. 2002. Male sterility thermal thresholds in *Drosophila*: *D. simulans* appears more cold-adapted than its sibling *D. melanogaster*. Genetica 114:195–205.

Chapin, F. S., III, K. Autumn, and F. Pugnaire. 1993. Evolution of suites of traits in response to environmental stress. American Naturalist 142:S78–S92.

Charnov, E. L. 1976. Optimal foraging: the marginal value theorem. Theoretical Population Biology 9:129–136.

Charnov, E. L. 1982. The Theory of Sex Allocation. Princeton University Press, Princeton, NJ.

Charnov, E. L. 1993. Life History Invariants. Oxford University Press, Oxford.

Chazdon, R. L., and R. W. Pearcy. 1991. The importance of sunflecks for forest understory plants. Bioscience 41:760–766.

Cheplick, G. P. 1995. Life history trade-offs in *Amphibromus scabrivalvis* (Poaceae): allocation to clonal growth, storage, and cleistogamous reproduction. American Journal of Botany 82:621–629.

Cheverud, J. M. 1984. Quantitative genetics and developmental constraints on evolution by selection. Journal of Theoretical Biology 110:155–171.

Cheverud, J. M., and E. J. Routman. 1995. Epistasis and its contribution to genetic variance components. Genetics 139:1455–1461.

Christiansen, F. B. 1975. Hard and soft selection in a subdivided population. American Naturalist 109:11–16.

Chuang, S.-E., D. L. Daniels, and F. R. Blattner. 1993. Global regulation of gene expression in *Escherichia coli*. Journal of Bacteriology 175:2026–2036.

Cipollini, D. F. 1999. Costs to flowering of the production of a mechanically hardened phenotype in *Brassica napus* L. International Journal of Plant Science 160:735–741.

Cipollini, D. F., and J. C. Schultz. 1999. Exploring cost constraints on stem elongation using phenotypic manipulation. American Naturalist 153:236–242.

Clark, A. G., D. J. Begin, and T. Prout. 1999. Female x male interactions in *Drosophila* sperm competition. Science 283:217–220.

Clark, C. W., and C. D. Harvell. 1992. Inducible defenses and the allocation of resources: a minimal model. American Naturalist 139:521–539.

Clark, M. A., N. A. Moran, P. Baumann, and J. J. Wernegreen. 2000. Cospeciation between bacterial endosymbionts (*Buchnera*) and a recent radiation of aphids (*Uroleucon*). Evolution 54:517–525.

Clarkson, D. T. 1985. Factors affecting mineral nutrient acquisition by plants. Annual Review of Plant Physiology 36:77–115.

Clausen, J. D., D. Keck, and W. M. Hiesey. 1940. Experimental studies on the nature of species. I. Effects of varied environments on Western North American plants. Carnegie Institute of Washington, Washington, DC, USA.

Clausen, J. D., D. Keck, and W. M. Hiesey. 1948. Experimental Studies on the Nature of Species. III. Environmental Responses of Climatic Races of *Achillea*. Carnegie Institution of Washington Publication no. 581. Washington, DC.

Coe, S. D. 1994. Theodosius Dobzhansky: a family story. Pages 13–28 in M. B. Adams, editor, The Evolution of Theodosius Dobzhansky: Essays on His Life and Thought in Russia and America. Princeton University Press, Princeton, NJ.

Cohen, D. 1968. A general model of optimal reproduction in a randomly varying environment. Journal of Ecology 56:219–228.

Cohen, D. 1976. Optimal timing of reproduction. American Naturalist 110:801–807.

Cohen, J. E. 1978. Food Webs and Niche Space. Princeton University Press, Princeton, NJ.

Coleman, J. S., K. D. M. McConnaughay, and D. D. Ackerly. 1994. Interpreting phenotypic variation in plants. Trends in Ecology and Evolution 9:187–190.

Collins, J. P. 1979. Intrapopulation variation in the body size at metamorphosis and timing of metamorphosis in the bullfrog, *Rana catesbeiana*. Ecology 60:738–749.

Collins, J. P., and J. R. Holomuzki. 1984. Intraspecific variation in diet within and between trophic morphs in larval tiger salamanders (*Ambystoma tigrinum nebulosum*). Canadian Journal of Zoology 62:168–174.

Conner, J. 1989. Density-dependent sexual selection in the fungus beetle, *Bolitotherus cornutus*. Evolution 43:1378–1386.

Conover, D. O., and E. T. Schultz. 1995. Phenotypic similarity and the evolutionary significance of countergradient variation. Trends in Ecology and Evolution 10:248–252.

Cowley, D. E. 1991. Prenatal effects on mammalian growth: embryo transfer results. Pages 762–779 in E. C. Dudley, editor, The Unity of Evolutionary Biology, Volume II. Proceedings of the Fourth International Congress of Systematic and Evolutionary Biology. Dioscorides Press, Portland, OR.

Cowley, D. E., and W. R. Atchley. 1992. Quantitative genetic models for development, epigenetic selection, and phenotypic evolution. Evolution 46:495–518.

Cowley, D. E., D. Pomp, W. R. Atchley, E. J. Eisen, and D. Hawkins-Brown. 1989. The impact of maternal uterine genotype on postnatal growth and adult body size in mice. Genetics 122:193–203.

Craig, T. P., J. K. Itami, and P. W. Price. 1989. A strong relationship between oviposition preference and larval performance in a shoot-gall sawfly. Ecology 70:1691–1699.

Crews, D. 1994. Temperature, steroids and sex determination. Journal of Endocrinology 142:1–8.

Crews, D. 1996. Temperature-dependent sex determination: the interplay of steroid hormones and temperature. Zoological Science 13:1–13.

Crnokrak, P., and D. A. Roff. 1995. Dominance variance: associations with selection and fitness. Heredity 75:530–540.

Crouse, D. T., L. B. Crowder, and H. Caswell. 1987. A stage-based population model for loggerhead sea turtles and implications for conservation. Ecology 68:1412–1423.

Crow, J. F. 1958. Some possibilities for measuring selection intensities in man. Human Biology 30:1–13.

Crowl, T. A., and A. P. Covich. 1990. Predator-induced life-history shifts in a freshwater snail. Science 247:949–951.

Crowley, P. H. 1992. Resampling methods for computation intensive data analysis in ecology and evolution. Annual Review of Ecology and Systematics 23:405–447.

Crowley, P. H., D. R. DeVries, and A. Sih. 1990. Inadvertent errors and error-constrained optimization: fallible foraging by bluegill sunfish. Behavioral Ecology and Sociobiology 27:135–144.

Curio, E. 1973. Towards a method of teleometry. Experientia 29:1045–1058.

Da Lage, J. L., P. Capy, and J. R. David. 1989. Starvation and desiccation tolerance in *Drosophila melanogaster* adults: effects of environmental temperature. Journal of Insect Physiology 35:453–457.

D'Antonio, C. M., and P. M. Vitousek. 1992. Biological invasions by exotic grasses, the grass/fire cycle, and global change. Annual Review of Ecology and Systematics 23:63–87.

David, J. R., and P. Capy. 1988. Genetic variation of *Drosophila melanogaster* natural populations. Trends in Genetics 4:106–111.

David, J. R., R. Allemand, J. Van Herrewege, and Y. Cohet. 1983. Ecophysiology: abiotic factors. Pages 105–170 in M. Ashburner, H. L. Carson, and J. N. Thompson, editors, The Genetics and Biology of *Drosophila*, Volume III. Academic Press, New York.

David, J. R., P. Capy, and Jean-P. Gauthier. 1990. Abdominal pigmentation and growth temperature in *Drosophila melanogaster*: similarities and differences in the norms of reaction of successive segments. Journal of Evolutionary Biology 3:429–445.

David, J. R., B. Moreteau, J. R. Gauthier, G. Pétavy, J. Stockel, and A. Imasheva. 1994. Reaction norms of size characters in relation to growth temperature in *Drosophila melanogaster*: an isofemale lines analysis. Genetics, Selection, Evolution 26:229–251.

David, J. R., P. Gibert, E. Gravot, G. Pétavy, J. P. Morin, D. Karan, and B. Moreteau. 1997. Phenotypic plasticity and developmental temperature in *Drosophila*: analysis and significance of reaction norms of morphometrical traits. Journal of Thermal Biology 22:441–451.

Day, T., and L. Rowe. 2002. Developmental thresholds and the evolution of reaction norms for age and size at life-history transitions. American Naturalist 159:338–350.

Day, T., and P. D. Taylor. 1997. Von Bertalanffy's growth equation should not be used to model age and size at maturity. American Naturalist 149:381–393.

de Jong, G. 1988. Consequences of a model of counter-gradient selection. Pages 264–277 in G. de Jong, editor, Population Genetics and Evolution. Springer-Verlag, Berlin.

de Jong, G. 1989. Phenotypically plastic characters in isolated populations. Pages 3–18 in A. Fontdevila, editor, Evolutionary Biology of Transient Unstable Populations. Springer-Verlag, Berlin.

de Jong, G. 1990a. Genotype-by-environment interaction and the genetic covariance between environments: multilocus genetics. Genetica 81:171–177.

de Jong, G. 1990b. Quantitative genetics of reaction norms. Journal of Evolutionary Biology 3:447–468.

de Jong, G. 1995. Phenotypic plasticity as a product of selection in a variable environment. American Naturalist 145:493–512.

de Jong, G. 1999. Unpredictable selection in a structured population leads to local genetic differentiation in evolved reaction norms. Journal of Evolutionary Biology 12:839–851.

de Jong, G., and N. Behera. 2002. The influence of life-history differences on the evolution of reaction norms. Evolutionary Ecology Research 4:1–25.

de Jong, G., and S. Gavrilets. 2000. Maintenance of genetic variation in phenotypic plasticity: the role of environmental variation. Genetical Research 76:295–304.

de Kroon, H., and M. J. Hutchings. 1994. Morphological plasticity in clonal plants: the foraging concept reconsidered. Journal of Ecology 83:113–122.

Delpuech, J.-M., B. Moreteau, J. Chiche, E. Pla, J. Vouidibio, and J. R. David. 1995. Phenotypic plasticity and reaction norms in temperate and tropical populations of *Drosophila melanogaster*: ovarian size and developmental temperature. Evolution 49:670–675.

DeLucia, E. H., W. H. Schlesinger, and W. D. Billings. 1989. Edaphic limitations to growth and photosynthesis in Sierran and Great Basin vegetation. Oecologia 78:184–190.

De Moed, G. H., G. de Jong, and W. Scharloo. 1997a. Environmental effects on body size variation in Drosophila melanogaster and its cellular basis. Genetical Research 70:35–43.

De Moed, G. H., G. de Jong, and W. Scharloo. 1997b. The phenotypic plasticity of wing size in *Drosophila melanogaster*: the cellular basis of its genetic variation. Heredity 79:260–267.

Dempster, E. R. 1955. Maintenance of genetic heterogeneity. Cold Spring Harbor Symposia on Quantitative Biology 20:25–32.

Denlinger, D. 1985. Diapause. Pages 353–412 in G. A. Kerkut and L. I. Gilbert, editors, Comprehensive Insect Physiology, Biochemistry, and Pharmacology. Pergamon Press, New York.

Denver, R. J. 1997a. Environmental stress as a developmental cue: corticotropin-releasing hormone is a proximate mediator of adaptive phenotypic plasticity in amphibian metamorphosis. Hormones and Behavior 31:169–179.

Denver, R. J. 1997b. Proximate mechanisms of phenotypic plasticity in amphibian metamorphosis. American Zoologist 37:172–184.
Denver, R. J. 1998a. Hormonal correlates of environmentally induced metamorphosis in the western spadefoot toad, *Scaphiopus hammondii*. General and Comparative Endocrinology 110:326–336.
Denver, R. J. 1998b. The molecular basis of thyroid hormone-dependent central nervous system remodeling during amphibian metamorphosis. Comparative Biochemistry and Physiology 119:219–228.
Denver, R. J., N. Mirhadi, and M. Phillips. 1998. Adaptive plasticity in amphibian metamorphosis: response of *Scaphiopus hammondii* tadpoles to habitat desiccation. Ecology 79:1859–1872.
De Vries, H. 1901. Die Mutationstheorie. Versuche und Beobachtungen über die Enstehung der Arten in Pflanzenreich, Volume 1. Veit and Co., Leipzig.
De Vries, H. 1903. Die Mutationstheorie. Versuche und Beobachtungen über die Enstehung der Arten in Pflanzenreich, Volume 2. Veit and Co., Leipzig.
DeWitt, T. J. 1996. Functional Tradeoffs and Phenotypic Plasticity in the Freshwater Snail *Physa*. Ph.D. diss., Binghamton University.
DeWitt, T. J. 1998. Costs and limits of phenotypic plasticity: tests with predator-induced morphology and life history in a freshwater snail. Journal of Evolutionary Biology 11:465–480.
DeWitt, T. J., and J. Yoshimura. 1998. The fitness threshold model: random environmental change alters adaptive landscapes. Evolutionary Ecology 12:615–626.
DeWitt, T. J., A. Sih, and D. S. Wilson. 1998. Costs and limits of phenotypic plasticity. Trends in Ecology and Evolution 13:77–81.
DeWitt, T. J., A. Sih, and J. A. Hucko. 1999. Trait compensation and cospecialization: size, shape, and antipredator behaviour. Animal Behavior 58:397–407.
DeWitt, T. J., and R. B. Langerhans. 2003. Multiple prey traits, multiple predators: keys to understanding complex community dynamics. Journal of Sea Research 49:143–155.
Dichtel, M. L., S. Louvet-Vallee, M. E. Viney, M. A. Felix, and P. W. Sternberg. 2001. Control of vulval cell division number in the nematode *Oscheius/Dolichorhabditis* sp. CEW1. Proceedings of the National Academy of Sciences of the United States of America 157:183–197.
Dickerson, G. E. 1955. Genetic slippage in response to selection for multiple objectives. Cold Spring Harbor Symposium on Quantitative Biology 20:213–224.
Diggle, P. K. 1993. Developmental plasticity, genetic variation, and the evolution of andromonoecy in *Solanum hirtum* (Solanaceae). American Journal of Botany 80:967–973.
Diggle, P. K. 1997. Ontogenetic constraints and floral morphology: the effects of architecture and resource limitation. International Journal of Plant Science 158:S99–S107.
Dixon, A. F. G., and B. K. Agarwala. 1999. Ladybird-induced life-history changes in aphids. Proceedings of the Royal Society of London, Series B. Biological Sciences 266:1549–1553.
Dixon, A. F. G., and R. Kundu. 1998. Resource tracking in aphids: programmed reproductive strategies anticipate seasonal trends in habitat quality. Oecologia 114:73–78.
Dobzhansky, T. 1937. Genetics and the Origin of Species. Columbia University Press, New York.
Dobzhansky, T. 1955a. Evolution, Genetics, and Man. John Wiley and Sons, New York.
Dobzhansky, T. 1955b. A review of some fundamental concepts and problems of population genetics. Cold Spring Harbor Symposia on Quantitative Biology 20:1–15.
Dobzhansky, T., and B. Spassky. 1963. Genetics of natural populations. XXXIV. Adaptive norm, genetic load and genetic elite in *Drosophila pseudoobscura*. Genetics 48:1467–1485.
Dodson, S. I. 1988. Cyclomorphosis in *Daphnia galeata mendotae* Birge and *D. retrocurva* Forbes as a predator-induced response. Freshwater Biology 19:109–114.
Dodson, S. I. 1989. Predator-induced reaction norms. Bioscience 39:447–452.
Donohue, K., and J. Schmitt. 1999. The genetic architecture of plasticity to density in *Impatiens capensis*. Evolution 53:1377–1386.
Donohue, K., D. Messiqua, E. Hammond-Pyle, M. S. Heschel, and J. Schmitt 2000. Evidence of adaptive divergence in plasticity: density- and site-dependent selection on shade avoidance responses in *Impatiens capensis*. Evolution 54:1956–1968.
Donovan, L. A., and J. R. Ehleringer. 1994. Potential for selection on plants for water use effi-

ciency as estimated by carbon isotope discrimination. American Journal of Botany 81:927–935.

Doughty, P. 1995. Testing the ecological correlates of phenotypically plastic traits within a phylogenetic framework. Acta Oecologica 16:519–524.

Doughty, P. 1996. Statistical analysis of natural experiments in evolutionary biology: comments on recent criticisms of the use of comparative methods to study adaptation. American Naturalist 148:943–956.

Doughty, P. 2002. Coevolution of developmental plasticity and large egg size in *Crinia georgiana* tadpoles. Copeia 2002:928–937.

Doughty, P., and R. Shine. 1998. Reproductive energy allocation and long-term energy stores in a viviparous lizard (*Eulamprus tympanum*). Ecology 79:1073–1083.

Drew, M. C. 1975. Comparison of effects of a localised supply of phosphate, nitrate, ammonium and potassium. New Phytologist 75:479–490.

Driver, E. C. 1931. Temperature and gene expression in *Drosophila*. Journal of Experimental Zoology 59:1–28.

Dudley, S. A. 1996. Differing selection on plant physiological traits in response to environmental water availability: a test of adaptive hypotheses. Evolution 50:92–102.

Dudley, S. A., and J. Schmitt. 1995. Genetic differentiation in morphological responses to simulated foliage shade between populations of *Impatiens capensis* from open and woodland sites. Functional Ecology 9:655–666.

Dudley, S. A., and J. Schmitt. 1996. Testing the adaptive plasticity hypothesis: density-dependent selection on manipulated stem length in *Impatiens capensis*. American Naturalist 147:445–465.

Duellman, W. E., and L. Trueb. 1986. Biology of the Amphibians. McGraw-Hill, New York.

Dugaktin, L. A., and H. K. Reeve. 1998. Game Theory and Animal Behavior. Oxford University Press, Oxford.

Eberhard, W. G. 1982. Beetle horn dimorphism: making the best of a bad lot. American Naturalist 119:420–426.

Ebert, D., L. Yampolsky, and A. J. Van Noordwijk. 1993. Genetics of life history in *Daphnia magna*: 2. Phenotypic plasticity. Heredity 70:344–352.

Edelstein-Keshet, L., and M. Rausher. 1989. The effects of inducible plant defenses on herbivore populations. I. Mobile herbivores in continuous time. American Naturalist 133:787–810.

Ehleringer, J. 1985. Annuals and perennials of warm deserts. Pages 162–180 in B. F. Chabot and H. A. Mooney, editors, Physiological Ecology of North American Plant Communities. Chapman and Hall, New York.

Ehleringer, J. R., and C. Clark. 1988. Evolution and adaptation in *Encelia* (Asteraceae). Pages 221–248 in L. D. Gottleib, editor, Plant Evolutionary Biology. Chapman and Hall, New York.

Ehrlich, P. R., and P. H. Raven. 1964. Butterflies and plants: a study in coevolution. Evolution 18:586–608.

Emlen, D. J. 1994. Environmental control of horn length dimorphism in the beetle *Onthophagus acuminatus* (Coleoptera: Scarabaeidae). Proceedings of the Royal Society of London, Series B. Biological Sciences 256:131–136.

Emlen, D. J. 1996. Artificial selection on horn length-body size allometry in the horned beetle *Onthophagus acuminatus* (Coleopter:Scarabaeidae). Evolution 50:1219–1230.

Emlen, D. J. 1997a. Alternative reproductive tactics and male-dimorphism in the horned beetle *Onthophagus acuminatus* (Coleoptera: Scarabaedidae). Behavioural Ecology and Sociobiology 41:335–341.

Emlen, D. J. 1997b. Diet alters male horn allometry in the beetle *Onthophagus acuminatus* (Coleoptera: Scarabaeidae). Proceedings of the Royal Society of London, Series B. Biological Sciences 264:567–574.

Emlen, D. J. 2000. Integrating development with evolution: a case study with beetle horns. Bioscience 50:403–418.

Emlen, D. J. 2001. Costs and the diversification of exaggerated animal structures. Science 291:1534–1536.

Emlen, D. J., and H. F. Nijhout. 1999. Hormonal control of male horn length dimorphism in the dung beetle *Onthophagus taursus* (Coleoptera: Scarabaeidae). Journal of Insect Physiology 45:45–53.

Emlen, D. J., and H. F. Nijhout. 2000. The development and evolution of exaggerated morphologies in insects. Annual Review of Entomology 45:661–708.

Emlen, D. J., and H. F. Nijhout. 2001. Hormonal control of male horn length dimorphism in *Onthophagus taurus* (Coleoptera:Scarabaeidae): a second critical period of sensitivity to juvenile hormone. Journal of Insect Physiology 47:1045–1054.

Endler, J. A. 1986. Natural Selection in the Wild. Princeton University Press, Princeton, NJ.

Endler, J. A. 1995. Multiple-trait coevolution and environmental gradients in guppies. Trends in Ecology and Evolution 10:22–29.

Endo, K., and S. Funatsu. 1985. Hormonal control of seasonal morph determination in the swallowtail butterfly, *Papilio xuthus* L. (Lepidoptera:Papilionidae). Journal of Insect Physiology 31:669–674.

Endo, K., and Y. Kamata. 1985. Hormonal control of seasonal-morph determination in the small copper butterfly, *Lycaena phlaeas daimio* Seitz. Journal of Insect Physiology 31:701–706.

Eshel, I., and C. Matessi. 1998. Canalization, genetic assimilation and preadaptation. A quantitative genetic model. Genetics 149:2119–2133.

Evans, J. P. 1995. A spatially explicit test of foraging behavior in a clonal plant. Ecology 76:1147–1155.

Fagen, R. 1987. Phenotypic plasticity and social environment. Evolutionary Ecology 1:263–271.

Fairbairn, D. J., and D. E. Yadlowski. 1997. Coevolution of traits determining migratory tendency: correlated response of a critical enzyme, juvenile hormone esterase, to selection on wing morphology. Journal of Evolutionary Biology 10:495–513.

Falconer, D. S. 1952. The problem of environment and selection. American Naturalist 86:293–298.

Falconer, D. S. 1960. Selection of mice for growth on high and low planes of nutrition. Genetical Research 1:91–113.

Falconer, D. S. 1990. Selection in different environments: effects on environmental sensitivity (reaction norm) and on mean performance. Genetical Research 56:57–70.

Falconer, D. S., and T. F. C. Mackay. 1996. Introduction to Quantitative Genetics, 4th edition. Longman, Essex, UK.

Feder, J. L., and K. E. Filchak. 1999. It's about time: the evidence for host plant-mediated selection in the apple maggot fly, *Rhagoletis pomonella*, and its implications for fitness trade-offs in phytophagous insects. Entomologia Experimentalis et Applicata 91:211–225.

Feldman, M. W., and R. C. Lewontin. 1975. The heritability hang-up. Science 190:1163–1168.

Filchak, K. E., J. L. Feder, J. B. Roethele, and U. Stolz. 1999. A field test for host-plant dependent selection on larvae of the apple maggot fly, *Rhagoletis pomonella*. Evolution 53:187–200.

Filchak, K. E., J. B. Roethele, and J. L. Feder. 2000. Natural selection and sympatric divergence in the apple maggot, *Rhagoletis pomonella*. Nature 407:739–742.

Fire, A., S. Xu, M. Montgomery, S. A. Kostas, S. E. Driver, and C. C. Mello. 1998. Potent and specific genetic interference by double-stranded RNA in *Caenorhabditis elegans*. Nature 391:806–811.

Fisher, R. A. 1918. The correlation between relatives on the supposition of Mendelian inheritance. Transactions of the Royal Society of Edinburgh Proceedings B 52:399–433.

Fisher, R. A. 1928a. The possible modification of the response of the wild type to recurrent mutations. American Naturalist 62:115–126.

Fisher, R. A. 1928b. Two further notes on the origin of dominance. American Naturalist 62:571–574.

Fisher, R. A. 1931. The evolution of dominance. Biological Reviews 6:345–368.

Fitter, A. H., and C. J. Ashmore. 1974. Response of two *Veronica* species to a simulated woodland light climate. New Phytologist 1974:997–1001.

Fitter, A. H., and R. K. M. Hay. 1987. Environmental Physiology of Plants. Academic Press, London.

Fox, C. W. 1993. A quantitative genetic analysis of oviposition preference and larval performance on two hosts in th bruchid beetle, *Callosobruchus maculatus*. Evolution 47:166–175.

Fox, C. W., and T. A. Mousseau. 1996. Larval host plant affects fitness consequences of egg size variation in the seed beetle *Stator limbatus*. Oecologia 107:541–548.

Fox, C. W., and T. A. Mousseau. 1998. Adaptive maternal effects and the evolution of transgeneration phenotypic plasticity. Pages 159–177 in T. A. Mousseau and C. W. Fox, editors, Maternal Effects as Adaptations. Oxford University Press, New York.

Fox, C. W., and U. M. Savalli. 2000. Maternal effects mediate diet expansion in a seed-feeding beetle. Ecology 81:3–7.

Fox, C. W., K. J. Waddell, and T. A. Mousseau. 1994. Host-associated fitness variation in a seed beetle (Coleoptera: Bruchidae): evidence for local adaptation to a poor quality host. Oecologia 99:329–336.

Fox, C. W., M. S. Thakar, and T. A. Mousseau. 1997. Egg size plasticity in a seed beetle: an adaptive effect. American Naturalist 149:149–163.

Fox, C. W., M. E. Czesak, T. A. Mousseau, and D. A. Roff. 1999. The evolutionary genetics of an adaptive maternal effect: egg size plasticity in a seed beetle. Evolution 53:552–560.

Frank, S. A. 1998. Foundations of Social Evolution. Princeton University Press, Princeton, NJ.

Freeling, M., and D. C. Bennett. 1985. Maize Adh1. Annual Review of Genetics 19:297–323.

Fricker, M. D., S. G. Gilroy, and A. J. Trewavas. 1990. Signal transduction in plant cells and the calcium message. Pages 89–102 in R. Ranjeva and A. M. Boudet, editors, Signal Perception and Transduction in Higher Plants. Springer Verlag, Berlin.

Fristrom, D., and J. W. Fristrom. 1993. The metamorphic development of the adult epidermis. Pages 843–897 in M. Bate and A. M. Arias, editors, The Development of *Drosophila melanogaster*. Cold Spring Harbor Laboratory Press, Plainview, NY.

Fry, J. D. 1992. The mixed-model analysis of variance applied to quantitative genetics: biological meaning of the parameters. Evolution 46:540–550.

Fry, J. D. 1996. The evolution of host specialization: are tradeoffs overrated? American Naturalist 148:S84–S107.

Futuyma, D. J. 1983. Evolutionary interactions among herbivorous insects and plants. Pages 207–231 in D. J. Futuyma and M. Slatkin, editors, Coevolution. Sinauer Associates, Inc., Sunderland, MA.

Futuyma, D. J., and G. Moreno. 1988. The evolution of ecological specialization. Annual Review of Ecology and Systematics 19:207–233.

Gabriel, W. 1998. Evolution of reversible plastic responses: inducible defenses and environmental tolerance. Pages 286–305 in R. Tollrian and C. D. Harvell, editors, The Ecology and Evolution of Inducible Defenses. Princeton University Press, Princeton, NJ.

Gabriel, W., and M. Lynch. 1992. The selective advantage of reaction norms for environmental tolerance. Journal of Evolutionary Biology 5:41–59.

Galis, F., and E. G. Drucker. 1996. Pharyngeal biting mechanics in centrarchid and cichlid fishes: insights into a key evolutionary innovation. Journal of Evolutionary Biology 9:641–670.

Galis, F., and J. A. J. Metz. 1998. Why are there so many cichlid species? Trends in Ecology and Evolution 13:1–2.

Galloway, L. F. 1995. Response to natural environmental heterogeneity: maternal effects and selection on life-history characters and plasticities in *Mimulus guttatus*. Evolution 49:1095–1107.

Galton, V. A. 1985. 3,5,3'-Triiodothyronine receptors and thyroxine 5'-monodeiodonating activity in thyroid hormone-insensitive amphibia. General and Comparative Endocrinology 57:465–741.

Galton, V. A. 1992a. The role of thyroid hormone in amphibian metamorphosis. Trends in Endocrinology and Metabolism 3:96–100.

Galton, V. A. 1992b. Thyroid hormone receptors and iodothyronine deiodinases in the developing Mexican axolotl, *Ambystoma mexicanum*. General and Comparative Endocrinology 85:62–70.

García-Dorado, A. 1990. The effect of soft selection on variability of a quantitative trait. Evolution 44:168–179.

Garland, T., Jr., and S. C. Adolph. 1994. Why not to do two-species comparative studies: limitations on inferring adaptation. Physiological Zoology 67:797–828.

Gause, G. F. 1941. The effect of natural selection in the acclimitization of *Euplotes* to different salinities of the medium. Journal of Experimental Zoology 87:85–100.

Gause, G. F. 1942. The relation of adaptability to adaptation. Quarterly Review of Biology 17:99–114.

Gause, G. F. 1947. Problems of evolution. Transactions of the Connecticut Academy of Sciences 37:17–68.

Gavrilets, S. Y. 1986. An approach to modeling the evolution of populations with consideration of genotype-environment interaction. Soviet Genetics 22:28–36.

Gavrilets, S. Y. 1988. Evolution of modificational variability in random environment. Journal of General Biology (USSR) 49:271–276.

Gavrilets, S., and S. M. Scheiner. 1993a. The genetics of phenotypic plasticity. V. Evolution of reaction norm shape. Journal of Evolutionary Biology 6:31–48.

Gavrilets, S., and S. M. Scheiner. 1993b. The genetics of phenotypic plasticity. VI. Theoretical predictions for directional selection. Journal of Evolutionary Biology 6:49–68.

Geber, M. A. 1989. Interplay of morphology and development on size inequality: a *Polygonum* greenhouse study. Ecological Monographs 59:267–288.

Gebhardt, M. D., and S. C. Stearns. 1988. Reaction norms for development time and weight at eclosion in *Drosophila mercatorum*. Journal of Evolutionary Biology 1:335–354.

Gedroc, J. J., K. D. M. McConnaughay, and J. S. Coleman. 1996. Plasticity in root/shoot partitioning: optimal, ontogenetic, or both? Functional Ecology 10:44–50.

Gerisch, B., C. Weitzel, C. Kober-Eisermann, V. Rottiers, and A. Antebi. 2001. A hormonal signaling pathway influencing *C. elegans* metabolism, reproductive development, and life span. Developmental Cell 1:841–851.

Gersani, M., Z. Abramsky, and O. Falik. 1998. Density-dependent habitat selection in plants. Evolutionary Ecology 12:223–234.

Getty, T. 1996. The maintenance of phenotypic plasticity as a signal detection problem. American Naturalist 148:378–385.

Gibert, P., and R. B. Huey. 2001. Chill-coma temperature in *Drosophila*: effects of developmental temperature, latitude and phylogeny. Physiological and Biochemical Zoology 74:429–434.

Gibert, P., B. Moreteau, Jean-C. Moreteau, and J. R. David. 1996. Growth temperature and adult pigmentation in two *Drosophila* sibling species: an adaptive convergence of reaction norms in sympatric populations? Evolution 50:2346–2353.

Gibert, P., B. Moreteau, J. C. Moreteau, and J. R. David. 1997. Genetic variability of quantitative traits in *Drosophila melanogaster* (fruit fly) natural populations: analysis of wild-living flies and of several laboratory generations. Heredity 80:326–335.

Gibert, P., B. Moreteau, J. R. David, and S. M. Scheiner. 1998a. Describing the evolution of reaction norm shape: body pigmentation in *Drosophila*. Evolution 52:1501–1506.

Gibert, P., B. Moreteau, J. C. Moreteau, R. Parkash, and J. R. David. 1998b. Light body pigmentation in Indian *Drosophila melanogaster*: a likely adaptation to a hot and arid climate. Journal of Genetics 77:13–20.

Gibert, P., B. Moreteau, S. M. Scheiner, and J. R. David. 1998c. Phenotypic plasticity of body pigmentation in *Drosophila*: correlated variation between segments. Genetics, Selection, Evolution 30:183–196.

Gibert, P., B. Moreteau, and J. R. David. 2000. Developmental constraints of an adaptive plasticity: reaction norms of pigmentation in adult segments of *Drosophila melanogaster*. Evolution and Development 2:249–260.

Gibert, P., R. B. Huey, and G. W. Gilchrist. 2001. Locomotor performance of *Drosophila melanogaster*: interactions among developmental and adult temperatures, age and geography. Evolution 55:205–209.

Gibert, P., P. Capy, A. Imasheva, B. Moreteau, J. P. Morin, G. Pétavy, and J. R. David. 2003. Comparative analysis of morphological traits among *Drosophila melanogaster* and *D. simulans*: genetic variability, clines and phenotypic plasticity. Genetica 118:(in press).

Gibson, G. 2002. Microarrays in ecology and evolution: a preview. Molecular Ecology 11:17–24.

Gibson, G., and D. S. Hogness. 1996. Effect of polymorphism in the *Drosophila* regulatory gene ultrabithorax on homeotic stability. Science 271:200–203.

Gilbert, S. 1994. Dobzhansky, Waddington, and Schmalhausen: embryology and the modern synthesis. Pages 143–154 in M. B. Adams, editor, The Evolution of Theodosius Dobzhansky. Princeton University Press, Princeton, NJ.

Gilbert, S. F. 2000. Developmental Biology. Sinauer Associates, Inc., Sunderland, MA.

Gilchrist, G. W. 1995. Specialists and generalists in changing environments . I. Fitness landscapes of thermal sensitivity. American Naturalist 146:252–270.

Gillespie, J. H., and M. Turelli. 1989. Genotype-environment interaction and the maintenance of polygenic variation. Genetics 121:129–138.

Givnish, T. J. 1982. On the adaptive significance of leaf height in forest herbs. American Naturalist 120:353–381.

Givnish, T. J. 1986. Optimal stomatal conductance, allocation of energy between leaves and roots, and the marginal cost of transpiration. Pages 171–214 in T. J. Givnish, editor, On the Economy of Plant Form and Function. Cambridge University Press, New York.

Godwin, J., and D. Crews. 1997. Sex differences in the nervous system of reptiles. Cellular and Molecular Neurobiology 17:649–669.

Golden, J. W., and D. L. Riddle. 1984. The *C. elegans* dauer larva, developmental effects of pheromone, food and temperature. Developmental Biology 102:368–378.

Goldschmidt, R. 1920. Mechanismus und Physiologie der Geschlechtsbestimmung. Gebrüder Borntraeger, Berlin.

Goldschmidt, R. 1928. Einführung in Die Vererbungswissenschaft, 5th edition. Julius Springer, Berlin.

Goldschmidt, R. 1938. Physiological Genetics. McGraw-Hill, New Haven, CT.

Gomulkiewicz, R., and M. Kirkpatrick. 1992. Quantitative genetics and the evolution of reaction norms. Evolution 46:390–411.

Goodnight, C. J. 1991. Intermixing ability in two-species communities of *Tribolium* flour beetles. American Naturalist 138:342–354.

Goodnight, C. J. 2000. Quantitative trait loci and gene interaction: the quantitative genetics of metapopulations. Heredity 84:589–600.

Gotthard, K. 1998. Life history plasticity in the satyrine butterfly *Lasiommata petropolitana*: investigating an adaptive reaction norm. Journal of Evolutionary Biology 11:21–39.

Gotthard, K., and S. Nylin. 1995. Adaptive plasticity and plasticity as an adaptation: a selective review of plasticity in animal morphology and life history. Oikos 74:3–17.

Gotthard, K., S. Nylin, and C. Wiklund. 1994. Adaptive variation in growth rate: life history costs and consequences in the speckled wood butterfly, *Pararge aegeria*. Oecologia 99:281–289.

Gottlieb, G. 1992. Individual Development and Evolution: The Genesis of Novel Behavior. Oxford University Press, New York.

Gould, S. J., and R. Lewontin. 1979. The spandrels of San Marco and the Panglossian paradigm. Proceedings of the Royal Society of London, Series B. Biological Sciences 205:581–598.

Grantham, T. A. 1995. Hierarchical approaches to macroevolution: recent work on species selection and the "effect hypothesis." Annual Review of Ecology and Systematics 26:301–321.

Grbi, M., and M. R. Strand. 1998. Shifts in the life history of parasitic wasps correlate with pronounced alterations in early development. Proceedings of the National Academy of Sciences, U.S.A. 95:1097–1101.

Greene, E. 1989a. A diet-induced developmental polymorphism in a caterpillar. Science 243:643–646.

Greene, E. 1989b. Effect of light quality and larval diet on morph induction in the polymorphic caterpillar *Nemoria arizonaria* (Lepidoptera: Geometridae). Biological Journal of the Linnean Society 58:277–285.

Gregor, J. W. 1956. Adaptation and ecotypic components. Proceedings of the Royal Society of London, Series B. Biological Sciences 145:333–337.

Griffing, B. 1977. Selection for populations of interacting phenotypes. Pages 413–434 in E. Pollak, O. Kempthorne, and T. B. Bailey, editors, Proceedings of the International Conference on Quantitative Genetics. Iowa State University Press, Ames, IA.

Griffing, B. 1989. Genetic analysis of plant mixtures. Genetics 122:943–956.

Grime, J. P. 1977. Evidence for the existence of three primary strategies in plants and its relevance to ecological and evolutionary theory. American Naturalist 111:1169–1194.

Gross, J., B. C. Husband, and S. C. Stewart. 1998. Phenotypic selection in a natural population of *Impatiens pallida* Nutt. (Balsaminaceae). Journal of Evolutionary Biology 11:589–609.

Gross, K. L., A. Peters, and K. S. Pregitzer. 1993. Fine root growth and demographic responses to nutrient patches in four old-field plant species. Oecologia 95:61–64.

Gross, K. L., K. S. Pregitzer, and A. J. Burton. 1995. Spatial variation in nitrogen availability in three successional plant communities. Journal of Ecology 83:357–367.

Gross, M. R. 1996. Alternative reproductive strategies and tactics: diversity within sexes. Trends in Ecology and Evolution 11:92–98.

Grunbaum, D. 1998. Hydromechanical mechanisms of colony organization and the cost of defense in an encrusting bryozoan, *Membranipora membranacea*. Limnology and Oceanography 42:741–752.

Guntrip, J., and R. M. Sibly. 1998. Phenotypic plasticity, genotype-by-environment interaction and the analysis of generalism and specialization in *Callosobruchus maculatus*. Heredity 81:198–204.

Guntrip, J., R. M. Sibly, and G. J. Holloway. 1997. The effect of novel environment and sex on the additive genetic variation and covariation in and between emergence body weight and development period in the cowpea weevil, *Callosobruchus maculatus* (Coleoptera, Bruchidae). Heredity 78:158–165.

Gupta, A. P., and R. C. Lewontin. 1982. A study of reaction norms in natural populations of *Drosophila pseudoobscura*. Evolution 36:934–948.

Gurganus, M. C., J. D. Fry, S. V. Nuzhdin, F. G. Pasyukova, R. F. Lyman, and T. F. C. Mackay. 1998. Genotype environment interaction at quantitative trait loci affecting sensory bristle number in *Drosophila melanogaster*. Genetics 149:1883–1898.

Gustafson, A. 1953. The cooperation of genotypes in barley. Hereditas 39:1–18.

Haag, E. S., and J. R. True. 2001. Perspective: from mutants to mechanisms? Assessing the candidate gene paradigm in evolutionary biology. Evolution 55:1077–1084.

Hacia, J. G., J. B. Fan, O. Ryder, L. Jin, K. Edgemon, G. Ghandour, R. A. Mayer, B. Sun, L. Hsie, C. M. Robbins, L. C. Brody, D. Wang, E. S. Lander, R. Lipshutz, S. P. Fodor, and F. S. Collins. 1999. Determination of ancestral alleles for human single-nucleotide polymorphisms using high-density oligonucleotide arrays. Nature Genetics 22:164–167.

Hairston, N. G., Jr., C. L. Holtmeier, W. Lampert, L. J. Weider, D. M. Post, J. M. Fischer, C. E. Cáceres, J. A. Fox, and U. Gaedke. 2001. Natural selection for grazer resistance to toxic cyanobacteria. Evolution 55:2203–2214.

Haldane, J. B. S. 1924. A mathematical theory of natural and artificial selection. Part I. Transactions of the Cambridge Philosophical Society 23:195–217.

Haldane, J. B. S. 1936. Some principles of causal analysis in genetics. Erkenntnis 6:346–357.

Haldane, J. B. S. 1946. The interaction of nature and nurture. Annals of Eugenics 13:197–205.

Halliday, K. J., M. Koornneef, and G. C. Whitelam. 1994. Phytochrome B and at least one other phytochrome mediate the accelerated flowering response of *Arabidopsis thaliana* L. to low red/far-red ratio. Plant Physiology 104:1311–1315.

Hammerstein, P. 1996. Darwinian adaptation, population genetics and the streetcar theory of evolution. Journal of Math Biology 34:511–532.

Hanson, A. D., and A. H. D. Brown. 1984. Three alcohol dehydrogenase genes in wild and cultivated barley: characterization of the products of variant alleles. Biochemical Genetics 22:495–515.

Hanson, A. D., and W. D. Hitz. 1982. Metabolic responses of mesophytes to plant water deficits. Annual Review of Plant Physiology 33:163–203.

Harberd, N. P., and K. J. R. Edwards. 1983. Further studies on the alcohol dehydrogenases in barley: evidence for a third alcohol dehydrogenase locus and data on the effect of alcohol dehydrogenase-1 null mutation in homozygous and in heterozygous condition. Genetical Research 41:109–116.

Hard, J. J., W. E. Bradshaw, and C. M. Holzapfel. 1992. Epistasis and the genetic divergence of photoperiodism between populations of the pitcher-plant mosquito, *Wyeomyia smithii*. Genetics 131:389–396.

Hard, J. J., W. E. Bradshaw, and C. M. Holzapfel. 1993. The genetic basis of photoperiodism and its evolutionary divergence among populations of the pitcher-plant mosquito, *Wyeomyia smithii*. American Naturalist 142:457–473.

Harris, R. N., R. D. Semlitsch, H. M. Wilbur, and J. E. Fauth. 1990. Local variation in the genetic basis of paedomorphosis in the salamander Ambysotma talpoideum. Evolution 44:1588–1603.

Harvell, C. D. 1994. The evolution of polymorphism in colonial invertebrates and social insects. Quarterly Review of Biology 69:155–185.

Harvey, P. H., and A. Purvis. 1999. Understanding the ecological and evolutionary reasons for life history variation: mammals as a case study. Pages 232–248 in J. McGlade, editor, Advanced Theoretical Ecology. Blackwell Scientific Publishing, Oxford.

Heckathorn, S. A., C. A. Downs, T. D. Sharkey, and J. S. Coleman. 1998. The small, methionine-rich chloroplast heat shock protein protects photosystem II electron transport during heat stress. Plant Physiology 116:439–444.

Hensley, F. R. 1993. Ontogenetic loss of phenotypic plasticity of age at metamorphosis in tadpoles. Ecology 74:2405–2412.

Hersh, A. H. 1930. The facet-temperature relation in the Bar series of *Drosophila*. Journal of Experimental Zoology 57:283–306.

Hersh, A. H. 1934. On Mendelian dominance and the serial order of phenotypic effects in the Bar series of *Drosophila melanogaster*. American Naturalist 68:186–189.

Hiesey, W. M., and H. W. Milner. 1965. Physiology of ecological races and species. Annual Review of Plant Physiology 16:203–216.

Hillesheim, E., and S. C. Stearns. 1991. The response of *Drosophila melanogaster* to artificial selection on body weight and its phenotypic plasticity in two larval food environments. Evolution 45:1909–1923.

Hinton, G. E., and S. J. Nowlan. 1987. How learning can guide evolution. Complex Systems 1:495–502.

Ho, M.-W., and P. T. Saunders. 1979. Beyond neo-Darwinism: an epigenetic approach to evolution. Journal of Theoretical Biology 78:573–591.

Ho, M.-W., E. Bolton, and P. T. Saunders. 1983a. The bithorax phenocopy and pattern formation. I. Spatiotemporal characteristics of the phenocopy response. Experimental Cell Biology 51:282–290.

Ho, M.-W., C. Tucker, D. Keeley, and P. T. Saunders. 1983b. Effects of successive generations of ether treatment on penetrance and expression of the bithorax phenocopy in *Drosophila melanogaster*. Journal of Experimental Zoology 225:357–368.

Hoang, A. 2001. Immune response to parasitism reduces resistance of *Drosophila melanogaster* to desiccation and starvation. Evolution 55:2353–2358.

Hochachka, P. W., and G. N. Somero. 1984. Biochemical Adaptation. Princeton University Press, Princeton, NJ.

Hoffmann, A. A., and J. Merila. 1999. Heritable variation and evolution under favourable and unfavourable conditions. Trends in Ecology and Evolution 14:96–101.

Hoffmann, A. A., and M. Schiffer. 1998. Changes in the heritability of five morphological traits under combined environmental stresses in *Drosophila melanogaster*. Evolution 52:1207–1212.

Hogben, L. 1933. Nature and Nurture. W. W. Norton, New York.

Holloway, G. J., H. J. Crocker, and A. Callaghan. 1997. The effects of novel and stressful environments on trait distribution. Functional Ecology 11:579–584.

Holt, R. D., and M. S. Gaines. 1992. Analysis of adaptation in heterogeneous landscapes: implications for the evolution of fundamental niches. Evolutionary Ecology 6:433–447.

Houle, D. 1992. Comparing evolvability and variability in quantitative traits. Genetics 130:195–204.

Houston, A. I., and J. M. McNamara. 1992. Phenotypic plasticity as a state-dependent life-history decision. Evolutionary Ecology 6:243–253.

Huey, R. B., and A. F. Bennett. 1987. Phylogenetic studies of coadaptation: preferred temperatures versus optimal performance temperatures of lizards. Evolution 41:1098–1115.

Huey, R. B., and D. Berrigan. 1996. Testing evolutionary hypotheses of acclimation. Pages 205–238 in I. A. Johnston and A. F. Bennett, editors, Phenotypic and Evolutionary Adaptation to Temperature. Cambridge University Press, Cambridge.

Huey, R., and M. Slatkin. 1976. Cost and benefits of lizard thermoregulation. Quarterly Review of Biology 51:363–384.

Huey, R. B., G. W. Gilchrist, M. L. Carlson, D. Berrigan, and L. Serra. 2000. Rapid evolution of a geographic cline in size in an introduced fly. Science 287:308–309.

Humphrey, L. D., and D. A. Pyke. 1997. Clonal foraging in perennial wheatgrasses: a strategy for exploiting patchy soil nutrients. Journal of Ecology 85:601–610.

Hunt, J., and L. W. Simmons. 1998. Patterns of parental provisioning covary with male morphology in a horned beetle (*Onthophagus taurus*) (Coleoptera: Scarabaeidae). Behavioural Ecology and Sociobiology 42:447–451.

Hunt, J., and L. W. Simmons. 2000. Maternal and paternal effects on offspring phenotype in the dung beetle *Onthophagus taurus*. Evolution 54:936–941.

Hunt, J., and L. W. Simmons. 2001. Status-dependent selection in the dimorphic beetle *Onthophagus taurus*. Proceedings of the Royal Society of London, Series B. Biological Sciences 268:2409–2414.

Hunt, J., and L. W. Simmons. 2002. The genetics of maternal care: direct and indirect genetic effects on phenotype in the dung beetle. *Onthophagus taurus*. Proceedings of the National Academy of Sciences, U.S.A., 99:6828–6832.

Hutchings, J. A., and R. A. Meyers. 1994. The evolution of alternative mating strategies in variable environments. Evolutionary Ecology 8:256–268.

Hutchinson, G. E. 1959. Homage to Santa Rosalia, or, Why are there so many animals? American Naturalist 93:145–159.

Huxley, J. S. 1942. Evolution: The Modern Synthesis. George Allen and Unwin, London.

Imasheva, A. G., V. Loeschcke, L. A. Zhivotovsky, and O. E. Lazenby. 1997. Effects of extreme temperatures on phenotypic variation and developmental stability in *Drosophila melanogaster* and *Drosophila buzzatii*. Biological Journal of the Linnean Society 61:117–126.

Imasheva, A. G., V. Loeschcke, and O. E. Lazenby. 1998. Stress temperatures and quantitative variation in *Drosophila melanogaster*. Heredity 81:246–253.

Imasheva, A. G., B. Moreteau, and J. R. David. 2000. Growth temperature and genetic variability of wing dimensions in *Drosophila*: opposite trends in two sibling species. Genetical Research 76:237–247.

Jablonka, E., and E. Szathmary. 1995. The evolution of information storage and heredity. Trends in Ecology and Evolution 10:206–211.

Jablonka, E., B. Oborny, I. Molnar, E. Kisdi, J. Hofbauer, and T. Czaran. 1995. The adaptive advantage of phenotypic memory in changing environments. Philosophical Transactions of the Royal Society of London, Series B. Biological Sciences 350:133–141.

Jablonski, D. 1987. Heritability at the species level: analysis of geographic ranges of Cretaceous mollusks. Science 238:360–363.

Jacquard, A. 1983. Heritability: one word, three concepts. Biometrics 39:465–477.

Jaenike, J. 1990. Factors maintaining genetic variation for host preference in *Drosophila*. Pages 195–207 in J. S. F. Barker, W. T. Starmer, and R. J. MacIntyer, editors, Ecological and Evolutionary Genetics of *Drosophila*. Plenum, New York.

Jain, S. K. 1978. Inheritance of phenotypic plasticity in soft chess, *Bromus mollis* L. (Gramineae). Experentia 34:835–836.

Jain, S. 1979. Adaptive strategies: polymorphism, plasticity, and homeostasis. Pages 160–187 in O. T. Solbrig, S. Jain, G. B. Johnson, and P. H. Raven, editors, Topics in Plant Population Biology. Columbia University Press, New York.

James, A. C., R. B. R. Azevedo, and L. Partridge. 1995. Cellular basis and developmental timing in a size cline of *Drosophila melanogaster*. Genetics 140:659–666.

Jelinsky, S. A., and L. D. Samson. 1999. Global response of *Saccharomyces cerevisiae* to an alkylating agent. Proceedings of the National Academy of Sciences of the United States of America 96:1486–1491.

Jeschke, J. M., and R. Tollrian. 2000. Density-dependent effects of prey defences. Oecologia 123:391–396.

Jinks, J. L., and H. S. Pooni. 1988. The genetic basis of environmental sensitivity. Pages 505–522 in B. S. Weir, E. J. Eisen, M. M. Goodman, and G. Namkoong, editors, Proceedings of the Second International Conference on Quantitative Genetics. Sinauer Associates, Inc., Sunderland, MA.

Jockusch, E. L. 1997. Geographic variation and phenotypic plasticity of number of trunk vertebrae in slender salamanders, *Batrachoseps* (Caudata: Plethodontidae). Evolution 51:1966–1982.

Johannsen, W. 1909. Elemente der exacten Erblichkeitslehre. Gustav Fischer, Jena.

Johannsen, W. 1911. The genotype conception of heredity. American Naturalist 45:129–159.

Johansson, F. 2002. Reaction norms and production costs of predator-induced morphological defences in a larval dragonfly (*Leucorrhinia dubia*: Odonata). Canadian Journal of Zoology 80:944–950.

Johnson, C. B. 1989. Signal transduction mechanisms in phytochrome action. Pages 229–247 in A. M. Ranjeva, editor, Signal Perception and Transduction in Higher Plants. Springer Verlag, Berlin.

Johnson, M. S. 1982. Polymorphism for direction of coil in *Partula surturalis*: behavioural isolation and positive frequency dependent selection. Heredity 49:145–151.

Jurik, T. W. 1991. Population distributions of plant size and light environment of giant ragweed (*Ambrosia trifida* L.) at three densities. Oecologia 87:539–550.

Kalisz, S. 1986. Variable selection on the timing of germination in *Collinsia verna* (Schrophulariaceae). Evolution 40:479–491.

Kalisz, S., and J. A. Teeri. 1986. Population-level variation in photosynthetic metabolism and growth in *Sedum wrightii*. Ecology 67:20–26.

Kaltenbach, J. C. 1996. Endocrinology of amphibian metamorphosis. Pages 403–423 in L. I. Gilbert, J. R. Tata, and B. G. Atkinson, editors, Metamorphosis: Postembryonic Reprogramming of Gene Expression in Amphibian and Insect Cells. Academic Press, San Diego.

Kanki, K., and M. Wakahara. 1999. Precocious testicular growth in metamorphosis-arrested larvae of a salamander *Hynobius retardatus*: role of thyroid-stimulating hormone. Journal of Experimental Biology 283:548–558.

Kaplan, R. H. 1989. Ovum size plasticity and maternal effects on the early development of the frog, *Bombina orientalis* Boulenger, in a field population in Korea. Functional Ecology 3:597–604.

Kaplan, R. H., and W. S. Cooper. 1984. The evolution of developmental plasticity in reproductive characteristics: an application of the "adaptive coin-flipping" principle. American Naturalist 123:393–410.

Karan, D., and J. R. David. 2000. Cold tolerance in *Drosophila*: adaptive variations revealed by the analysis of starvation survival reaction norms. Journal of Thermal Biology 25:345–351.

Karan, D., J. P. Morin, B. Moreteau, and J. R. David. 1998. Body size and developmental temperature in *Drosophila melanogaster*: analysis of body weight reaction norm. Journal of Thermal Biology 23:301–309.

Karan, D., B. Moreteau, and J. R. David. 1999a. Growth temperature and reaction norms of morphometrical traits in a tropical drosophilid: *Zaprionus indianus*. Heredity 83:398–407.

Karan, D., J. P. Morin, E. Gravot, B. Moreteau, and J. R. David. 1999b. Temporal stability of body size reaction norms in a natural population of *Drosophila melanogaster*. Genetics, Selection, Evolution 31:491–508.

Karan, D., J. P. Morin, P. Gibert, B. Moreteau, S. M. Scheiner, and J. R. David. 2000. The genetics of phenotypic plasticity. IX. Genetic architecture and sex differences in *Drosophila melanogaster*. Evolution 54:1035–1040.

Karban, R., and I. T. Baldwin. 1997. Induced Responses to Herbivory. University of Chicago Press, Chicago.

Kawecki, T. J. 1995a. Demography of source-sink populations and the evolution of ecological niches. Evolutionary Ecology 9:38–44.

Kawecki, T. J. 1995b. Expression of genetic and environmental variation for life history characters on the usual and novel hosts in *Callosobruchus maculatus* (Coleoptera:Bruchidae). Heredity 75:70–76.

Kawecki, T. J., and S. C. Stearns. 1993. The evolution of life histories in spatially heterogeneous environments: optimal reaction norms revisited. Evolutionary Ecology 7:155–174.

Kemp, P., and M. D. Bertness. 1984. Snail shape and growth rates: evidence for plastic shell allometry in *Littorina littorea*. Proceedings of the National Academy of Sciences of the United States of America 81:811–813.

Kennerdell, J. R., and R. W. Carthew. 1998. Use of dsRNA-mediated genetic interference to demonstrate that frizzled and frizzled 2 act in the wingless pathway. Cell 95:1017–1026.

Kikuyama, S., K. Kawamura, S. Tanaka, and K. Yamamoto. 1993. Aspects of amphibian metamorphosis: hormonal control. International Review of Cytology 145:105–148.

King, J. 1991. The Genetic Basis of Plant Physiological Processes. Oxford University Press, Oxford.

Kingsolver, J. G., and R. B. Huey. 1998. Evolutionary analyses of morphological and physiological plasticity in thermally variable environments. American Zoologist 38:545–560.

Kirkpatrick, M. 1996. Genes and adaptation: a pocket guide to the theory. Pages 123–146 in M. R. Rose and G. V. Lauder, editors, Adaptation. Academic Press, San Diego.

Kirkpatrick, M., and N. Heckman. 1989. A quantitative genetic model for growth, shape, reaction norms, and other infinite-dimensional characters. Journal of Mathematical Biology 27:429–450.

Kirkpatrick, M., and D. Lofsvold. 1992. Measuring selection and constraint in the evolution of growth. Evolution 46:954–971.

Kirkpatrick, M., D. Lofsvold, and M. Bulmer. 1990. Analysis of the inheritance, selection and evolution of growth trajectories. Genetics 124:979–993.

Klingenberg, C. P., and H. F. Nijhout. 1998. Competition among growing organs and developmental control of morphological asymmetry. Proceedings of the Royal Society of London, Series B. Biological Sciences 265:1135–1139.

Koch, P. B., and D. Bückmann. 1987. Hormonal control of seasonal morphs by the timing of ecdysteroid release in *Araschnia levana* L. (Nymphalidae: Lepidoptera). Journal of Insect Physiology 33:823–929.

Koch, P. B., and N. Kaufmann. 1995. Pattern specific melanin synthesis and DOPA decarboxylase activity in a butterfly wing of *Precis coenia* Hübner. Insect Biochemistry and Molecular Biology 25:73–82.

Kohler, R. E. 1994. Lords of the Fly. University of Chicago Press, Chicago.

Kooi, R. E., and P. M. Brakefield. 1999. The critical period for wing pattern induction in the polyphenic tropical butterfly *Bicyclus anynana* (Satyrinae). Journal of Insect Physiology 45:201–212.

Kooi, R. E., P. M. Brakefield, and W. E. M. T. Rossie. 1996. Effects of food plant on phenotypic plasticity in the tropical butterfly *Bicyclus anynana*. Entomologia Experimentalis et Applicata 80:149–151.

Koorneef, M., and R. E. Kendrick. 1986. A genetic approach to photomorphogenesis. Pages 521–546 in R. E. Kendrick and G. H. M. Kronenberg, editors, Photomorphogenesis in Plants. Martinus Nijhoff Publishers, Dordrecht.

Kopp, A., I. Duncan, and S. B. Carroll. 2000. Genetic control and evolution of sexually dimorphic characters in *Drosophila*. Nature 408:553–559.

Kozlowski, J., and R. G. Wiegert. 1986. Optimal allocation of energy to growth and reproduction. Theoretical Population Biology 29:16–37.

Kozlowski, J., and R. G. Wiegert. 1987. Optimal age and size at maturity in annuals and perennials with determinate growth. Evolutionary Ecology 1:231–244.

Krafka, J. 1920. The effect of temperature upon facet number in the bar-eyed mutant of *Drosophila*. Journal of General Physiology 2:409–464.

Krebs, J. R., and N. B. Davies. 1996. Behavioural Ecology: An Evolutionary Approach. Blackwell Scientific Publication, Oxford.

Krebs, R. A., and M. E. Feder. 1998. Experimental manipulation of the cost of thermal acclimation in *Drosophila melanogaster*. Biological Journal of the Linnean Society 63:593–601.

Krebs, R. A., M. E. Feder, and J. Lee. 1998. Heritability of expression of the 70KD heat-shock protein in *Drosophila melanogaster* and its relevance to the evolution of thermotolerance. Evolution 52:841–847.

Kudoh, H., Y. Ishiguri, and S. Kawano. 1996. Phenotypic plasticity in age and size at maturity and its effects on the integrated phenotypic expressions of life history traits of *Cardamine flexiosa* (Cauciferal). Journal of Evolutionary Biology 9:541–570.

Kwesiga, F., and J. Grace. 1986. The role of the red/far-red ratio in the response of tropical tree seedlings to shade. Annals of Botany 57:283–290.

Lande, R. 1979. Quantitative genetic analysis of multivariate evolution, applied to brain:body size allometry. Evolution 33:402–416.

Lande, R. 1984. The genetic correlation between characters maintained by selection, linkage and inbreeding. Genetical Research 44:309–320.

Lande, R., and S. J. Arnold. 1983. The measurement of selection on correlated characters. Evolution 37:1210–1226.

Lande, R., and S. Shannon. 1996. The role of genetic variation in adaptation and population presistence in a changing enviromnent. Evolution 50:434–437.

Langerhans, R. B., and T. J. DeWitt. 2002. Plasticity constrained: overgeneralized environmental cues induce phenotype errors in a freshwater snail. Evolutionary Ecology Research 4:857–870.

Larson, A., and J. B. Losos. 1996. Phylogenetic systematics of adaptation. Pages 187–220 in M. R. Rose and G. V. Lauder, editors, Adaptation. Academic Press, San Diego.

Lauder, G. V. 1996. The argument from design. Pages 55–91 in M. R. Rose and G. V. Lauder, editors, Adaptation. Academic Press, San Diego.

Layzer, D. 1974. Heritability analyses of IQ scores: science or numerology? Science 183:1259–1266.

Leather, S. R. 1985. Oviposition preferences in relation to larval growth rates and survival in the pine beauty moth, *Panolis flammea*. Ecological Entomology 10:213–217.

Lechowicz, M. J., D. J. Schoen, and G. Bell. 1988. Environmental correlates of habitat distribution and fitness components in *Impatiens capensis* and *Impatiens pallida*. Journal of Ecology 76:1043–1054.

Leips, J., and J. Travis. 1994. Metamorphic responses to changing food levels in two species of hylid frogs. Ecology 75:1345–1356.

Leon, J. A. 1993. Plasticity in fluctuating environments. Pages 105–121 in J. Yoshimura and C. W. Clark, editors, Lecture Notes in Biomathematics, Volume 98. Adaptation in Stochastic Environments. Springer Verlag, Berlin.

Lerner, I. M. 1950. Population Genetics and Animal Improvement. Cambridge University Press, Cambridge.

Lerner, I. M. 1954. Genetic Homeostasis. Oliver and Boyd, London.

Leroi, A. M., R. E. Lenski, and A. F. Bennett. 1994. Evolutionary adaptation to temperature. III. Adaptation of *Escherichia coli* to a temporally varying environment. Evolution 48:1222–1229.

Levin, D. A. 1988. Plasticity, canalization and evolutionary stasis in plants. Pages 35–45 in A. J. Davy, M. J. Hutchings, and A. R. Watkinson, editors, Plant Population Ecology. Blackwell Scientific Publications, Oxford.

Levins, R. 1962. Theory of fitness in a heterogeneous environment I. The fitness set and adaptive function. American Naturalist 96:361–378.

Levins, R. 1963. Theory of fitness in a heterogeneous environment II. Developmental flexibility and niche selection. American Naturalist 97:75–90.

Levins, R. 1968. Evolution in Changing Environments. Princeton University Press, Princeton, NJ.

Lewontin, R. 1955. The effects of population density and composition on viability in *Drosophila melanogaster*. Evolution 9:27–41.

Lewontin, R. C. 1974. The analysis of variance and the analysis of causes. American Journal of Human Genetics 26:400–411.

Lewontin, R. C., and D. Cohen. 1969. On population growth in a randomly varying environment. Proceedings of the National Academy of Sciences of the United States of America 62:1056.

Lewontin, R., and Y. Matsuo. 1963. Interaction of genotypes determining the viability in *Drosophila busckii*. Proceedings of the National Academy of Sciences of the United States of America 49:270–278.

Liem, K. F. 1990. Key evolutionary innovations, differential diversity, and symecomorphosis. Pages 147–170 in M. Nitecki, editor, Evolutionary Innovations. University of Chicago Press, Chicago.

Lilliendahl, K. 1997. The effect of predator presence on body mass in captive greenfinches. Animal Behavior 53:75–81.
Lima, S. L., and L. M. Dill. 1990. Behavioral decisions made under the risk of predation: a review and prospectus. Canadian Journal of Zoology 68:619–640.
Linhart, Y. B., and M. C. Grant. 1966. Evolutionary significance of local genetic differentiation in plants. Annual Review of Ecology and Systematics 27:237–277.
Lively, C. M. 1986a. Canalization versus developmental conversion in a spatially variable environment. American Naturalist 128:561–572.
Lively, C. M. 1986b. Predator-induced shell dimorphism in the acorn barnacle *Chthamalus anisopoma*. Evolution 40:232–242.
Lively, C. M. 1996. Host-parasite coevolution and sex. Bioscience 46:107–114.
Lively, C. M. 1999a. Developmental strategies in spatially variable environments: barnacle shell dimorphism and strategic models of selection. Pages 245–258 in R. Tollrian and C. D. Harvell, editors, The Ecology and Evolution of Inducible Defenses. Princeton University Press, Princeton, NJ.
Lively, C. M. 1999b. Migration, virulence, and the geographic mosaic of adaptation by parasites. American Naturalist 153:S34–S47.
Lively, C. M., W. N. Hazel, M. J. Schellenberger, and K. S. Michelson. 2000. Predator-induced defense: variation for inducibility in an intertidal barnacle. Ecology 81:1240–1247.
Lloyd Morgan, C. 1896. Habit and Instinct. E. Arnold, London.
Lloyd Morgan, C. 1900. Animal Behaviour. E. Arnold, London.
Loeb, M. L. G., J. P. Collins, and T. J. Maret. 1994. The role of prey in controlling expression of a trophic polymorphism in *Ambystoma tigrinum nebulosum*. Functional Ecology 8:151–158.
Loeschcke, V., and R. A. Krebs. 1996. Selection for heat-shock resistance in larval and in adult *Drosophila buzzatii*: comparing direct and indirect responses. Evolution 50:2354–2359.
Lundholm, J. T., and L. W. Aarssen. 1994. Neighbour effects on gender variation in *Ambrosia artemisiifolia*. Canadian Journal of Botany 72:794–800.
Lush, J. L. 1943. Animal Breeding Plans, 2nd edition. Collegiate Press, Ames, IA.
Lynch, M., and W. Gabriel. 1987. Enviromental tolerance. American Naturalist 129:283–303.
Lynch, M., and B. Walsh. 1997. Genetics and Analysis of Quantitative Traits. Sinauer Associates, Inc., Sunderland, MA.
Lyons, S. K., and M. R. Willig. 1999. A hemispheric assessment of scale dependence in latitudinal gradients of species richness. Ecology 80:2483–2491.
MacArthur, R. H. 1957. On the relative abundance of species. Proceedings of the National Academy of Sciences of the United States of America 43:293–295.
MacArthur, R. H. 1960. On the relative abundance of species. American Naturalist 94:25–36.
Macdonald, S. E., and C. C. Chinnappa. 1989. Population differentiation for phenotypic plasticity in the *Stellaria longipes* complex. American Journal of Botany 76:1627–1637.
MacKenzie, A. 1996. A trade-off for host plant utilization in the black bean aphid *Aphis fabae*. Evolution 50:155–162.
Malhó, R. 1999. Coding information in plant cells: the multiple roles of Ca^{2+} as a second messenger. Plant Biology 1:487–494.
Maliakal, S., K. McDonnell, S. A. Dudley, and J. Schmitt. 1999. Effects of red to far-red ratio and plant density on biomass allocation and gas exchange in *Impatiens capensis*. International Journal of Plant Science 160:723–733.
Mangel, M., and C. W. Clark. 1986. Dynamic Modeling in Behavioral Ecology. Princeton University Press, Princeton, NJ.
Marko, P. B., and A. R. Palmer. 1991. Responses of a rocky shore gastropod to the effluents of predatory and non-predatory crabs: avoidance and attraction. Biological Bulletin 181:363–370.
Marshall, D. R., and S. K. Jain. 1968. Phenotypic plasticity of *Avena fatua* and *A. barbata*. American Naturalist 102:457–467.
Martin, M. M., and J. Harding. 1981. Evidence for the evolution of competition between two species of annual plants. Evolution 35:975–987.
Matsuda, R. 1982. The evolutionary process in talitrid amphipods and salamanders in changing environments, with a discussion of "genetic assimilation" and some other evolutionary concepts. Canadian Journal of Zoology 60:733–749.

Maynard Smith, J. 1982. Evolution and the Theory of Games. Cambridge University Press, Cambridge.

McCollum, S. A., and J. D. Leimberger. 1997. Predator-induced morphological changes in an amphibian: predation by dragonflies affects tadpole color, shape, and growth rate. Oecologia 109:615–621.

McCollum, S. A., and J. Van Buskirk. 1996. Costs and benefits of a predator-induced polyphenism in the gray treefrog Hyla chrysoscelis. Evolution 50:583–593.

McLaren, A. 1999. Too late for the midwife toad: stress, variability and Hsp90. Trends in Genetics 15:169–171.

McLeod, L. 1968. Controlled environment experiments with *Precis octavia* Cram. (Nymphalidae). Journal of Research on the Lepidoptera 7:1–18.

McNamara, J. M., and A. I. Houston. 1986. The common currency for behavioral decisions. American Naturalist 127:358–378.

McNamara, J. M., and A. I. Houston. 1990. State-dependent ideal free distribution. Evolutionary Ecology 4:298–311.

Meffert, L. M. 1995. Bottleneck effects on genetic variance for courtship repertoire. Genetics 139:365–374.

Meffert, L. M. 2000. The evolutionary potential of morphology and mating behavior: the role of epistasis in bottlenecked populations. Pages 177–193 in J. B. Wolf, E. D. Brodie III, and M. J. Wade, editors, Epistasis and the Evolutionary Process. Oxford University Press, New York.

Meffert, L. M., J. L. Regan, and B. W. Brown. 1999. Convergent evolution of the mating behaviour of founder-flush populations of the housefly. Journal of Evolutionary Biology 12:859–868.

Metzger, J. D. 1996. A physiological comparison of vernalization and dormancy chilling requirement. Pages 147–156 in G. A. Lang, editor, Plant Dormancy, Physiology, Biochemistry and Molecular Biology. CAB International, Oxfordshire, UK.

Meyer, A. 1987. Phenotypic plasticity and heterochrony in *Cichlasoma managuense* (Pisces, Cichlidae) and their implications for speciation in Cichlid fishes. Evolution 41:1357–1369.

Milinski, M., and G. A. Parker. 1991. Competition for resources. Pages 137–168 in J. R. Krebs and N. B. Davies, editors, Behavioural Ecology: An Evolutionary Approach, 3rd edition. Blackwell Scientific Publications, Oxford.

Miller, R. E., J. M. Ver Hoef, and N. L. Fowler. 1995. Spatial heterogeneity in eight central Texas grasslands. Journal of Ecology 83:919–928.

Miller, T. E. 1995. Evolution of *Brassica rapa* L. (Cruciferae) populations in intra- and interspecific competition. Evolution 49:1125–1133.

Miranda, J. R., M. Hemmat, and P. Eggleston. 1991. The competition diallel and the exploitation and interference components of larval competition in *Drosophila melanogaster*. Heredity 66:333–342.

Mitchell, P. L., and F. I. Woodward. 1988. Responses of three woodland herbs to reduced photosynthetically active radiation and low red to far-red ratios in shade. Journal of Ecology 76:807–825.

Mitchell-Olds, T., and R. G. Shaw. 1987. Regression analysis of natural selection: statistical inference and biological interpretation. Evolution 41:1149–1161.

Mitter, C., and D. J. Futuyma. 1983. An evolutionary-genetic view of host-plant utilization by insects. Pages 427–459 in R. F. Denno and M. S. McClure, editors, Variable Plants and Herbivores in Natural and Managed Systems. Academic Press, New York.

Moczek, A. P. 1998. Horn polyphenism in the beetle *Othophagus tarusus*: larval diet quality and plasticity in parental investment determine adult body size and male horn morphology. Behavioral Ecology 9:636–641.

Moczek, A. P., and D. J. Emlen. 2000. Male horn dimorphism in the scarab beetle, *Onthophagus taurus*: do alternative reproductive tactics favor alternative phenotypes? Animal Behaviour 59:459–466.

Monterio, A., P. M. Brakefield, and V. French. 1997. Butterfly eyespots: the genetics and development of the color rings. Evolution 51:1207–1216.

Moore, A. J., and C. R. B. Boake. 1994. Optimality and evolutionary genetics: complementary procedures for evolutionary analysis in behavioural ecology. Trends in Ecology and Evolution 9:69–72.

Moore, A. J., E. D. Brodie III, and J. B. Wolf. 1997. Interacting phenotypes and the evolutionary process. I. Direct and indirect genetic effects of social interactions. Evolution 51:1352–1362.

Moran, N. A. 1988. The evolution of host-plant alteration in aphids: evidence for specialization as a dead end. American Naturalist 132:681–706.

Moran, N. A. 1992. The evolutionary maintenance of alternative phenotypes. American Naturalist 139:971–989.

Moran, P. A. P. 1973. A note on heritability and the correlation between relatives. Annals of Human Genetics 37:217.

Moreteau, B., J. P. Morin, P. Gibert, G. Pétavy, E. Pla, and J. R. David. 1997. Evolutionary changes of nonlinear reaction norms according to thermal adaptation: a comparison of two *Drosophila* species. Comptes Rendus de l'Academie des Sciences 320:833–841.

Moreteau, B., A. G. Imasheva, J. P. Morin, and J. R. David. 1998. Wing shape and developmental temperature in two *Drosophila* sibling species: different regions exhibit different norms of reaction. Russian Journal of Genetics 34:183–192.

Moreteau, B., P. Gibert, J.-M. Delpuech, G. Pétavy, and J. R. David. 2003. Phenotypic plasticity of sternopleural bristle number in temperate and tropical populations of *Drosophila melanogaster*. Genetical Research, Cambridge 81:25–35.

Morey, S., and D. Reznick. 2000. A comparative analysis of plasticity in larval development in three species of spadefoot toads. Ecology 81:1736–1749.

Morgan, D. C., and H. Smith. 1979. A systematic relationship between phytochrome-controlled development and species habitat, for plants grown in simulated natural radiation. Planta 145:253–258.

Morin, J. P., B. Moreteau, G. Pétavy, A. G. Imasheva, and J. R. David. 1996. Body size and developmental temperature in *Drosophila simulans*: comparison of reaction norms with sympatric *Drosophila melanogaster*. Genetics, Selection, Evolution 28:415–436.

Morin, J. P., B. Moreteau, G. Pétavy, and J. R. David. 1999. Divergence of reaction norms of size characters between tropical and temperate populations of *Drosophila melanogaster* and *D. simulans*. Journal of Evolutionary Biology 12:329–339.

Mousseau, T. A., and D. A. Roff. 1995. Genetic and environmental contributions to geographic variation in the ovipositor length of a cricket. Ecology 76:1473–1482.

Moxon, E. R., and C. F. Higgins. 1997. A blueprint for life. Nature 389:120–121.

NRC (National Research Council, United States). 1985. Models for biomedical research, a new perspective. A report by the Committee on Models for Biomedical Research. National Academy Press, Washington, D.C.

Newman, R. A. 1987. Effects of density and predation on *Scaphiopus couchii* tadpoles in desert ponds. Oecologia 71:301–307.

Newman, R. A. 1988a. Adaptive plasticity in development of *Scaphiopus couchii* tadpoles in desert ponds. Evolution 42:774–783.

Newman, R. A. 1988b. Genetic variation for larval anuran (*Scaphiopus couchii*) development time in an uncertain environment. Evolution 42:763–773.

Newman, R. A. 1989. Developmental plasticity of *Scaphiopus couchii* tadpoles in an unpredictable environment. Ecology 70:1775–1787.

Newman, R. A. 1992. Adaptive plasticity in amphibian metamorphosis. Bioscience 42:671–678.

Newman, R. A. 1994a. Effects of changing density and food levels on metamorphosis of a desert amphibian, *Scaphiopus couchii*. Ecology 75:1085–1096.

Newman, R. A. 1994b. Genetic variation for phenotypic plasticity in the larval life history of spadefoot toads (*Scaphiopus couchii*). Evolution 48:1773–1785.

Newman, R. A., and A. E. Dunham. 1994. Size at metamorphosis and water loss in a desert anuran (*Scaphiopus couchii*). Copeia 1994:372–381.

Niewiarowski, P. H., and W. Roosenburg. 1993. Reciprocal transplant reveals sources of variation in growth rates of the lizard *Sceloporus undulatus*. Ecology 74:1992–2002.

Nijhout, H. F. 1980. Pattern formation on lepidopteran wings: determination of an eyespot. Developmental Biology 80:267–274.

Nijhout, H. F. 1991. The Development and Evolution of Butterfly Wing Patterns. Smithsonian Institution, Washington, DC.

Nijhout, H. F. 1994. Insect Hormones. Princeton University Press, Princeton, NJ.

Nijhout, H. F. 1999a. Control mechanisms of polyphenic development in insects. Bioscience 49:181–192.

Nijhout, H. F. 1999b. Hormonal control in larval development and evolution—insects. Pages 217–254 in B. K. Hall and M. H. Wake, editors, The Origin and Evolution of Larval Forms. Academic Press, San Diego.

Nijhout, H. F., and D. J. Emlen. 1998. Competition among body parts in the development and evolution of insect morphology. Proceedings of the National Academy of Sciences of the United States of America 95:3685–3689.

Nilsson-Ehle, H. 1914. Vilka erfarenheter hava hittills vunnits rörande möjligheten av växters acklimatisering. Kgl Landtbruks-Akad Handl Tidskr 53:537–572.

Noach, E. J. K., G. de Jong, and W. Scharloo. 1996. Phenotypic plasticity in morpholgical traits in two populations of *Drosophila melanogaster*. Journal of Evolutionary Biology 9:831–844.

Norris, R. D. 1991. Biased extinction and evolutionary trends. Paleobiology 17:388–399.

Novoplansky, A., D. Cohen, and T. Sachs. 1990. How portulaca seedlings avoid their neighbours. Oecologia 82:490–493.

Nylin, S. 1992. Seasonal plasticity in life history traits: growth and development in *Polygonia c-album* (Lepidoptera: Nymphalidae). Biological Journal of the Linnean Society 47:301–323.

Nylin, S., and K. Gotthard. 1998. Plasticity in life-history traits. Annual Reviews of Entomology 43:63–83.

Nylin, S., K. Gotthard, and C. Wiklund. 1996. Reaction norms for age and size at maturity in *Lasiommata* butterflies: predictions and tests. Evolution 50:1351–1358.

Oliver, S. G. 1996. From DNA sequence to biological function. Nature 379:597–600.

Olvido, A., and T. A. Mousseau. 1995. The effect of rearing environment on calling song plasticity in the striped ground cricket. Evolution 49:1271–1277.

O'Neill, S. D. 1997. Pollination regulation of flower development. Annual Review of Plant Physiology and Plant Molecular Biology 48:547–574.

Orr, H. A., and J. A. Coyne. 1992. The genetics of adaptation: a reassessment. American Naturalist 140:725–742.

Orzack, S. H. 1985. Population dynamics in variable environments V. The genetics of homeostasis revisited. American Naturalist 125:550–572.

Osborn, H. F. 1897a. The limits of organic selection. American Naturalist 31:944–951.

Osborn, H. F. 1897b. Organic selection. Science 15:583–587.

Paddison, P. J., A. A. Caudy, and G. J. Hannon. 2002. Stable suppression of gene expression by RNAi in mammalian cells. Proceedings of the National Academy of Sciences of the United States of America 99:1443–1448.

Padilla, D. K., and S. C. Adolph. 1996. Plastic inducible morphologies are not always adaptive: the importance of time delays in a stochastic environment. Evolutionary Ecology 10:105–117.

Pagel, M. 1994. The adaptationist wager. Pages 29–51 in P. Eggleton and R. Vane-Wright, editors, Phylogenetics and Ecology. Academic Press, London.

Pál, C., and I. Miklos. 1999. Epigenetic inheritance, genetic assimilation and speciation. Journal of Theoretical Biology 200:19–37.

Palmer, A. R. 1990. Effect of crab effluent and scent of damaged conspecifics on feeding, growth, and shell morphology of the Atlantic dogwhelk *Nucella lapillus* (L.). Hydrobiologia 193:155–182.

Palmer, M. W., and P. S. White. 1994. Scale dependence and the species-area relationship. American Naturalist 144:717–740.

Parejko, K. 1992. Embryology of *Chaoborus*-induced spines in *Daphnia pulex*. Hydrobiologia 231:77–84.

Parejko, K., and S. I. Dodson. 1991. The evolutionary ecology of an antipredator reaction norm: *Daphnia pulex* and *Chaoborus americanus*. Evolution 45:1665–1674.

Parichy, D. M., and R. H. Kaplan. 1995. Maternal investment and developmental plasticity: functional consequences for locomotor performance of hatchling frog larvae. Functional Ecology 9:606–617.

Park, T. 1936. Studies in population physiology. VI. The effects of differentially conditioned flour upon fecundity and fertility of *Tribolium confusum*. Journal of Experimental Zoology 73:393–404.

Parker, G., and J. Maynard Smith. 1990. Optimality theory in evolutionary biology. Nature 348:27–33.

Patel, M. N., C. G. Knight, C. Karageorgi, and A. M. Leroi. 2002. Evolution of germ-line signals that regulate growth and aging in nematodes. Proceedings of the National Academy of Sciences of the United States of America 99:769–774.

Paterson, A. H., S. Damon, J. D. Hewett, D. Zamir, H. D. Rabinowitch, S. E. Lincoln, E. S. Lander, and S. D. Tanksley. 1991. Mendelian factors underlying quantitative traits in tomato: comparison across species, generations, and environments. Genetics 127:181–197.

Peichel, C. L., K. S. Nereng, K. A. Ohgi, B. L. Cole, P. F. Colosimo, C. A. Buerkle, D. Schluter, and D. M. Kingsley. 2001. The genetic architecture of divergence between threespine stickleback species. Nature 414:901–905.

Perrin, N., and J. F. Rubin. 1990. On dome-shaped norms of reaction for size-to-age at maturity in fishes. Functional Ecology 4:53–57.

Pétavy, G., J. P. Morin, B. Moreteau, and J. R. David. 1997. Growth temperature and phenotypic plasticity in two *Drosophila* sibling species: probable adaptive changes in flight capacities. Journal of Evolutionary Biology 10:875–887.

Pfennig, D. W. 1990. The adaptive significance of an environmentally cued developmental switch in an anuran tadpole. Oecologia 85:101–107.

Pfennig, D. W. 1992. Polyphenism in spadefoot toad tadpoles as a locally adjusted evolutionarily stable strategy. Evolution 46:1408–1420.

Pfennig, D. W., and P. J. Murphy. 2000. Character displacement in polyphenic tadpoles. Evolution 54:1738–1749.

Philippi, T. and J. Seger. 1989. Hedging one's evolutionary bets, revisited. Trends in Ecology and Evolution 4:41–44.

Phillips, P. C. 1998. The language of gene interaction. Genetics 149:1167–1171.

Phillips, P. C., and S. J. Arnold. 1989. Visualizing multivariate selection. Evolution 43:1209–1222.

Pieau, C., M. Girondot, N. Richard-Mercier, G. Desvages, M. Dorizzi, and P. Zaborski. 1994. Temperature sensitivity of sexual differentiation of gonads in the European pond turtle: hormonal involvement. Journal of Experimental Zoology 270:86–94.

Piersma, T., and A. Linstrom. 1997. Rapid reversible changes in organ size as a component of adaptive behaviour. Trends in Ecology and Evolution 12:134–138.

Pigliucci, M. 1996a. How organisms respond to environmental changes: from phenotypes to molecules (and vice versa). Trends in Ecology and Evolution 11:168–173.

Pigliucci, M. 1996b. Modelling phenotypic plasticity. II. Do genetic correlations matter? Heredity 77:453–460.

Pigliucci, M., and C. D. Schlichting. 1995. Reaction norms of *Arabidopsis* (Brassicaceae). III. Response to nutrients in 26 populations from a worldwide collection. American Journal of Botany 82:1117–1125.

Pigliucci, M., C. D. Schlichting, and J. Whitton. 1995a. Reaction norms of *Arabidopsis*. II. Response to stress and unordered environmental variation. Functional Ecology 9:537–547.

Pigliucci, M., J. Whitton, and C. D. Schlichting. 1995b. Reaction norms of *Arabidopsis*. I. Plasticity of characters and correlations across water, nutrient and light gradients. Journal of Evolutionary Biology 8:421–438.

Pigliucci, M., K. Cammell, and J. Schmitt. 1999. Evolution of phenotypic plasticity: a comparative approach in the phylogenetic neighbourhood of *Arabidopsis thaliana*. Journal of Evolutionary Biology 12:779–791.

Pletcher, S. D., S. J. Macdonald, R. Marguerie, U. Certa, S. C. Stearns, D. B. Goldstein, and L. Partridge. 2002. Genome-wide transcript profiles in aging and calorically restricted *Drosophila melanogaster*. Current Biology 12:712–723.

Precht, H., J. Christophersen, and H. Hensel. 1955. Temperature and Leben. Springer Verlag, Berlin.

Price, G. R. 1970. Selection and covariance. Nature 227:520–521.

Price, T., M. Turelli, and M. Slatkin. 1993. Peak shifts produced by correlated response to selection. Evolution 47:280–290.

Provine, W. B. 1971. The Origins of Theoretical Population Genetics. University of Chicago Press, Chicago.

Quail, P. H. 1991. Phytochrome: a light-activated molecular switch that regulates plant gene expression. Annual Review of Genetics 25:389–409.

Rachootin, S. P., and K. S. Thomson. 1981. Epigenetics, paleontology, and evolution. Pages 181–193 in G. G. E. Scudder and J. L. Reveal, editors, Evolution Today. Hunt Institute for Botanical Documentation, Pittsburgh, PA.

Raff, R. 1996. The Shape of Life: Genes, Development, and the Evolution of Animal Form. University of Chicago Press, Chicago.

Ramachandran, V. S., C. W. Tyler, R. L. Gregory, D. Rogers-Ramachandran, S. Duensing, C. Pillsbury, and C. Ramachandran. 1996. Rapid adaptive camouflage in tropical flounders. Nature 379:815–818.

Rast, J. P., G. Amore, C. Calestani, C. B. Livi, A. Ransick, and E. H. Davidson. 2000. Recovery of developmentally defined gene sets from high-density cDNA macroarrays. Developmental Biology 228:270–286.

Rausher, M. D. 1983. Ecology of host-race behavior in phytophagous insects. Pages 223–257 in R. F. Denno and M. S. McClure, editors, Variable Plants and Herbivores in Natural and Managed Ecosystems. Academic Press, New York.

Real, L., and T. Caraco. 1986. Risk and foraging in stochastic environments. Annual Review of Ecology and Systematics 17:371–390.

Reboud, X., and G. Bell. 1997. Experimental evolution of *Chlamydomonas*. III. Evolution of specialist and generalist types in environments that vary in space and time. Heredity 78:507–514.

Reed, J. W., A. Nagatani, T. D. Elich, M. Fagan, and J. Chory. 1994. Phytochrome A and phytochrome B have overlapping but distinct functions in *Arabidopsis* development. Plant Physiology 104:1139–1149.

Reekie, J. Y. C., and P. R. Hiclenton. 1994. Effects of elevated CO_2 on time of flowering in four short-day and four long-day species. Canadian Journal of Botany 72:533–538.

Reeve, H. K., and P. W. Sherman. 1993. Adaptation and the goals of evolutionary research. Quarterly Review of Biology 68:1–32.

Reinhardt, R. 1969. Über den Einfluss der Temperatur auf den Saisondimorphismus von *Araschnia levana* L. (Lepidopt. Nymphalidae) nach photoperiodischer Diapause-Induktion. Zoologische Jahrbücher (Physiologie) 75:41–75.

Relyea, R. A. 2001. Morphological and behavioural plasticity of larval anurans in response to different predators. Ecology 82:523–540.

Relyea, R. A. 2002. Costs of phenotypic plasticity. American Naturalist 159:272–282.

Ren, P., C.-S. Lim, R. Johnsen, P. S. Albert, D. Pilgrim, and D. L. Riddle. 1996. Control of *C. elegans* larval development by neuronal expression of a TGF-b homolog. Science 274:1389–1391.

Reznick, D. N. 1983. The structure of guppy life histories: the tradeoff between growth and reproduction. Ecology 64:862–873.

Reznick, D. N. 1990. Plasticity in age and size at maturity in male guppies (*Poecilia reticulata*): an experimental evaluation of alternative models of development. Journal of Evolutionary Biology 3:185–203.

Reznick, D. N., and J. Travis. 1996. The empirical study of adaptation in natural populations. Pages 243–289 in M. R. Rose and G. V. Lauder, editors, Adaptation. Academic Press, San Diego.

Reznick, D. N., and A. P. Yang. 1993. The influence of fluctuating resources on life history: patterns of allocation and plasticity in female guppies. Ecology 71:2011–2019.

Rhen, T., and J. W. Lang. 1994. Temperature-dependent sex determination in the snapping turtle: manipulation of the embryonic sex steroid environment. General and Comparative Endocrinology 96:243–254.

Rice, S. H. 1998. The evolution of canalization and the breaking of Von Baers laws: modeling the evolution of development with epistasis. Evolution 52:647–656.

Rice, W. R. 1996. Sexually antagonistic male adaptation triggered by experimental arrest of female evolution. Nature 381:232–234.

Rice, W. R. 1998. Intergenomic conflict, interlocus antagonistic coevolution, and the evolution of reproductive isolation. Pages 261–270 in D. J. Howard and S. H. Berlocher, editors, Endless Forms: Species and Speciation. Oxford University Press, Oxford.

Rice, W. R., and B. Holland. 1997. The enemies within: intergenomic conflict, interlocus contest evolution (ICE) and the intraspecific red queen. Behavioral Ecology and Sociobiology 41:1–10.

Ridley, M. 1997. Evolution, 2nd edition. Blackwell Scientific Publishers, Cambridge.

Riedel, H. 1935. Der Einfluss der Entwicklungstemperatur auf Flügel und Tibialänge von *Drosophila melanogaster* (wild, vestigial und der reziproken Kreuzungen). Wilhelm Roux Archiv für Entwicklungsmechanik der Organismen 132:463–503.

Riessen, H. P. 1992. Cost-benefit model for the induction of an antipredator defense. American Naturalist 140:349–362.

Robertson, F. W. 1959. Studies in quantitative inheritance. XII. Cell size and relation to genetic and environmental variation of body size in *Drosophila*. Genetics 44:869–896.

Robertson, F. W. 1964. The ecological genetics of growth in *Drosophila*. 7. The role of canalization in the stability of growth relations. Genetical research 5: 107–126. .

Robinson, B. W., and R. Dukas. 1999. The influence of phenotypic modifications on evolution: the Baldwin effect and modern perspectives. Oikos 85:582–589.

Rodd, F. H., and D. N. Reznick. 1997. Variation in the demography of natural populations of guppies: the importance of predation and life histories. Ecology 78:405–418.

Rodd, F. H., D. N. Reznick, and M. B. Sokolowski. 1997. Phenotypic plasticity in the life history traits of guppies: responses to social environment. Ecology 78:419–433.

Roff, D. A. 1992. The Evolution of Life Histories: Theory and Analysis. Chapman and Hall, London.

Roff, D. A. 1996. The evolution of threshold traits in animals. Quarterly Review of Biology 71:3–35.

Roff, D. A. 1997. Evolutionary Quantitative Genetics. Chapman and Hall, New York.

Roff, D. A., G. Stirling, and D. J. Fairbairn. 1997. The evolution of threshold traits: a quantitative genetic analysis of the physiological and life-historical correlates of wind dimorphism in the sand cricket. Evolution 51:1910–1919.

Rollo, C. D. 1994. Phenotypes: Their Epigenetics, Ecology and Evolution. Chapman and Hall, London.

Romaschoff, D. D. 1925. Über die Variabilität in der Manifestierung eines erblichen Merkmales (Abdomen abnormalis) bei *Drosophila funebris* F. Journal für Psychologie und Neurologie 31:323–325.

Rose, C. S. 1996. An endocrine based model for developmental and morphogenetic diversification in metamorphic and paedomorphic urodeles. Journal of Zoology 239:253–284.

Rose, C. S. 1999. Hormonal control in larval development and evolution—amphibians. Pages 167–205 in B. K. Hall and M. H. Wake, editors, The Origin and Evolution of Larval Forms. Academic Press, San Diego.

Rountree, D. B., and H. F. Nijhout. 1995a. Genetic control of a seasonal morph in *Precis coenia*. Journal of Insect Physiology 41:1141–1145.

Rountree, D. B., and H. F. Nijhout. 1995b. Hormonal control of a seasonal polyphenism in *Precis coenia*. Journal of Insect Physiology 41:987–992.

Rowe, L., and D. Ludwig. 1991. Size and timing of metamorphosis in complex life cycles: time constraints and variation. Ecology 72:413–427.

Roy, S. K. 1960. Interaction between rice varieties. Journal of Genetics 57:137–145.

Ruan, Y., J. Gilmore, and T. Conner. 1998. Towards *Arabidopsis* genome analysis: monitoring expression profiles of 1400 genes using cDNA microarrays. Plant Journal 15:821–833.

Rundle, S. D., and C. Brönmark. 2001. Inter- and intraspecific trait compensation of defence mechanisms in freshwater snails. Proceedings of the Royal Society, London (B) 268:1463–1468.

Rutherford, S. L., and S. Lindquist. 1998. Hsp90 as a capacitor for morphological evolution. Nature 396:336–342.

Sage, R. F., and R. W. Pearcy. 1987. The nitrogen use efficiency of C_3 and C_4 plants. II. Leaf nitrogen effects on the gas exchange characteristics of *Chenopodium album* (L.) and *Amaranthus retroflexus* (L.). Plant Physiology 84:959–963.

Salisbury, F. B., and C. W. Ross. 1992. Plant Physiology. Wadsworth Publishing Company, Belmont, CA.

Salzman, A. G., and M. Parker. 1985. Neighbors ameliorate local salinity stress for a rhizomatous plants in heterogeneous environment. Oecologia 65:273–277.

Samson, D. A., and K. S. Werk. 1986. Size-dependent effects in the analysis of reproductive effort in plants. American Naturalist 127:667–680.

Sapp, J. 1987. Beyond the Gene: Cytoplasmic Inheritance and the Struggle for Authority in Genetics. Oxford University Press, New York.

Sarà, M. 1996a. The problem of adaptations: an holistic approach. Rivista de Biologia 82:75–101.

Sarà, M. 1996b. A "sensitive" cell system. Its role in a new evolutionary paradigm. Rivista de Biologia 89:139–156.

Sarkar, S. 1998. Genetics and Reductionism. Cambridge University Press, New York.

Sarkar, S. 1999. From the Reaktionsnorm to the adaptive norm: the norm of reaction, 1909–1960. Biology and Philosophy 14:235–252.

Sasaki, A., and G. de Jong. 1999. Density dependence and unpredictable selection in a heterogeneous environment: compromise and polymorphism in the ESS reaction norm. Evolution 53:1329–1342.

Satou, S. 1988. A fitness function for optimal life history in relation to density effects, competition, predation and stability of the environment. Ecological Research 3:145–161.

Schaal, B. A. 1975. Population structure and local differentiation in *Liatris cylindracea*. American Naturalist 109:511–528.

Schaal, B. A., and W. J. Leverich. 1987. Genetic constraints on plant adaptive evolution. Pages 173–184 in V. Loeschcke, editor, Genetic Constraints on Adaptive Evolution. Springer Verlag, Berlin.

Scheiner, S. M. 1993a. Genetics and evolution of phenotypic plasticity. Annual Review of Ecology and Systematics 24:35–68.

Scheiner, S. M. 1993b. Plasticity as a selectable trait: reply to Via. American Naturalist 142:372–374.

Scheiner, S. M. 1998. The genetics of phenotypic plasticity. VII. Evolution in a spatially structured environment. Journal of Evolutionary Biology 11:303–320.

Scheiner, S. M. 2002. Selection experiments and the study of phenotypic plasticity. Journal of Evolutionary Biology 15:889–898.

Scheiner, S. M., and D. Berrigan. 1998. The genetics of phenotypic plasticity. VIII. The cost of plasticity in *Daphnia pulex*. Evolution 52:368–378.

Scheiner, S. M., and H. S. Callahan. 1999. Measuring natural selection on phenotypic plasticity. Evolution 53:1704–1713.

Scheiner, S. M., and C. J. Goodnight. 1984. The comparison of phenotypic plasticity and genetic variation in populations of the grass *Danthonia spicata*. Evolution 38:845–855.

Scheiner, S. M., and R. F. Lyman. 1989. The genetics of phenotypic plasticity. I. Heritability. Journal of Evolutionary Biology 2:95–107.

Scheiner, S. M., and R. Lyman. 1991. The genetics of phenotypic plasticity. II. Response to selection. Journal of Evolutionary Biology 4:23–50.

Scheiner, S. M., and L. Y. Yampolsky. 1998. The evolution of *Daphnia pulex* in a temporally varying environment. Genetical Research 72:25–37.

Scheiner, S. M., R. L. Caplan, and R. F. Lyman. 1991. The genetics of phenotypic plasticity. III. Genetic correlations and fluctuating asymmetries. Journal of Evolutionary Biology 4:51–68.

Scheiner, S. M., S. B. Cox, M. R. Willig, G. G. Mittelbach, C. Osenberg, and M. Kaspari. 2000a. Species richness, species-area curves, and Simpson's paradox. Evolutionary Ecology Research 2:791–802.

Scheiner, S. M., R. J. Mitchell, and H. Callahan. 2000b. Using path analysis to measure natural selection. Journal of Evolutionary Biology 13:423–433.

Scheiner, S. M., K. Donohue, L. A. Dorn, S. J. Mazer, and L. M. Wolfe. 2002. Reducing environmental bias when measuring natural selection. Evolution 56:2156–2167.

Schlichting, C. D. 1986. The evolution of phenotypic plasticity in plants. Annual Review of Ecology and Systematics 17:667–693.

Schlichting, C. D. 1989. Phenotypic plasticity in *Phlox* II. Plasticity of character correlations. Oecologia 78:496–501.

Schlichting, C. D., and D. A. Levin. 1984. Phenotypic plasticity of annual phlox: tests of some hypotheses. American Journal of Botany 71:252–260.

Schlichting, C. D., and D. A. Levin. 1986. Phenotypic plasticity: an evolving plant character. Biological Journal of the Linnean Society 29:37–47.

Schlichting, C. D., and D. A. Levin. 1988. Phenotypic plasticity in *Phlox*. I. Wild and cultivated populations of *P. drummondii*. American Journal of Botany 75:161–169.

Schlichting, C. D., and D. A. Levin. 1990. Phenotypic plasticity in *Phlox*. II. Variation among natural populations of *P. drummondii*. Journal of Evolutionary Biology 3:411–428.

Schlichting, C. D., and M. Pigliucci. 1993. Control of phenotypic plasticity via regulatory genes. American Naturalist 142:366–370.

Schlichting, C. D., and M. Pigliucci. 1995. Gene regulation, quantitative genetics and the evolution of reaction norms. Evolutionary Ecology 9:154–168.

Schlichting, C. D., and M. Pigliucci. 1998. Phenotypic Evolution: A Reaction Norm Perspective. Sinauer Associates, Inc., Sunderland, MA.

Schluter, D. 1995. Adaptive radiation along genetic lines of least resistance. Evolution 50:1766–1774.

Schmalhausen, I. I. 1949. Factors of Evolution. Blakiston, Philadelphia, PA.

Schmitt, J. 1993. Reaction norms of mophological and life-history traits to light availability in *Impatiens capensis*. Evolution 47:1654–1668.

Schmitt, J., and S. E. Gamble. 1990. The effect of distance from the parental site on offspring performance and inbreeding depression in *Impatiens capensis*: test of the local adaptation hypothesis. Evolution 44:2022–2030.

Schmitt, J., and R. D. Wulff. 1993. Light spectral quality, phytochrome and plant competition. Trends in Ecology and Evolution 8:47–51.

Schmitt, J., J. Ecclestas, and D. W. Ehrhardt. 1987. Density-dependent flowering phenology, outcrossing, and reproduction in *Impatiens capensis*. Oecologia 72:341–347.

Schmitt, J., A. C. McCormac, and H. Smith. 1995. A test of the adaptive plasticity hypothesis using transgenic and mutant plant disabled in phytochrome-mediated elongation responses to neighbors. American Naturalist 146:937–953.

Schmitt, J., S. A. Dudley, and M. Pigliucci. 1999. Manipulation approaches to testing adaptive plasticity: phytochrome-mediated shade-avoidance responses in plants. American Naturalist 154:S43–S54.

Schroeder, J. I. 1990. Function of ion channels and cytosolic Ca^{2+} in the mediation of higher plant cell ion transport. Pages 103–116 in R. Ranjeva and A. M. Boudet, editors, Signal Perception and Transduction in Higher Plants. Springer Verlag, Berlin.

Seger, J., and H. J. Brockmann. 1987. What is bet-hedging? Oxford Surveys in Evolutionary Biology 4:182–211.

Semlitsch, R. D., and J. W. Gibbons. 1986. Phenotypic variation in metamorphosis and paedomorphosis in the salamander Ambystoma talpoideum. Ecology 66:1123–1130.

Semlitsch, R. D., and H. M. Wilbur. 1989. Artificial selection for paedomorphosis in the salamander *Ambystoma talpoideum*. Evolution 43:943–945.

Semlitsch, R. D., R. N. Harris, and H. M. Wilbur. 1990. Paedomorphosis in *Ambystoma talpoideum*: maintenance of population variation and alternative life-history pathways. Evolution 44:1604–1613.

Serna, L., and C. Fenoll. 1997. Tracing the ontogeny of stomatal clusters in *Arabidopsis* with molecular markers. Plant Journal 12:747–755.

Service, P. M., and M. R. Rose. 1985. Genetic covariation among life-history components: the effect of novel environments. Evolution 39:943–945.

Sgrò, C. M., and A. A. Hoffmann. 1998a. Effects of stress combinations on the expression of additive genetic variation for fecundity in *Drosophila melanogaster*. Genetical Research 72:13–18.

Sgrò, C. M., and A. A. Hoffmann. 1998b. Effects of temperature extremes on genetic variances for life history traits in *Drosophila melanogaster* as determined from parent-offspring comparisons. Journal of Evolutionary Biology 11:1–20.

Shaffer, H. B., and R. Voss. 1996. Phylogenetic and mechanistic analysis of a developmentally integrated character complex: alternate life history modes in ambystomatid salamanders. American Zoologist 36:24–35.

Shapiro, A. M. 1976. Seasonal polyphenism. Evolutionary Biology 9:259–333.

Shapiro, A. M. 1980. Convergence in pierine polyphenisms (Lepidoptera). Journal of Natural History 14:781–802.

Sharrock, R. A., and P. H. Quail. 1989. Novel phytochrome sequences in *Arabidopsis thaliana*: structure, evolution, and differential expression of a plant regulatory photoreceptor family. Genes and Development 3:1745–1757.

Shaw, R. G., and G. A. J. Plantenkamp. 1993. Quantitative genetics of response to competitors in *Nemophila menziesii*: a greenhouse study. Evolution 47:801–812.

Shi, Y. 1994. Molecular biology of amphibian metamorphosis. Trends in Endocrinology and Metabolism 5:14–20.

Sibly, R. M. 1995. Life-history evolution in spatially heterogeneous environments, with and without phenotypic plasticity. Evolutionary Ecology 9:242–257.

Sievers, A. 1990. Transduction of the gravity signal in plants. Pages 297–306 in R. Ranjeva and A. M. Boudet, editors, Signal Perception and Transduction in Higher Plants. Springer Verlag, Berlin.

Sih, A. 1980. Optimal foraging: partial consumption of prey. American Naturalist 116:281–290.

Sih, A. 1982. Optimal patch use: variations in selective pressure for efficient foraging. American Naturalist 120:666–685.

Sih, A. 1984. Optimal behavior and density-dependent predation. American Naturalist 123:314–326.

Sih, A. 1987. Predators and prey lifestyles: an evolutionary and ecological overview. Pages 203–224 in W. C. Kerfoot and A. Sih, editors, Predation: Direct and Indirect Impacts on Aquatic Communities. University Press of New England, Hanover, NH.

Sih, A. 1992. Forager uncertainty and the balancing of antipredator and feeding needs. American Naturalist 139:1052–1069.

Sih, A. 1994. Predation risk and the evolutionary ecology of reproductive behavior. Journal of Fish Biology 45A:111–130.

Sih, A. 1998. Game theory and predator-prey response races. Pages 221–238 in L. A. Dugatkin and H. K. Reeve, editors, Game Theory and Animal Behavior. Oxford University Press, Oxford.

Sih, A., and R. D. Moore. 1993. Delayed hatching of salamander eggs in response to enhanced larval predation risk. American Naturalist 142:947–960.

Sih, A., G. Englund, and D. E. Wooster. 1998. Emergent impacts of multiple predators on prey. Trends in Ecology and Evolution 13:350–355.

Simm, G. 1998. Genetic Improvement of Cattle and Sheep. Farming Press, Ipswich, UK.

Simms, E. L., and M. D. Rausher. 1989. The evolution of resistance to herbivory in *Ipomoea purpurea*. II. Natural selection by insects and the costs of resistance. Evolution 43:573–585.

Simpson, G. G. 1953. The Baldwin effect. Evolution 7:110–117.

Sinervo, B., and A. L. Basalo. 1996. Testing adaptations using phenotypic manipulations. Pages 149–185 in M. R. Rose and G. V. Lauder, editors, Adaptation. Academic Press, San Diego.

Sinervo, B., and P. Doughty. 1996. Interactive effects of offspring size and timing of reproduction on offspring reproduction: experimental, maternal, and quantitative genetic aspects. Evolution 50:1314–1327.

Sinnott, E. W., L. C. Dunn, and T. Dobzhansky. 1950. Principles of Genetics, 4 ed. McGraw-Hill, New York.

Skávlová, H., and F. Krahulec. 1992. The response of three *Festuca rubra* clones to changes in light quality and plant density. Functional Ecology 6:282–290.

Skelly, D. K. 1994. Activity level and susceptibility of anuran larvae to predation. Animal Behaviour 48:465–468.

Skelly, D. K., and E. E. Werner. 1990. Behavioral and life-historical responses of larval American toads to an odonate predator. Ecology 71:2313–2322.

Skelly, D. K. 1992. Field evidence for a cost of behavioral antipredator response in a larval amphibian. Ecology 73:704–708.

Skulason, S., and T. B. Smith. 1995. Resource polymorphisms in vertebrates. Trends in Ecology and Evolution 10:366–370.

Smith, D. C. 1987. Adult recruitment in chorus frogs: effects of size and date at metamorphosis. Ecology 68:344–350.

Smith, H. 1982. Light quality, photoreception, and plant strategy. Annual Review of Plant Physiology 33:481–518.

Smith, H. 1990. Signal perception, differential expression within multigene families and the molecular basis of phenotypic plasticity. Plant, Cell and Environment 13:585–594.

Smith, H. 1995. Physiological and ecological function within the phytochrome family. Annual Review of Plant Physiology and Plant Molecular Biology 46:289–315.

Smith, H., and G. C. Whitelam. 1990. Phytochrome, a family of photoreceptors with multiple physiological roles. Plant, Cell and Environment 13:695–707.

Smith, H., J. J. Casal, and G. M. Jackson. 1990. Reflection signals and the perception by phytochrome of the proximity of neighbouring vegetation. Plant, Cell and Environment 13:73–78.

Smith, L. D., and A. R. Palmer. 1994. Effects of manipulated diet on size and performance of brachyuran crab claws. Science 264:710–712.

Smith, V., K. N. Chou, D. Lashkari, D. Botstein, and P. O. Brown. 1996. Functional analysis of the genes of yeast chromosome V by genetic footprinting. Science 274:2069–2074.

Smith-Gill, S. J. 1983. Developmental plasticity: developmental conversion versus phenotypic modulation. American Zoologist 23:47–55.

Smits, J. D., F. Witte, and G. D. E. Povel. 1996a. Differences between inter- and intraspecific architectonic adaptations to pharyngeal mollusc crushing in cichlid fishes. Biological Journal of the Linnean Society 59:367–387.

Smits, J. D., F. Witte, and F. G. Van Veen. 1996b. Functional changes in the anatomy of the pharyngeal jaw apparatus of *Astatoreochromis alluaudi* (Pisces, Cichlidae), and their effects on adjacent structures. Biological Journal of the Linnean Society 59:389–409.

Snaydon, R. W. 1970. Rapid population differentiation in a mosaic environment I. The response of *Anthoxanthum odoratum* populations to soils. Evolution 24:257–269.

Solangaarachchi, S. M., and J. L. Harper. 1987. The effect of canopy filtered light on growth of white clover *Trifolium repens*. Oecologia 72:372–376.

Solomon, B. P. 1985. Environmentally influenced changes in sex expression in an andromonoecious plant. Ecology 66:1321–1332.

Southwood, T. R. E. 1961. A hormonal theory of the mechanism of wing polymorphism in Heteroptera. Proceedings of the Royal Entomological Society of London 36:63–66.

Sparkes, T. C. 1996. The effects of size-dependent predation risk on the interaction between behavioral and life history traits in a stream-dwelling isopod. Behavioral Ecology and Sociobiology 39:411–417.

Spieth, H. R. 1995. Change in photoperiodic sensitivity during larval development of *Pieris brassicae*. Journal of Insect Physiology 41:77–83.

Spitze, K. 1992. Predator-mediated plasticity of prey life history and morphology: *Chaoborus americanus* predation on *Daphnia pulex*. American Naturalist 139:229–247.

Spitze, K., and T. D. Sadler. 1996. Evolution of a generalist genotype: multivariate analysis of the adaptiveness of phenotypic plasticity. American Naturalist 148:S108–S123.

Sprules, W. G. 1974. The adaptive significance of paedogenesis in the North American species of *Amybstoma* (Amphibia: Caudata): an hypothesis. Canadian Journal of Zoology 52:393–400.

Stalker, H. D. 1980. Chromosome studies in wild populations of *Drosophila melanogaster*. II. Relationship of inversion frequencies to latitude, season, wing-loading and flight activity. Genetics 95:211–223.

Stanley, S. M. 1975. A theory of evolution above the species level. Proceedings of the National Academy of Sciences of the United States of America 72:646–650.

Starnecker, G. 1996. Color preference for pupation sites of the butterfly larvae of *Inachis io* and the significance of the pupal melanization reducing factor. Naturwiss 83:474–476.

Starnecker, G., and W. Hazel. 1999. Convergent evolution of neuroendocrine control of phenotypic plasticity in pupal colour in butterflies. Proceedings of the Royal Society of London, Series B. Biological Sciences 266:2409–2412.

Stearns, S. C. 1976. Life-history tactics: a review of the ideas. Quarterly Review of Biology 51:3–47.

Stearns, S. C. 1982. The role of development in the evolution of life histories. Pages 237–258 in J. T. Bonner, editor, Evolution and Development. Springer Verlag, New York.

Stearns, S. C. 1983. The evolution of life-history traits in mosquitofish since their introduction to Hawaii in 1905: rates of evolution, heritabilities, and developmental plasticity. American Zoologist 23:65–75.

Stearns, S. C. 1989. The evolutionary significance of phenotypic plasticity. Bioscience 7:436–445.

Stearns, S. C. 1992. The Evolution of Life Histories. Oxford University Press, Oxford.

Stearns, S. C., and J. C. Koella. 1986. The evolution of phenotypic plasticity in life-history traits: predictions of reaction norms for age and size at maturity. Evolution 40:893–913.

Stephens, D. W., and J. R. Krebs. 1986. Foraging Theory. Princeton University Press, Princeton, NJ.

Stern, D. L. 1998. A role of Ultrabithorax in morphological differences between *Drosophila* species. Nature 396:463–466.

Stern, D. L. 2000. Evolutionary developmental biology and the problem of variation. Evolution 54:1079–1091.

Stern, D. L., and D. J. Emlen. 1999. The developmental basis of allometry in insects. Development 126:1091–1101.

Stevens, D. J., M. H. Hansell, J. A. Freel, and P. Monaghan. 1999. Developmental trade-offs in caddis flies: increased investment in larval defence alters adult resource allocation. Proceedings of the Royal Society of London, Series B. Biological Sciences 266:1049–1054.

Stewart, S. C., and D. J. Schoen. 1987. Pattern of phenotypic variability and fecundity selection in a natural population of *Impatiens pallida*. Evolution 41:1290–1301.

Stratton, D. A. 1994. Genotype-by-environment interactions for fitness of *Erigeron annuus* show fine-scale selective heterogeneity. Evolution 48:1607–1618.

Stratton, D. A. 1995. Spatial scale of variation in fitness of *Erigeron annuus*. American Naturalist 146:608–624.

Stratton, D. A., and C. C. Bennington. 1996. Measuring spatial variation in natural selection using randomly-sown seeds of *Arabidopsis*. Journal of Evolutionary Biology 9:215–228.

Sturmbauer, C. M. W., and R. Dallinger. 1992. Ecophysiology of Aufwuchs-eating cichlids in Lake Tanganyika—niche separation by trophic specialization. Environmental Biology of Fishes 35:283–290.

Sucena, E., and D. L. Stern. 2000. Divergence of larval morphology between *Drosophila schellia* and its sibling species due to cis-regulatory evolution of ovo/shaven-baby. Proceedings of the National Academy of Sciences of the United States of America 97:4530–4534.

Sultan, S. E. 1987. Evolutionary implications of phenotypic plasticity in plants. Evolutionary Biology 21:127–178.

Sultan, S. E. 1992. Phenotypic plasticity and the Neo-Darwinian legacy. Evolutionary Trends in Plants 6:61–71.

Sultan, S. E., and F. A. Bazzaz. 1993a. Phenotypic plasticity in *Polygonum persicaria*. I. Diversity and uniformity in genotypic norms of reaction to light. Evolution 47:1009–1031.

Sultan, S. E., and F. A. Bazzaz. 1993b. Phenotypic plasticity in *Polygonum persicaria*. II. Norms of reaction to soil moisture and the maintenance of genetic diversity. Evolution 47:1032–1049.

Sultan, S. E., and F. A. Bazzaz. 1993c. Phenotypic plasticity in *Polygonum persicaria*. III. The evolution of ecological breadth for nutrient environment. Evolution 47:1050–1071.

Sultan, S. E., and H. G. Spencer. 2002. Metapopulation structure favors plasticity over local adaptation. American Naturalist 160:271–283.

Summerton, J. 1999. Morpholino antisense oligomers: the case for an RNase H-independent structural type. Biochimica et Biophysica Acta – Gene Structure and Expression 1489:141–158.

Suvanto, L., J. O. Liimatainen, and A. Hoikkala. 1999. Variability and evolvability of male song characters in *Drosophila montana* populations. Hereditas 130:13–18.

Swenson, W., D. S. Wilson, and R. Elias. 2000a. Artificial ecosystem selection. Proceedings of the National Academy of Sciences of the United States of America 97:9110–9114.

Swenson, W., J. Arendt, and D. S. Wilson. 2000b. Artificial selection of microbial ecosystems for 3-chloroaniline biodegradation. Environmental Microbiology 2:564–571.

Taiz, L., and E. Zeiger. 1998. Plant Physiology, 2nd edition. Sinauer Associates, Inc., Sunderland, MA.

Takeda, K., R. Kubota, and C. Yagioka. 1985. Blueing of sepal color of *Hydrangea macrophylla*. Phytochemistry 24:2251–2254.

Tallis, G. M., and P. Leppard. 1988. The joint effects of selection and assortative mating on multiple polygenic characters. Theoretical and Applied Genetics 75:278–281.

Tata, J. R. 1993. Gene expression during metamorphosis: an ideal model for post-embryonic development. BioEssays 15:239–248.

Tauber, C. A., and M. J. Tauber. 1992. Phenotypic plasticity in *Chrysoperla*: genetic variation in the sensory mechanism and in correlated reproductive traits. Evolution 46:1754–1773.

te Velde, J. H., and W. Scharloo. 1988. Natural and artificial selection on a deviant character of the anal papillae in *Drosophila melanogaster* and their significance for salt adaptation. Journal of Evolutionary Biology 1:155–164.

te Velde, J. H., J. H. Gordens, and W. Scharloo. 1987. The genetic fixation of phenotypic response of an ultrastructural character in the anal papillae of *Drosophila melanogaster*. Heredity 61:47–53.

Thoday, J. M. 1955. Balance, heterozygosity and developmental stability. Cold Spring Harbor Symposium on Quantitative Biology 20:318–326.

Thoday, J. M. 1958. Homeostasis in a selection experiment. Heredity 12:401–415.

Thomas, B. 1991. Phytochrome and photoperiodic induction. Physiologia Plantarum 81:571–577.

Thomas, C. D., and M. C. Singer. 1998. Scale-dependent evolution of specialization in a checkerspot butterfly: from individuals to metapopulations and ecotypes. Pages 343–374 in S. Mopper and S. Y. Strauss, editors, Genetic Structure and Local Adaptation in Natural Insect Populations. Chapman and Hall, New York.

Thompson, D. B. 1999. Genotype-environment interaction and the ontogeny of diet-induced phenotypic plasticity in size and shape of *Melanopus femurrubrum* (Orthoptera: Acrididae). Journal of Evolutionary Biology 12:38–48.

Thompson, J. D. 1991. Phenotypic plasticity as a component of evolutionary change. Trends in Ecology and Evolution 6:246–249.

Thompson, J. N. 1982. Interaction and Coevolution. John Wiley and Sons, New York.

Thompson, J. N. 1994. The Coevolutionary Process. University of Chicago Press, Chicago.

Thompson, J. N. 1997. Evaluating the dynamics of coevolution among geographically structured populations. Ecology 78:1619–1623.

Thompson, J. N. 1999a. The evolution of species interactions. Science 284:2116–2118.

Thompson, J. N. 1999b. The raw material for coevolution. Oikos 84:1–11.

Thompson, J. N. 1999c. Specific hypotheses on the geographic mosaic of coevolution. American Naturalist 153:S1–S13.

Thompson, J. N., and O. Pellmyr. 1991. Evolution of oviposition behavior and host preference in lepidoptera. Annual Review of Entomology 36:65–89.

Thompson, L., and J. L. Harper. 1988. The effect of grasses on the quality of transmitted radiation and its influence on the growth of white clover *Trifolium repens*. Oecologia 75:343–347.

Timoféeff-Ressovsky, H. A., and N. W. Timoféeff-Ressovsky. 1926. Über das phänotypische Manifestieren des Genotyps. II. Über idio-somatische Variationsgruppen bei *Drosophila funebris*. Wilhelm Roux' Archiv für Entwicklungsmechanik der Organismen 108:146–170.

Timoféeff-Ressovsky, N. W. 1925. Über den Einfluss des Genotypus auf das phänotypen Auftreten eines einzelnes Gens. Journal für Psychologie und Neurologie 31:305–310.

Tollrian, R., and C. D. Harvell. 1999. The Ecology and Evolution of Inducible Defenses. Princeton University Press, Princeton, NJ.

Travis, J. 1994. Evaluating the adaptive role of morphological plasticity. Pages 99–122 in P. C. Wainwright and S. M. Reilly, editors, Ecological Morphology. University of Chicago Press, Chicago.

Trexler, J. C., and J. Travis. 1990. Phenotypic plasticity in the sailfin molly, *Poecilia latipinna* (Pisces: Poeciliidae). I. Field experiment. Evolution 44:143–156.

Trexler, J. C., J. Travis, and M. Trexler. 1990. Phenotypic plasticity in the sailfin molly, *Poecilia latipinna* (Pisces: Poeciliidae). II. Laboratory experiment. Evolution 44: 157–167.

Truman, J. W. 1971. Hour-glass behavior of the circadian clock controlling eclosion of the silkmoth *Antheraea pernyi*. Proceedings of the National Academy of Sciences of the United States of America 68:595–599.

Trussell, G. C. 1996. Phenotypic plasticity in an intertidal snail: the role of a common crab predator. Evolution 50:448–454.

Trussell, G. C. 2000. Phenotypic clines, plasticity, and morphological trade-offs in an intertidal snail. Evolution 54:151–166.

Tufto, J. 2000. The evolution of plasticity and nonplastic spatial and temporal adaptations in the presence of imperfect environmental cues. American Naturalist 156:121–130.

Turesson, G. 1922. The genotypical response of the plant species to the habitat. Hereditas 3:211–350.

Turkington, R., and J. L. Harper. 1979. The growth, distribution, and neighbour relationships of *Trifolium repens* in a permanent pasture. IV. Fine scale biotic differentiation. Journal of Ecology 67:245–254.

Turner, A. M. 1996. Freshwater snails alter habitat use in response to predation. Animal Behaviour 51:747–756.

Van Buskirk, J., and R. A. Relyea. 1998. Selection for phenotypic plasticity in *Rana sylvatica* tadpoles. Biological Journal of the Linnean Society 65:301–328.

Van der Have, T. M., and G. de Jong. 1996. Adult size in ectotherms: temperature effects on growth and differentiation. Journal of Theoretical Biology 183:329–340.

Van der Heijden, M. G. A., J. N. Klironomos, M. Ursic, P. Moutoglis, R. Streitwolf-Engel, T. Boller, A. Wiemken, and I. R. Sanders. 1998. Mycorrhizal fungal diversity determines plant biodiversity, ecosystem variability and production. Nature 396:69–72.

Van Hinsberg, A. 1996. On phenotypic plasticity in *Plantago lanceolata*: light quality and plant morphology. Ph.D. Dissertation. Universiteit Utrecht, Utrecht, The Netherlands.

Van Hinsberg, A. 1997. Morphological variation in *Plantago lanceolata* L.: effects of light quality and growth regulators on sun and shade population. Journal of Evolutionary Biology 10:687–701.

Van Noordwijk, A. J. 1989. Reaction norms in genetical ecology. Bioscience 39:453–458.

Van Noordwijk, A. J., and G. de Jong. 1986. Acquisition and allocation of resources: their influence on variation in life history tactics. American Naturalist 128:137–142.

Van Noordwijk, A. J., J. H. Van Balen, and W. Scharloo. 1988. Heritability of body size in a natural population of the great tit (*Parus major*) and its relation to age and environmental conditions during growth. Genetical Research 51:149–162.

Van Tienderen, P. H. 1990. Morphological variation in *Plantago lanceolata*: limits of plasticity. Evolutionary Trends in Plants 4:35–43.

Van Tienderen, P. H. 1991. Evolution of generalists and specialists in spatially heterogeneous environments. Evolution 45:1317–1331.

Van Tienderen, P. H. 1997. Generalists, specialists, and the evolution of phenoptypic plasticity in sympatric populations of distinct species. Evolution 51:1372–1380.

Van Tienderen, P. H., and H. P. Koelewijn. 1994. Selection on reaction norms, genetic correlations and constraints. Genetical Research 64:115–125.

Vavrek, M. C., J. B. McGraw, and H. S. Yang. 1997. Within-population variation in demography of *Taraxacum officinale*: season- and size-dependent survival, growth and reproduction. Journal of Ecology 85:277–287.

Vaz Nunes, M., R. D. Lewis, and D. S. Saunders. 1991a. A coupled oscillator feedback system as a model for the photoperiodic clock in insects and mites. I. The basic control system as a model for circadian rhytms. Journal of Theoretical Biology 152:287–298.

Vaz Nunes, M., D. S. Saunders, and R. D. Lewis. 1991b. A coupled oscillator feedback system as a model for the photoperiodic clock in insects and mites. II. Simulation of photoperiodic responses. Journal of Theoretical Biology 152:299–317.

Vermeij, G. J. 1987. Evolution and Escalation: An Ecological History of Life. Princeton University Press, Princeton, NJ.

Vermeij, G. J. 1994. The evolutionary interaction among species: selection, escalation, and coevolution. Annual Review of Ecology and Systematics 25:219–236.

Via, S. 1986. Genetic covariance between oviposition preference and larval performance in an insect herbivore. Evolution 49:778–785.

Via, S. 1987. Genetic constraints on the evolution of phenotypic plasticity. Pages 47–71 in V. Loeschcke, editor, Genetic Constraints on Adaptive Evolution. Springer-Verlag, Berlin.

Via, S. 1991. The genetic structure of host plant adaptation in a spatial patchwork: demographic variability among reciprocally transplanted pea aphid clones. Evolution 45:827–852.

Via, S. 1993a. Adaptive phenotypic plasticity: target or byproduct of selection in a variable environment. American Naturalist 142:352–365.

Via, S. 1993b. Regulatory genes and reaction norms. American Naturalist 142:374–378.

Via, S. 1994. The evolution of phenotypic plasticity: what do we really know? Pages 35–57 in L. A. Real, editor, Ecological Genetics. Princeton University Press, Princeton, NJ.

Via, S., and R. Lande. 1985. Genotype-environment interaction and the evolution of phenotypic plasticity. Evolution 39:505–522.

Via, S., and R. Lande. 1987. Evolution of genetic variability in a spatially heterogeneous environment: effects of genotype-environment interaction. Genetical Research 49:147–156.

Via, S., R. Gomulkiewicz, G. de Jong, S. M. Scheiner, C. D. Schlichting, and P. Van Tienderen. 1995. Adaptive phenotypic plasticity: consensus and controversy. Trends in Ecology and Evolution 10:212–217.

Vogt, O. 1926. Psychiatrisch wichtige Tatsachen der zoologisch-botanischen Systematik. Journal für Psychologie und Neurologie 101:805–832.

Von Arnim, A. G., and X.-W. Deng. 1996. Light control of seedling development. Annual Review of Plant Physiology and Plant Molecular Biology 47:215–243.

Voss, S. R., and H. B. Shaffer. 1997. Adaptive evolution via a major gene effect: paedomorphosis in the Mexican axolotl. Proceedings of the National Academy of Sciences of the United States of America 94:14185–14189.

Waddington, C. H. 1938. An Introduction to Modern Genetics. George Allen and Unwin, London.

Waddington, C. H. 1940a. The genetic control of wing development in *Drosophila*. Journal of Genetics 41:75–139.

Waddington, C. H. 1940b. Organisers and Genes. Cambridge University Press, Cambridge.

Waddington, C. H. 1942. Canalization of development and the inheritance of acquired characters. Nature 150:563–565.

Waddington, C. H. 1953. Genetic assimilation of an acquired character. Evolution 7:118–126.

Waddington, C. H. 1956. Genetic assimilation of the bithorax phenotype. Evolution 10:1–13.

Waddington, C. H. 1959. Canalization of development and the genetic assimilation of acquired characters. Nature 183:1654–1655.

Waddington, C. H. 1961. Genetic assimilation. Advances in Genetics 10:257–290.

Waddington, C. H. 1975. The Evolution of an Evolutionist. Cornell University Press, Ithaca, NY.

Wade, M. J. 1980. An experimental study of kin selection. Evolution 34:844–855.

Wade, M. J. 1988. Group selection for increased and decreased competitive ability in subdivided populations of flour beetles. Bulletin of the Society of Population Ecology 44:3–11.

Wade, M. J. 1990. Genotype-environment interaction for climate and competition in a natural population of flour beetles, *Tribolium castaneum*. Evolution 44:2004–2011.

Wade, M. J. 1992. Epistasis. Pages 87–91 in E. F. Keller and E. A. Lloyd, editors, Key Words in Evolutionary Biology. Harvard University Press, Cambridge, MA.

Wade, M. J. 1994. The biology of the willow leaf beetle, *Plagiodera versicolora* (Laicharting). Pages 541–547 in P. Joliviet and M. Cox, editors, Novel Aspects of the Biology of Chrysomelidae. Kluwer Academic Publishers, Dordrecht.

Wade, M. J. 1998. The evolutionary genetics of maternal effects. Pages 5–21 in T. A. Mousseau and C. W. Fox, editors, Maternal Effects as Adaptations. Oxford University Press, New York.

Wade, M. J. 2000. Epistasis as a genetic constraint within populations and an accelerant of adaptive divergence among them. Pages 213–231 in J. B. Wolf, E. D. Brodie III, and M. J. Wade, editors, Epistasis and the Evolutionary Process. Oxford University Press, New York.

Wade, M. J., and S. Kalisz. 1990. The causes of natural selection. Evolution 44:1947–1955.

Wagner, G. P., G. Booth, and H. BagheriChaichian. 1997. A population genetic theory of canalization. Evolution 51:329–347.

Wahlsten, D. 1990. Insensitivity of the analysis of variance to heredity-environment interaction. Behavioral and Brain Sciences 13:109–120.

Wake, D. B. 1987. Foreword. Pages v–xii in Schmalhausen, I. I., Factors of Evolution: The Theory of Stabilizing Selection. University of Chicago Press, Chicago.

Walls, S. C., and A. R. Blaustein. 1995. Larval marbled salamanders, *Ambystoma opacum*, eat their kin. Animal Behaviour 50:537–545.

Ward, J. M., and J. I. Schroeder. 1997. Roles of ion channels in initiation of signal transduction in higher plants. Pages 1–22 in P. Aducci, editor, Signal Transduction in Plants. Birkhauser Verlag, Basel.

Warkentin, K. M. 1995. Adaptive plasticity in hatching age: a response to predation risk tradeoffs. Proceedings of the National Academy of Sciences of the United States of America 92:3507–3510.

Wcislo, W. T. 1989. Behavioral environments and evolutionary change. Annual Review of Ecology and Systematics 20:137–170.

Weiner, J. 1985. Size hierarchies in experimental populations of annual plants. Ecology 66:743–752.

Weinig, C., M. C. Ungerer, L. A. Dorn, N. C. Kane, Y. Toyonaga, S. S. Halldorsdottir, T. F. C. Mackay, M. D. Purugganan, and J. Schmitt. 2003. Novel loci control variation in reproductive timing in *Arabidopsis thaliana* in natural environments. Genetics 162:1875–1884.

Weis, A. E., and W. L. Gorman. 1990. Measuring selection on reaction norms: an exploration of the *Eurosta-Solidago* system. Evolution 44:820–831.

Weisbrot, D. R. 1966. Genotypic interactions among competing strains and species of *Drosophila*. Genetics 53:427–435.

Weisser, W. W., C. G. Braendle, and N. Minoretti. 1999. Predator-induced morphological shift in the pea aphid. Proceedings of the Royal Society of London, Series B. Biological Sciences 266:1175–1181.

Werner, E. E. 1986. Amphibian metamorphosis: growth rate, predation risk, and the optimal size at transformation. American Naturalist 128:319–341.

West, K., A. Cohen, and M. Baron. 1991. Morphology and behavior of crabs and gastropods from Lake Tanganyika, Africa: implications for lacustrine predator-prey coevolution. Evolution 45:589–607.

West-Eberhard, M. J. 1986. Alternative adaptations, speciation and phylogeny. Proceedings of the National Academy of Sciences of the United States of America 83:1388–1392.

West-Eberhard, M. J. 1989. Phenotypic plasticity and the origins of diversity. Annual Review of Ecology and Systematics 20:249–278.

West-Eberhard, M. J. 1992. Behavior and evolution. Pages 57–75 in P. R. Grant and H. S. Horn, editors, Molds, Molecules and Metazoa. Princeton University Press, Princeton, NJ.

West-Eberhard, M. J. 2003. Developmental Plasticity and Evolution. Oxford University Press, New York.

White, B. A., and C. S. Nicoll. 1981. Hormonal control of amphibian metamorphosis. Pages 363–397 in L. I. Gilbert and E. Frieden, editors, Metamorphosis: A Problem in Developmental Biology. Plenum Press, New York.

Whitelam, G. C., and P. F. Devlin. 1997. Roles of different phytochromes in *Arabidopsis* photomorphogenesis. Plant, Cell and Environment 20:752–758.

Whiteman, H. H. 1994. Evolution of facultative paedomorphosis in salamanders. Quarterly Review of Biology 69:205–221.

Whitlock, M. C. 1996. The red queen beats the jack-of-all-trades: the limitations on the evolution of phenotypic plasticity and niche breadth. American Naturalist 148:S65–S77.

Wiebe, G. A., F. C. Petr, and H. Stevens. 1963. Interplant competition between barley genotypes. Pages 546–557 in W. D. Hanson and H. F. Robinson, editors, Statistical Genetics and Plant Breeding. National Academy of Sciences and theNational Research Council, Washington, DC.

Wigglesworth, V. B. 1961. Insect polymorphism—a tentative synthesis. Pages 103–113 in J. S. Kennedy, editor, Insect Polymorphism. Royal Entomological Society, London.

Wilbur, H. M., and J. P. Collins. 1973. Ecological aspects of amphibian metamorphosis. Science 182:1305–1314.

Wilkins, A. S. 1997. Canalization: a molecular genetic perspective. BioEssays 19:257–262.

Williams, G. C. 1966. Adaptation and Natural Selection. Princeton University Press, Princeton, NJ.

Williams, G. C. 1992. Natural Selection: Domains, Levels, and Challenges. Oxford University Press, Oxford.

Wilson, D. S. 1980. The Natural Selection of Populations and Communities. Benjamin/Cummings Publishing Company, Menlo Park, CA.

Wilson, D. S., and J. Yoshimura. 1994. On the coexistence of specialists and generalists. American Naturalist 144:692–707.

Wilson, R. S., and C. E. Franklin. 2002. Testing the beneficial acclimation hypothesis. Trends in Ecology and Evolution 17:66–70.

Wimberger, P. H. 1991. Plasticity of jaw and skull morphology in the neotropical cichlids *Geophagus brasiliensis* and *G. steindachneri*. Evolution 45:1545–1563.

Wimberger, P. H. 1992. Plasticity of fish body shape. The effects of diet, development, family and age in two species of *Geophagus* (Pisces: Cichlidae). Biological Journal of the Linnean Society 45:197–218.

Wimberger, P. H. 1994. Trophic polymorphisms, plasticity, and speciation in vertebrates. Pages 19–43 in D. J. Stouder, K. Fresh, and R. J. Feller, editors, Theory and Application in Fish Feeding Ecology. University of South Carolina Press, Columbia.

Windig, J. J. 1992. Seasonal polyphenism in *Bicyclus safitza* a continuous reaction norm. Netherlands Journal of Zoology 42:583–594.

Windig, J. J. 1994a. Genetic correlations and reaction norms in wing pattern of the tropical butterfly *Bicyclus anynana*. Heredity 73:459–470.

Windig, J. J. 1994b. Reaction norms and the genetic basis of phenotypic plasticity in the wing pattern of the butterfly *Bicyclus anynana*. Journal of Evolutionary Biology 7:665–695.

Windig, J. J. 1997. The calculation and significance testing of genetic correlations across environments. Journal of Evolutionary Biology 10:853–874.

Windig J. J. 1999. Evolutionary genetics of seasonal polyphenism in the map butterfly *Araschnia levana*. Evolutionary Ecology Research 1:875–894.

Winn, A. A. 1996a. Adaptation to fine-grained environmental variation: an analysis of within-individual variation in an annual plant. Evolution 50:1111–1118.

Winn, A. A. 1996b. The contributions of programmed developmental change and phenotypic plasticity to within-individual variation in leaf traits in *Dicerandra linearifolia*. Journal of Evolutionary Biology 9:737–752.

Winn, A. A. 1999. Is seasonal variation in leaf traits adaptive for the annual plant *Dicerandra linearifolia*? Journal of Evolutionary Biology 12:306–313.

Winn, A. A., and A. S. Evans. 1991. Variation among populations of *Prunella vulgaris* L. in plastic responses to light. Functional Ecology 5:562–571.

Witte, F., C. D. N. Barel, and R. J. C. Hoogerhoud. 1990. Phenotypic plasticity of anatomical structures and its ecomorphological significance. Netherlands Journal of Zoology 40:278–298.

Wodicka, L., H. Dong, M. Mittmann, M.-H. Ho, and D. J. Lockhart. 1997. Genome-wide expression monitoring in *Saccharomyces cerevisiae*. Nature Biotechnology 15:1359–1367.

Wolf, J. B. 2000. Indirect genetic effects and gene interactions. Pages 158–176 in J. B. Wolf, E.

D. Brodie III, and M. J. Wade, editors, Epistasis and the Evolutionary Process. Oxford University Press, New York.

Wolf, J. B. 2001. Gene interactions from maternal effects. Evolution 54:1882–1898.

Wolf, J. B., and E. D. Brodie III. 1998. Co-adaptation of parental and offspring characters. Evolution 52:535–544.

Wolf, J. B., E. D. Brodie III, J. M. Cheverud, A. J. Moore, and M. J. Wade. 1998. Evolutionary consequences of indirect genetic effects. Trends in Ecology and Evolution 13:64–69.

Wolf, J. B., E. D. Brodie III, and M. J. Wade. 2000. Epistasis and the Evolutionary Process. Oxford University Press, New York.

Wolf, J. B., W. A. Frankino, A. F. Agrawal, E. D. Brodie III, and A. J. Moore. 2001. Developmental interactions and the constituents of quantitative variation. Evolution 55:232–245.

Wolpert, L. 1994. The evolutionary origin of development: cycles, patterning, privilege and continuity. Development (Supplement):79–84.

Woltereck, R. 1909. Weitere experimentelle untersüchungen über artveränderung, speziell über das wesen quantitativer artunterschiede bei daphniden. Verhandlungen der Deutschen Zooligischen Gesellschaft 19:110–172.

Woods, H. A. 1999. Egg-mass size and cell size: effects of temperature on oxygen distribution. American Zoologist 39:244–252.

Wootton, J. T. 1994. The nature and consequences of indirect effects in ecological communities. Annual Review of Ecology and Systematics 25:443–466.

Wright, S. 1920. The relative importance of heredity and environment in determining the piebald pattern of guinea pigs. Proceedings of the National Academy of Sciences of the United States of America 6:320–332.

Wright, S. 1921. Systems of mating. Genetics 6:111–178.

Wright, S. 1931. Evolution in Mendelian populations. Genetics 16:97–159.

Wright, S. 1932. The roles of mutation, inbreeding, crossbreeding and selection in evolution. Proceedings of the Sixth International Congress of Genetics 1:356–366.

Yampolsky, L. Y., and D. Ebert. 1994. Variation and plasticity of biomass allocation in *Daphnia*. Functional Ecology 8:435–440.

Yeap, K. L., R. Black, and M. S. Johnson. 2001. The complexity of phenotypic plasticity in the intertidal snail *Nodilittorina australis*. Biological Journal of the Linnean Society 72:63–76.

Yoshimura, J., and W. M. Shields. 1987. Probabilistic optimization of phenotype distributions: a general solution for the effects of uncertainty on natural selection? Evolutionary Ecology 1:125–138.

Yoshimura, J., and W. M. Shields. 1992. Components of uncertainty in clutch size optimization. Bulletin of Mathematical Biology 54:445–464.

Yoshimura, J., and W. M. Shields. 1995. Probabilistic optimization of body size: a discrepancy between genetic and phenotypic optima. Evolution 49:375–378.

Zangerl, A. R., and F. A. Bazzaz. 1983. Plasticity and genotypic variation in photosynthetic behavior of an early and a late successional species of *Polygonum*. Oecologia 57:270–273.

Zera, A. J. 1999. The endocrine genetics of wind polymorphism in *Gryllus*: critique of recent studies and state of the art. Evolution 53:973–977.

Zera, A. J., and R. F. Denno. 1997. Physiology and ecology of dispersal polymorphism in insects. Annual Review of Entomology 42:207–231.

Zera, A. J., and Y. Huang. 1999. Evolutionary endocrinology of juvenile hormone esterase: functional relationship with wing polymorphism in the cricket, *Gryllus firmus*. Evolution 53:837–847.

Zera, A. J., and C. Zhang. 1995. Evolutionary endocrinology of juvenile hormone esterase in *Gryllus assimili*: direct and correlated responses to selection. Genetics 141:1125–1134.

Zera, A. J., J. Potts, and K. Kobus. 1998a. The physiology of life-history trade-offs: experimental analysis of a hormonally induced life-history trade-off in *Gryllus assimilis*. American Naturalist 152:7–23.

Zera, A. J., T. Sanger, and G. L. Cisper. 1998b. Direct and correlated responses to selection on JHE activity in adult and juvenile *Gryllus assimilis*: implications for stage-specific evolution of insect endocrine traits. Heredity 80:300–309.

Zhang, H., and B. G. Forde. 1998. *Arabidopsis* MADS box gene that controls nutrient-induced changes in root architecture. Science 279:407–409.

Zhivotovsky, L., and S. Gavrilets. 1992. Quantitative variability and multilocus polymorphism under epistatic selection. Theoretical Population Biology 42:254–283.

Zhivotovsky, L. A., M. W. Feldman, and A. Bergman. 1996. On the evolution of phenotypic plasticity in a spatially heterogeneous environment. Evolution 50:547–558.

Index